Catalytic Energy Conversion and Catalytic Environmental Purification

Catalytic Energy Conversion and Catalytic Environmental Purification

Guest Editors

Yongjun Ji
Yuxi Liu
Shaomian Liu

MDPI

Basel • Beijing • Wuhan • Barcelona • Belgrade • Novi Sad • Cluj • Manchester

Guest Editors

Yongjun Ji
School of Light Industry and Science
Beijing Technology and Business University
Beijing
China

Yuxi Liu
School of Materials Science and Engineering
Beijing University of Technology
Beijing
China

Shaomian Liu
College of Chemistry and Chemical Engineering
Hebei Normal University for Nationalities
Chengde
China

Editorial Office
MDPI AG
Grosspeteranlage 5
4052 Basel, Switzerland

This is a reprint of the Special Issue, published open access by the journal *Catalysts* (ISSN 2073-4344), freely accessible at: https://www.mdpi.com/journal/catalysts/special_issues/YCZOA062Q5.

For citation purposes, cite each article independently as indicated on the article page online and as indicated below:

Lastname, A.A.; Lastname, B.B. Article Title. *Journal Name* **Year**, *Volume Number*, Page Range.

ISBN 978-3-7258-6754-7 (Hbk)
ISBN 978-3-7258-6755-4 (PDF)
https://doi.org/10.3390/books978-3-7258-6755-4

Contents

catalysts

MDPI

Editorial

Energy Catalytic Conversion and Environmental Catalytic Purification

Zhijin Zhang [1], Shaomian Liu [2,*], Yuxi Liu [3,*] and Yongjun Ji [1,*]

1 School of Light Industry and Science, Beijing Technology and Business University, Beijing 100048, China; 2451041022@st.btbu.edu.cn
2 College of Chemistry and Chemical Engineering, Hebei Minzu Normal University, Chengde 067000, China
3 School of Materials Science and Engineering, Beijing University of Technology, Beijing 100124, China
* Correspondence: shaomianliu@163.com (S.L.); yxliu@bjut.edu.cn (Y.L.); yjji@btbu.edu.cn (Y.J.)

Energy shortages, environmental pollution, and climate change have become critical global issues, presenting significant challenges to the sustainable development of human society [1,2]. In this context, advancing a low-energy-consumption and low-carbon economy, reducing dependence on fossil fuels, and restoring ecological environments have emerged as collective priorities [3]. Within this framework, catalytic conversion technology plays a crucial role in addressing the interconnected challenges of energy and environmental sustainability [4].

Advancements in catalytic technology face significant challenges. From a scientific perspective, current catalytic systems are constrained by three fundamental issues [5,6]. First, achieving an optimal balance among catalytic activity, selectivity, and stability remains a persistent challenge [7]. Second, there is a limited understanding of the fundamental catalytic mechanisms, which perpetuates reliance on empirical trial-and-error methods in catalyst development, thereby reducing research efficiency and extending innovation timelines [8–10]. Third, rigid reaction conditions impose critical constraints on catalytic applications. From industrial viewpoints, basic research tends to prioritize theoretical breakthroughs while insufficiently addressing engineering problems, such as large-scale catalyst synthesis and ensuring long-term stability under industrial operating conditions [11]. This Special Issue aims to provide a comprehensive overview of recent advances in catalysis for environmental and energy conversion applications, emphasizing both heterogeneous and homogeneous catalysis. The scope includes key topics of both scientific and industrial significance.

Lian et al. provided a comprehensive review of the applications of Mn-based catalysts, including both bulk and supported MnO_x catalysts as well as bulk and supported Mn-based composite oxide catalysts, in the CO-selective catalytic reduction of NO_x (CO-SCR) technology. Compared to other transition metal-based catalysts, Mn-based catalysts demonstrate superior catalytic efficiency under low-temperature and low-oxygen conditions, underscoring their considerable potential for practical applications [Contribution 1].

Gomha et al. synthesized three 2-cyano-N′-(2-cyano-3-azobutyl-2-enyl)-3-azobutyl-2-enyl hydrazides employing L-proline as an organic catalyst. This synthetic approach is notable for the catalyst's reusability, mild reaction conditions, environmentally benign process, operational simplicity, high product purity, short reaction time, and elevated yield. Consequently, it holds significant importance in the domain of green drug synthesis [Contribution 2].

Another contribution by the Gomha team involves the use of homophenyl tetramethylimide benzoyl thiourea cross-linked chitosan (PIBTUCS) hydrogel as a green biocatalyst for the efficient synthesis of novel thiazole derivatives. A series of new thi-

 https://doi.org/10.3390/catal16010067

azole derivatives were synthesized through the catalysis of the reaction between 2-(4-((2-aminoformylhydrazyl) methyl) phenoxy)-N-(4-chlorophenyl) acetamide and various hydrazone acyl halides as well as α-halogenated ketones [Contribution 3].

Hu et al. discussed the current challenges confronted by biogas dry reforming technology and outlined the future development trends, providing valuable insights and suggestions for subsequent research and industrial applications [Contribution 4].

Adriana et al. highlighted that iron oxides and hydroxides (Fe-OH) extracted from natural sources have attracted extensive attention in a range of catalytic applications, such as advanced oxidation processes (AOPs), electrochemical applications, catalytic cracking, and biodiesel production [Contribution 5].

Carlos et al. synthesized CeO_2 photocatalytic supports via a precipitation method. Their study demonstrated that sulfation, characterized by bidentate coordination between sulfate groups and semiconductor metal ions within the material, results in an increased specific surface area and a decreased average grain size. The surface modification of CeO_2 was confirmed to be feasible, with sulfate groups acting as electron traps during the photocatalytic process. This function reduces the recombination rate of charge carriers and increases the availability of vacancies, thereby enhancing the degradation of pollutants. This mechanism was supported by the material's high photocatalytic activity in degrading the herbicide 2,4-dichlorophenoxyacetic acid [Contribution 6].

Baaloudj et al. examined the synthesis and properties of a novel silicate material, $Bi_{12}SnO_{20}$. Due to its unique characteristics and the distinctive structure associated with the I_{23} space group, this silicate has attracted considerable research attention in recent years. In their study, tetracycline, bisphenol A, and Rhodamine B were utilized as model pollutants, and performance evaluations were conducted under visible light irradiation. The results indicated that, at a pollutant concentration of 20 mg/L, the degradation efficiency reached nearly 100% within 2 h [Contribution 7].

Li et al. investigated the effects of calcination temperature and hydrogen pretreatment on the structural characteristics and catalytic performance of Pt/CeO_2 catalysts. Hydrogen pretreatment induces the formation of oxygen vacancies on the catalyst surface, which is attributed to the reduction of PtO_x to metallic Pt. This facilitates the generation of reactive oxygen species, thereby improving the catalytic oxidation of NO. Experimental results indicated that calcination temperature significantly influences the redox properties of the catalyst, which in turn affects its efficiency in NO oxidation reactions. This research provides important insights for the design and optimization of Pt/CeO_2 catalysts in environmental applications, particularly in the development of exhaust gas post-treatment technologies [Contribution 8].

Almotiry et al. utilized hydrothermally synthesized UiO-67 Metal–Organic Framework (MOF) to evaluate the efficacy of various adsorption parameters in the removal of antiviral drugs, specifically ritonavir (RTV) and lopinavir (LPV), from aqueous solutions. Diffusion kinetic studies revealed that the adsorption process on the surface of UiO-67 MOF is predominantly catalytic. Thermodynamic analyses, including calculations of Gibbs free energy change (ΔG), enthalpy change (ΔH), and entropy change (ΔS), indicated that the adsorption of RTV and LPV onto UiO-67 from aqueous solutions is spontaneous. These results underscore the potential application of UiO-67 MOF as an effective adsorbent for antiviral drugs [Contribution 9].

Guo et al. addressed the problems of low mineralization rates and limited durability in the ozonated catalytic oxidation (OCO) of volatile organic compounds (VOCs) by enhancing the oxygen mobility and low-temperature reducibility of transition metal oxides. They achieved this by incorporating highly dispersed Ag into Mn_3O_4 via a co-precipitation method utilizing oxalic acid. The resulting Ag/Mn_3O_4 catalyst exhibited superior mineralization rates and stability relative to pure Mn_3O_4 during the catalytic ozonation of benzene

https://doi.org/10.3390/catal16010067

at ambient temperature. This study is expected to contribute to the improvement of VOC mineralization efficiency through ozonated catalytic oxidation [Contribution 10].

These research directions are interrelated and mutually reinforcing, collectively forming a framework for catalytic technologies aimed at addressing energy and environmental challenges. It is anticipated that the academic exchanges facilitated by this Special Issue will stimulate novel ideas and significant breakthroughs, thereby advancing innovations in catalytic materials, principles, methodologies, and technologies. Moreover, such progress is expected to accelerate the industrial implementation of catalytic technologies in areas such as clean energy conversion and environmental pollution control. Additionally, this Special Issue is envisioned to act as a catalyst for establishing a virtuous cycle encompassing fundamental research, technological development, and industrial application within the global catalysis community. This cycle will foster the widespread adoption of catalytic technologies across the energy, environmental, and chemical engineering sectors, ultimately contributing scholarly expertise toward the realization of a low-carbon, clean, and sustainable future.

Funding: This research was funded by the National Natural Science Foundation of China (Nos. 22301012 and 21978299), the Shccig-Qinling Program (No. SMYJY202400134C), the Research Foundation for Youth Scholars of Beijing Technology and Business University (No. QNJJ2022-23), the R&D Program of Beijing Municipal Education Commission (No. KM202310011005), the Research Foundation for Advanced Talents of Beijing Technology and Business University (No. 19008020159), ITF–RTH-Global STEM Professorship (9446006), and JC STEM lab of Advanced CO_2 Upcycling (9228005).

Conflicts of Interest: The authors declare no conflicts of interest.

List of Contributions:

1. Lian, D.; Chen, M.; Wang, H.; Li, C.; Liu, B.; Dai, G.; Hou, S.; Liu, Y.; Ji, Y. Mn-Based Catalysts in the Selective Reduction of NO_x with CO: Current Status, Existing Challenges, and Future Perspectives. *Catalysts* **2024**, *14*, 462. https://doi.org/10.3390/catal14070462.
2. Gomha, S.M.; Abolibda, T.Z.; Alruwaili, A.H.; Farag, B.; Boraie, W.E.; Al-Hussain, S.A.; Zaki, M.E.A.; Hussein, A.M. Efficient Green Synthesis of Hydrazide Derivatives Using L-Proline: Structural Characterization, Anticancer Activity, and Molecular Docking Studies. *Catalysts* **2024**, *14*, 489. https://doi.org/10.3390/catal14080489.
3. Gomha, S.M.; El-Ghany, N.A.A.; Ebaid, M.S.; Abolibda, T.Z.; Zaki, M.E.A.; Alhilal, M.; Alhilal, S.; Mohamed, N.A. Eco-Friendly Synthesis of Thiazole Derivatives Using Recyclable Cross-Linked Chitosan Hydrogel Biocatalyst Under Ultrasonic Irradiation as Anti-Hepatocarcinogenic Agents. *Catalysts* **2024**, *14*, 840. https://doi.org/10.3390/catal14120840.
4. Hu, W.; Wu, J.; Huang, Z.; Tan, H.; Tang, Y.; Feng, Z.; Deng, R.; Zhang, H.; Zairov, R.; Pan, Z. Catalyst Development for Biogas Dry Reforming: A Review of Recent Progress. *Catalysts* **2024**, *14*, 494. https://doi.org/10.3390/catal14080494.
5. Jiménez-Vázquez, A.; Jaimes-López, R.; Morales-Bautista, C.M.; Pérez-Rodríguez, S.; Gochi-Ponce, Y.; Estudillo-Wong, L.A. Catalytic Applications of Natural Iron Oxides and Hydroxides: A Review. *Catalysts* **2025**, *15*, 236. https://doi.org/10.3390/catal15030236.
6. Rodríguez, C.; Castañeda, C.; Sosa, E.; Martínez, J.J.; Mancipe, S.; Rojas, H.; Tzompantzi, F.; Gómez, R. Enhanced Photocatalytic Degradation of Herbicide 2,4-Dichlorophenoxyacetic Acid Using Sulfated CeO_2. *Catalysts* **2024**, *14*, 594. https://doi.org/10.3390/catal14090594.
7. Baaloudj, O.; Kenfoud, H.; Brienza, M.; El Jery, A.; Aldrdery, M.; Assadi, A.A. Exploring the Synthesis of Novel Sillenite $Bi_{12}SnO_{20}$: Effect of Calcination Temperature on the Phase Formation and Catalytic Performance. *Catalysts* **2024**, *14*, 650. https://doi.org/10.3390/catal14090650.
8. Li, X.; Li, S.; Cheng, Y.; Zheng, L.; Song, L.; Zi, X.; Dai, H. The Activation of Oxygen Species on the Pt/CeO_2 Catalyst by H_2 for NO Oxidation. *Catalysts* **2024**, *14*, 778. https://doi.org/10.3390/catal14110778.

9. Almotiry, S.; Alhogbi, B.G.; Abdel Salam, M.; Jaremko, M. UiO-67 Metal–Organic Framework as Advanced Adsorbent for Antiviral Drugs from Water Environment. *Catalysts* **2024**, *14*, 573. https://doi.org/10.3390/catal14090573.
10. Guo, H.; Cen, L.; Deng, K.; Mo, W.; Hajime, H.; Hu, D.; Zhang, P.; Shangguan, W.; Huang, H.; Einaga, H. Boosting Benzene's Ozone Catalytic Oxidation at Mild Temperatures over Highly Dispersed Ag-Doped Mn_3O_4. *Catalysts* **2024**, *14*, 554. https://doi.org/10.3390/catal14090554.

References

1. Wei, Y.M.; Liu, L.C.; Kang, J.N.; Zhang, Y.L.; Peng, S.; Liao, H.; Xu, S.; Zhao, L.T.; Yan, H.B.; Qian, X.Y.; et al. Navigating energy transition solutions for climate targets with minerals constraint. *Nat. Clim. Chang.* **2025**, *15*, 833–841. [CrossRef]
2. Zhao, J.; Tang, Y.; Zhu, X.; Zhu, J. National environmental monitoring and local enforcement strategies. *Nat. Cities* **2025**, *2*, 58–69. [CrossRef]
3. Shi, H.T.; Heng, J.N.; Duan, H.B.; Li, H.J.; Chen, W.Q.; Wang, P.; Cui, L.B.; Wang, S.Y. Critical mineral constraints pressure energy transition and trade toward the Paris Agreement climate goals. *Nat. Commun.* **2025**, *16*, 833–841. [CrossRef] [PubMed]
4. Skalska, K.; Miller, J.S.; Ledakowicz, S. Trends in NO_x abatement: A review. *Sci. Total Environ.* **2010**, *408*, 3976–3989. [CrossRef] [PubMed]
5. Granger, P.; Parvulescu, V.I. Catalytic NO_x Abatement Systems for Mobile Sources: From Three-Way to Lean Burn after-Treatment Technologies. *Chem. Rev.* **2011**, *111*, 3155–3207. [PubMed]
6. Lu, K.; Guo, S.; Tan, Z.; Wang, H.; Shang, D.; Liu, Y.; Li, X.; Wu, Z.; Hu, M.; Zhang, Y. Exploring atmospheric free-radical chemistry in China: The self-cleansing capacity and the formation of secondary air pollution. *Natl. Sci. Rev.* **2018**, *6*, 579–594. [CrossRef] [PubMed]
7. Wang, C.; Wang, W.; Sardans, J.; An, W.; Zeng, C.; Abid, A.A.; Peñuelas, J. Effect of simulated acid rain on CO_2, CH_4 and N_2O fluxes and rice productivity in a subtropical Chinese paddy field. *Environ. Pollut.* **2018**, *243*, 1196–1205. [CrossRef] [PubMed]
8. Nurunnabi, M.; Mukainakano, Y.; Kado, S.; Li, B.; Kunimori, K.; Suzuki, K.; Fujimoto, K.I.; Tomishige, K. Additive effect of noble metals on NiO-MgO solid solution in oxidative steam reforming of methane under atmospheric and pressurized conditions. *Appl. Catal. A Gen.* **2006**, *299*, 145–156. [CrossRef]
9. He, Z.H.; Li, C.C.; Yang, S.Y.; Liu, J.; Cao, H.H.; Wang, K.; Wang, W.; Yang, Y.; Liu, Z.T. Electrocatalytic CO_2 reduction to ethylene over CuO_x boosting CO_2 adsorption by lanthanide neodymium. *Catal. Sci. Technol.* **2023**, *13*, 6675–6684. [CrossRef]
10. Zhao, X.; Luo, B.; Long, R.; Wang, C.M.; Xiong, J.Y. Composition-dependent activity of Cu-Pt alloy nanocubes for electrocatalytic CO_2 reduction. *J. Mater. Chem. A* **2015**, *3*, 4134–4138. [CrossRef]
11. Nguyen, T.V.; Wu, J.C.S. Photoreduction of CO_2 in an optical-fiber photoreactor: Effects of metals addition and catalyst carrier. *Appl. Catal. A Gen.* **2008**, *335*, 112–120. [CrossRef]

https://doi.org/10.3390/catal16010067

catalysts

MDPI

Review

Mn-Based Catalysts in the Selective Reduction of NO_x with CO: Current Status, Existing Challenges, and Future Perspectives

Dianxing Lian [1], Mohaoyang Chen [1], Huanli Wang [1], Chenxi Li [1], Botao Liu [1], Guiyao Dai [1], Shujun Hou [1], Yuxi Liu [2,*] and Yongjun Ji [1,*]

1 College of Light Industry Science and Engineering, Beijing Technology and Business University, Beijing 100048, China
2 College of Environmental and Energy Engineering, Beijing University of Technology, Beijing 100124, China
* Correspondence: yxliu@bjut.edu.cn (Y.L.); yjji@btbu.edu.cn (Y.J.)

Abstract: The technology for the selective catalytic reduction of NO_x by CO (CO-SCR) has the capability to simultaneously eliminate CO and NO_x from industrial flue gas and automobile exhaust, thus making it a promising denitrification method. The advancement of cost-effective and high-performing catalysts is crucial for the commercialization of this technology. Mn-based catalysts demonstrate enhanced catalytic efficiency under conditions of low temperature and low oxygen content when compared to other transition metal-based catalysts, indicating significant potential for practical applications. This review outlines the diverse Mn-based catalysts, including bulk or supported MnO_x catalysts, bulk or supported Mn-based composite oxide catalysts, and the use of MnO_x as dopants. Subsequently, the synthesis methods and catalytic mechanism employed by Mn-based catalysts are presented. The following section examines the impact of O_2, H_2O, and SO_2 on the catalytic performance. Finally, the potential and implications of this reaction are deliberated. This work aims to offer theoretical guidance for the rational design of highly efficient Mn-based catalysts in the CO-SCR reaction for industrial applications.

Keywords: CO-SCR; reaction mechanism; Mn-based catalysts; O_2; H_2O; SO_2

1. Introduction

NO_x, including NO and NO_2, constitutes a significant portion of atmospheric pollutants, which is primarily derived from vehicle exhaust and industrial emissions, notably from coke oven and coal-fired power plant emissions. NO_x can cause severe environmental consequences, such as acid rain, photochemical smog, and ozone depletion [1–3], which present substantial hazards to human health. Currently, the prevalent industrial denitrification technique involves the selective reduction of NO_x by NH_3, known as NH_3-SCR [4,5]. However, this technology presents several concerns, including the potential for ammonia storage leakage, the formation of by-products that can lead to equipment blockage and corrosion, and the inherent toxicity of ammonia [6]. Given the incomplete combustion of fuel, flue gas also contains a certain amount of CO. The use of CO as the reducing agents for the selective catalytic reduction of NO_x (CO-SCR) obviates the necessity for extra reducing agents, thereby reducing costs. Moreover, this technology assists in the removal of NO_x and CO emissions, enabling the concurrent elimination of both pollutants. Hence, CO-SCR technology shows promising potential for practical application [7].

The ideal CO-SCR reaction is presented in Equation (1). However, in real flue gas, there exists a high concentration of O_2 and H_2O, along with a minor quantity of SO_2. The presence of oxygen can lead to the preferential oxidation of NO and CO (Equations (2) and (3)), while the existence of H_2O and SO_2 may cause catalyst poisoning; meanwhile, concurrent side reactions may occur, producing N_2O (Equation (4)), consequently diminishing the selectivity for N_2.

$$CO + NO \rightarrow CO_2 + N_2 \quad (1)$$

$$2CO + O_2 \rightarrow 2CO_2 \quad (2)$$

$$2NO + O_2 \rightarrow 2NO_2 \quad (3)$$

$$2NO + CO \rightarrow N_2O + CO_2 \quad (4)$$

Currently, the catalysts employed for the CO-SCR reaction are predominantly categorized into two groups: noble metal catalysts and transition metal catalysts. Noble metal catalysts, such as those based on Pd [8], Pt [9], and Ir [10], offer certain benefits such as resistance to oxygen and high temperatures. However, some challenges exist, such as the scarcity of precious metal, elevated costs, high reduction temperatures, and susceptibility to sulfur- and water-induced deactivation [11]. For instance, in three-way catalyst (TWC) systems, the CO-SCR reaction occurs concurrently with other reactions in the exhaust pipe of a vehicle, which operates at high temperatures and low oxygen concentrations. In this instance, the typical catalyst comprises Pt, Pd, and Rh. In contrast, transition metal catalysts like Co-based [12], Mn-based [13], Cu-based [14] ones are preferred due to their lower cost and superior catalytic activity at lower temperatures compared to noble metal catalysts. Therefore, employing transition metal catalysts in CO-SCR is a favorable approach. Nevertheless, in oxygen-rich environments, transition metals tend to be oxidized easily, leading to diminished oxygen resistance, N_2 selectivity, and stability. Therefore, it is imperative to develop cost-effective and efficient transition metal catalysts, which should exhibit low-temperature catalytic reactivity, a broad temperature range, high O_2 tolerance, resilience to SO_2 and H_2O poisoning, and long-lasting stability.

In prior reviews, researchers have provided a summary of catalysts based on Ir [15], Cu [16,17], and various transition metals [18]. Additionally, they have categorized and deliberated on various noble catalyst types [19–21]. It is known that Mn-based catalysts exhibit multiple oxidation states (Mn^{2+}, Mn^{3+}, and Mn^{4+}), excellent redox capacity, and abundant reactive oxygen species. Furthermore, Mn has the capability to be integrated into various functional materials. More importantly, Mn-based catalysts have demonstrated superior performances at low temperatures compared to other transition metal catalysts, showing promising potential for use in the CO-SCR reaction. Thus, in recent years, Mn-based catalysts have garnered increasing attention. However, they still encounter issues related to inadequate oxygen resistance and instability. Given the scarcity of literature reviews dedicated to Mn-based catalysts in the CO-SCR reaction, this review provides a comprehensive examination of Mn-based catalysts, in terms of their current status, existing challenges, and future perspectives, aiming to facilitate their industrial applications.

As illustrated in Figure 1, this review commences with an exploration of the various classifications of Mn-based catalysts, followed by an examination of the synthesis methods and a discussion of the reaction mechanism. It also offers a summary of the impact of O_2, H_2O, and SO_2 on the catalytic efficiency of Mn-based catalysts. This discussion is followed by an examination of the challenges faced by these catalysts and the strategies developed to address them. Finally, this review concludes by suggesting innovative strategies for the advancement of highly effective Mn-based catalysts to enhance their future prospects.

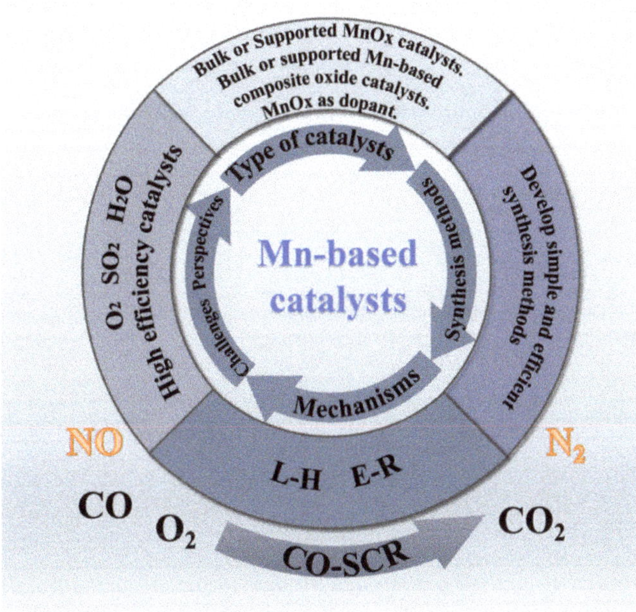

Figure 1. An overview of Mn-based catalysts.

2. Types of Mn-Based Catalysts

2.1. Bulk or Supported MnO_x Catalysts

Manganese oxides, which are present in various crystal phases, are compounds characterized by multiple oxidation states (Mn^{2+}, Mn^{3+}, and Mn^{4+}), excellent redox capabilities, and a rich supply of active oxygen. The efficiency of pure manganese oxides in reducing NO_x to N_2 is ranked MnO_2 > Mn_5O_8 > Mn_2O_3 > Mn_3O_4 > MnO. Highly oxidized manganese oxides have higher CO-SCR activity, and oxygen vacancies (OVs) produced in the reduction process serve as the adsorption site for NO_x [22]. These properties empower them to actively engage in and promote catalytic reactions. Therefore, manganese oxides hold great potential in the CO-SCR reaction (Table 1). The surface manganese oxide phases and redox properties of Mn oxide catalysts are crucial factors that significantly influence the overall catalytic activity [23]. For the α-MnO_2 nanorod catalyst, upon pretreatment with H_2, Mn^{4+} is first reduced to Mn^{2+}. Subsequently, these Mn^{2+} sites undergo rapid oxidation to Mn^{3+} in the presence of NO. In situ studies showed that the α-MnO_2 surface experiences significant recombination, forming a new surface active phase. Mn^{3+} and OVs were the active sites [24].

Compared with other transition metal oxides, the low temperature activity of manganese oxides on support for CO-SCR is higher than that of other transition metal oxides. Boningari et al. [25] studied different titanium-supported transition metal-based catalysts and found that MnO_x/TiO_2 had better CO-SCR performance at low temperatures. In situ infrared spectroscopy showed that the Lewis acid site on the catalyst surface was responsible for the reaction, not the Brønsted acid site. The reduced Mn site served as the active site for activating NO. In Mn/TiO_2 catalysts, the lattice oxygen proximal to Mn atoms facilitates the oxidation of CO to CO_2, leading to the creation of OVs. On Mn/TiO_2, the notable intermediate N_2O can be reduced to N_2, with an associated energy barrier of 0.51 eV, indicating a pronounced N_2 selectivity. The Mn–OV pair present on the catalyst surface serves as the active site for the adsorption of NO and the catalysis of N-O bond cleavage [26]. Moreover, the comparison of the activities of different transition metal oxides

(Cu, Ni, Fe, Mn, and Cr) showed that the TiO_2-supported manganese oxide has good NO reduction activity at 200 °C. The MnO_x/TiO_2 catalyst showed highly N_2 selectivity even in the presence of oxygen. The results indicate that the high surface area and the reducibility of manganese oxide play an important role in the high activity for CO-SCR [27].

Table 1. Comparison of catalytic performances of the Mn-based catalysts in the literature.

Catalyst	Gas Composition (%)					GHSV or WHSV	Temperature (°C)	NO Conversion (%)	CO Conversion (%)	N_2 Selectivity (%)	Ref.
	NO	CO	O_2	SO_2	H_2O						
α-MnO_2 nanorods	5.0	5.0	---	---	---	30,000 mL h^{-1} g^{-1}	400	85	---	80	[24]
MnO_2/TiO_2	0.4	0.4	2.0	---	---	50,000 h^{-1}	200	95	---	---	[25]
Mn/TiO_2	3.0	3.0	---	---	---	20,000 h^{-1}	450	100	---	90	[26]
MnO_x/TiO_2	0.4	0.4	2.0	---	---	50,000 h^{-1}	175	100	---	100	[27]
$2MnO_x$-Fe_3O_4	1.0	2.0	---	---	---	23,000 h^{-1}	400	90	---	70	[28]
$Cu_{1.5}Mn_{1.5}O_4$	5.0	10.0	---	---	---	24,000 mL h^{-1} g^{-1}	250	100	45	95	[29]
CuO-Mn_2O_3/γ-Al_2O_3	5.0	10.0	---	---	---	24,000 h^{-1}	300	70	---	85	[30]
NiMn-MOF-74	0.5	1.0	---	0.1	5.0	30,000 h^{-1}	200	100	---	---	[31]
$Co_{0.3}$-OMS-2	0.5	0.5	---	1.0	---	24,000 mL h^{-1} g^{-1}	150	95	---	80	[32]
$Cu_xMn_{3-x}O_4$	1.0	2.0	---	---	---	30,000 h^{-1}	200	100	---	55	[33]
Cu-Ce-Fe-Mn/TiO_2	0.02	0.02	1.0	---	---	10,000 h^{-1}	200	100	82	---	[34]
0.010_{MnFeCu}	5.0	10.0	---	1.0	10.0	15,000 mL h^{-1} g^{-1}	300	100	---	100	[35]
Mn@La3-Fe1/AC	0.25	5.0	10.0	---	---	26,000 h^{-1}	400	93.8	---	---	[36]
$Mn_xCo_{3-x}O4$	1.0	2.0	5.0	0.05	5.0	20,000 h^{-1}	180	100	---	100	[37]
Mn-CeO_2@Co_3O_4	1.0	2.0	5.0	0.05	10.0	24,000 h^{-1}	200	82	100	78	[38]
Cu-Mn_2	0.6	1.6	5.0	---	---	15,000 h^{-1}	400	90	100	87	[39]
$CuMnO_2$	1.0	2.0	---	---	---	13,000 h^{-1}	300	100	---	80	[40]
V_x-OMS-2	0.5	0.5	---	1.0	---	24,000 mL h^{-1} g^{-1}	300	95	---	---	[41]
$Sb_{0.2}$-OMS-2	0.05	0.05	---	---	---	---	300	100	90	---	[42]
Ho_x-OMS-2	0.05	0.05	0.05	---	---	15,000 h^{-1}	225	100	99	---	[43]

2.2. Bulk or Supported Mn-Based Composite Oxide Catalysts

When Mn forms a composite oxide with other metals, it frequently generates a synergistic effect, enhancing its catalytic performance beyond that of an individual Mn oxide (Table 1) [28]. The presence of surface synergistic oxygen vacancies (SSOVs) within composite oxides facilitates the adsorption and activation of gas molecules, thereby enhancing catalytic performance. In comparison to pure MnO_x, Cu-Mn catalysts demonstrate a higher propensity for oxidation and the generation of activated O species on their surface, thus promoting the adsorption of oxygen molecules. Mn-based catalysts doped with copper oxide comprise not only MnO_x but also a linear Cu-CO active substance, which enhances the adsorption of NO_x. An intriguing phenomenon is that the surface dispersed Cu^{x+}-O^{2-}-Mn^{y+} substance can be reduced to Cu^+-□-$Mn^{(4-x)+}$ active species with an elevation in adsorption temperature. The synergistic effect between Cu and Mn (Cu^{x+}-O^{2-}-Mn^{y+}) plays a crucial role in the CO-SCR reaction (Figure 2a) [29]. Yao et al. [44] synthesized CuO-MnO_2/CeO_2 using the co-impregnation method, which exhibited a stronger interaction and an excellent reduction behavior. This process facilitated the generation of low-valence copper species (Cu^+/Cu^0) and increased oxygen vacancies, particularly SSOVs (Cu^+-□-$Mn^{(4-x)+}$), during the reaction, which are conducive to the adsorption of CO and the dissociation of NO, respectively. In another study [30], it was found that the catalytic efficiency of CuO-Mn_2O_3/γ-Al_2O_3 catalysts, after being pretreated with CO, exhibited superior performance compared to CuO/γ-Al_2O_3 and Mn_2O_3/γ-Al_2O_3. The formation of SSOVs in the CuO-Mn_2O_3/γ-Al_2O_3 catalyst occurs following CO pretreatment, leading to distinct adsorption properties compared to SOVs in CuO/γ-Al_2O_3 and Mn_2O_3/γ-Al_2O_3 catalysts. Exposed Cu^+ and Mn^{2+} ions act synergistically as CO and NO adsorption sites, respectively, and the SSOVs function as a crucial bridge, facilitating the interaction between adjacent adsorbed CO and NO to promote the surface reaction. Therefore, SSOVs on the CO-CuMnAl catalyst exhibit superior catalytic activity, facilitating the reduction of NO by CO (Figure 2b) [30]. Due to the difference in cation radius, the incorporation of Mn and

Ni into MOF-74 leads to an improvement in the specific surface area, which is conducive to the formation of unsaturated coordination metal sites and Lewis acid sites for CO-SCR. In addition, the electron interaction between Ni and Mn in NiMn-MOF-74 facilitates a significant synergistic reaction ($Mn^{3+} + Ni^{3+} \leftrightarrow Mn^{4+} + Ni^{2+}$). This synergistic effect not only enhances the formation of more active Mn^{4+} components, but also promotes the generation of SSOVs, which exhibit higher catalytic activity compared to surface oxygen vacancies (SOVs) [31].

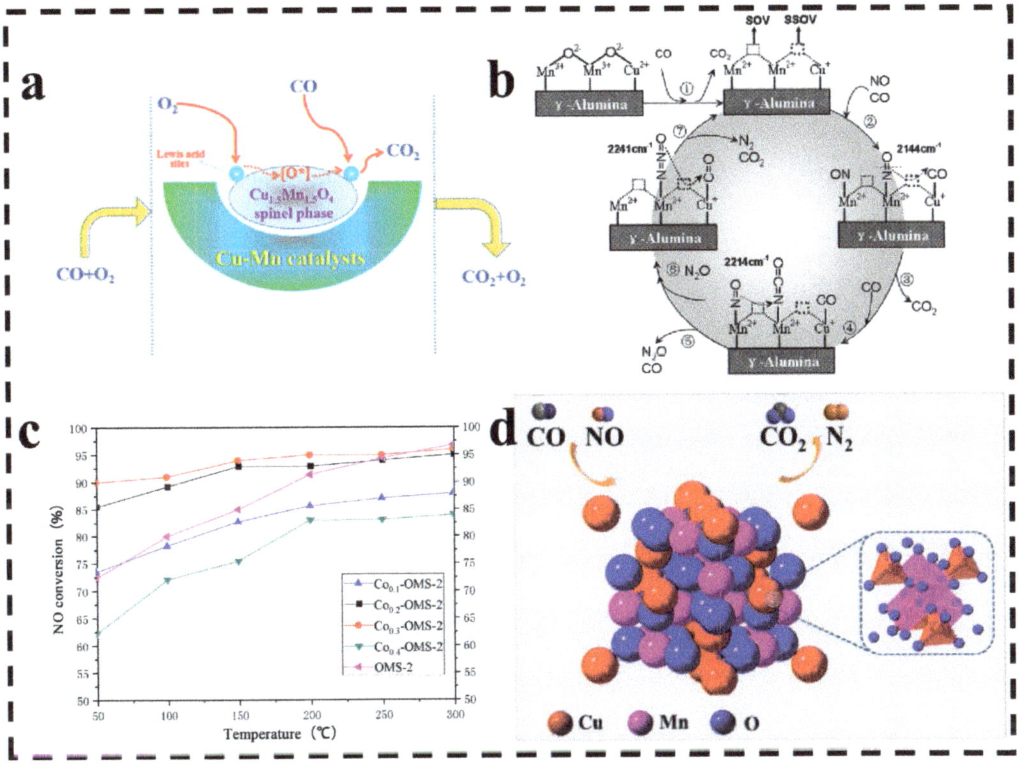

Figure 2. (**a**) Schematic illustration of the proposed mechanism for the catalytic oxidation of CO over Cu−Mn catalysts [29]; (**b**) proposed reaction mechanism for NO reduction by CO over CO−CuMnAl catalyst [30]; (**c**) NO conversion curves of OMS−2 prepared by Co doping method [32]; (**d**) schematic illustration of the proposed mechanism for the catalytic CO-SCR over $Cu_xMn_{3-x}O_4$ catalysts [33].

By adjusting the electronic configuration of manganese oxides, it is possible to alter their redox characteristics, subsequently impacting their catalytic efficiency. During the doping modification process, Co occupies the Mn site in the [MnO6] cell in the OMS-2 catalyst, forming a new [CoO6] crystal structure. The crystallinity of Co-doped OMS-2 decreases as the Co doping amount increases. Co substitutes for Mn, creating a novel active site. The active sites comprising Co^{2+}, Co^{3+}, Mn^{3+}, and Mn^{4+} adsorb CO and NO, catalyzing the conversion of these compounds into CO_2 and N_2 (Equations (9) and (10)). Compared with undoped OMS-2, the conversion of NO in the flue gas of $Co_{0.3}$-OMS-2 reaches approximately 95% at 100–300 °C (Figure 2c) [32]. In the case of the $Cu_xMn_{3-x}O_4$ catalyst, the introduction of copper ions into Mn_2O_3 results in the formation of a spinel structure with a high density of lattice defects and oxygen vacancies. This modification increases the concentration of Mn^{4+}, which has the capability to adsorb reactant molecules. Consequently, this process enhances the redox performance of the catalyst, increases the mobility of reactive oxygen species, and ultimately boosts its catalytic activity. Compared with CuO and Mn_2O_3, the catalytic activity of the $Cu_xMn_{3-x}O_4$ catalyst is enhanced. The strong synergy between binary metal oxides also results in its improved stability (Figure 2d) [33]. Mn demonstrates a propensity to exhibit synergistic interactions with other metals, resulting in increased efficacy of catalysts operating at low temperatures. When Ce, Fe and Co were introduced into a Mn-based catalyst and supported on TiO_2, the catalytic performance exhibited an enhancement compared to the performance of Mn alone. Under anaerobic conditions, the Mn-Ce-Fe-Co/TiO_2 catalyst achieved complete NO conversion at 200 °C. The enhanced performance of the modified Mn-based catalyst can be attributed to the increased oxygen mobility and the formation of Mn^{4+}, resulting in an elevated rate of NO conversion [34].

2.3. MnO_x as Dopants

Mn is also a frequently utilized dopant in CO-SCR (Table 1). The introduction of Mn doping has been found to facilitate the generation of oxygen vacancies on the catalyst surface. This, in turn, enhances the adsorption and activation of NO, leading to an overall improvement in the catalyst's performance. When Cu^{2+} and Fe^{3+} combine to form a Fe-based composite oxide in an alkaline solution, the addition of a specific quantity of Mn facilitates its bonding with Fe. This process weakens the interaction between Cu and Fe, leading to the precipitation of the CuO phase. The Cu^{2+} ions then selectively develop along the surface of CuO(110), leading to enhanced catalytic activity. At high temperatures (300–1000 °C), the presence of Mn^{3+} can impede the conversion of γ-Fe_2O_3 to α-Fe_2O_3, thus achieving stable catalytic efficacy. The existence of Mn^{4+} leads to the generation of reactive oxygen species, which can be readily reduced to create SSOVs during the reaction. The resulting SSOVs can promote the dissociation of NO and facilitate the oxygen transfer (Equation (5)). These factors ultimately result in the catalyst exhibiting good catalytic performance across a wide temperature range [35]. Different metal cations M (M = Zr, Cr, Mn, Fe, Co, and Sn) were doped into a CuO/$Ce_{20}M_1O_x$ catalyst to explore their impact on the catalyst's structure and activity in the CO-SCR reaction. It was found that the doping of Mn^{4+} into the CeO_2 lattice promoted the generation of oxygen vacancies and surface unsaturated metal cation. Hence, in comparison to Cu/CeO_2 and other metal cation-doped samples, Cu/$CeMnO_x$ demonstrated superior performance [45]. When MnO_x was doped into $CuCeO_x$/γ-Al_2O_3, it significantly enhanced the catalytic activity at low temperatures, with maximum conversions of NO_x and CO of 98% and 96%, respectively, at 200 °C. The robust interaction among Mn, Cu, and Ce oxides results in higher dispersibility and more coordination-unsaturated ions. In addition, this interaction enhances the formation of oxygen vacancies and acid sites, thereby enhancing the adsorption capacity for CO and NO_x [36]. The rise in the Fe^{2+}/Fe^{3+} ratio subsequent to the addition of an accelerant to the La-Fe/AC catalyst could be attributed to the supplementary synergistic impact of M (Mn and Ce) and Fe via the redox equilibrium ($M^{3+} + Fe^{3+} \leftrightarrow M^{4+} + Fe^{2+}$). This phenomenon improves the redox cycle, promotes the formation of SSOVs, aids in the decomposition of NO, and accelerates the CO-SCR process. The presence of O_2 promotes the formation of C(O) complexes and enhances the activation of metal sites. The NO conversion of a Mn@La3-Fe1/AC catalyst reached 93.8% at 400 °C with 10% O_2 [37]. Our group reported that the hierarchically interconnected porous (HIP) $Mn_xCo_{3-x}O_4$ spinel, synthesized through a citric acid-assisted sol–gel method, demonstrated high catalytic efficiency for CO-SCR. The NO removal efficiency of the synthesized $Mn_{0.3}Co_{2.7}O_4$ was 87% at 180 °C, exhibiting a wide active temperature range (100–400 °C). Its superior activity can be attributed to the following factors: (i) the high oxidation states of Mn^{3+}, Mn^{4+}, and Co^{3+} in the Co-O-Mn structure facilitates an effective redox cycle, creating adsorption sites for NO and CO (Equations (9) and (10)), and (ii) the HIP structure significantly enhances gas diffusion (Figure 3a) [46].

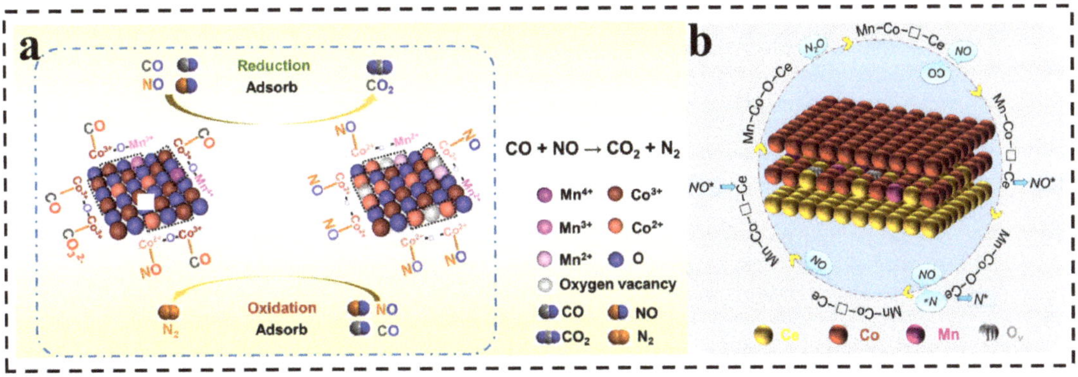

Figure 3. (**a**) Schematic illustration of proposed CO-SCR reaction mechanism on $Mn_{0.3}Co_{2.7}O_4$ [46]; (**b**) schematic illustration of proposed CO-SCR reaction mechanism on Mn−CeO_2@Co_3O_4 [38].

Moreover, the catalyst's electronic structure can be modified through Mn doping to enhance its catalytic performance. Doping Mn into CeO_2@Co_3O_4 generates additional active sites and modulates the electronic structure of the catalyst (Figure 3b). When the O_2 concentration was 5 vol% and the temperature was 200 °C, the Mn-CeO_2@Co_3O_4 catalyst exhibited an NO conversion of 82%, a CO conversion of 100%, and an N_2 selectivity of 78% [38]. In the Mn_2O_3-modified CuO/γ-Al_2O_3 catalyst prepared by Wan et al. [47], it was found that Mn_2O_3 can promote the dispersion of CuO on γ-Al_2O_3, forming a monomolecular layer. This enhances the reducibility of CuO (Cu^{2+} + Mn^{3+} ↔ Cu^{+} + Mn^{4+}), thus boosting the adsorption capacity of CO on the CuO/γ-Al_2O_3 catalyst (Equation (9)). As a result, both NO conversion and N_2 selectivity are enhanced.

3. Synthesis Methods

Various synthesis methods yield distinct structures, influencing factors like the particle size, the creation of specific morphologies, the generation of OVs, or the regulation of valence states of the active center. These factors, in turn, impact the catalytic activity [39].

In order to implement the catalyst in industrial settings, it is imperative to devise a synthesis approach that is easy to execute, amenable to upscaling, and cost-effective. Currently, alternative preparation methods distinct from conventional synthesis techniques have been reported in the literature. For instance, smooth spherical Fe_xMn_yO catalysts with varying Fe contents were synthesized using the solvothermal method, and their CO-SCR activity was tested. The results indicated that the introduction of FeO_x enhances the Fe^{2+} + Mn^{4+} ↔ Fe^{3+} + Mn^{3+} conversion, thereby facilitating oxygen cavitation and the production of reactive oxygen species [40]. Also, a series of spherical Mn_x-Fe_2O_3/C catalysts were synthesized using MnFe-MOF-74 as a precursor through the sacrif-membrane plate method, and they were used in the CO-SCR process. The characterization results revealed the formation of a redox reaction involving Fe^{2+} and Mn^{4+} ions at the Fe-O-Mn site, leading to the generation of Fe^{3+} + Mn^{3+} ions. This reaction significantly enhanced the adsorption and activation of NO molecules. DFT calculations revealed that the incorporation of Mn substantially enhances the local electron density at Fe-Mn sites, resulting not only in the formation of robust C-Fe/Mn and N-Fe/Mn bonds, but also in the effective weakening of N-O bonds [48]. While these approaches enhance the catalytic performance of Mn-based catalysts in the CO-SCR reaction, they are associated with complex operational procedures and high expenses.

Several researchers have employed coprecipitation as a method to synthesize La-Cu-Mn-O catalysts with varying La concentrations. The incorporation of La has been shown to facilitate the diminution of the particle size of copper–manganese oxides, inhibit its agglomeration, improve the reducibility of the catalyst, and stimulate the augmentation of the exposed active sites and the reactivity between the reactants (NO and CO) [49]. Although the co-precipitation method is straightforward, the utilization of La is cost-prohibitive, rendering it impractical for industrial applications.

Due to the complexity or high cost associated with the current synthesis methods, there is a need to establish a straightforward and effective synthesis approach that achieves catalytic activity for the preparation of the catalyst.

4. Reaction Mechanism

Generally, there exist three mechanisms for the CO-SCR reaction. The mechanisms are categorized into the Langmuir–Hinshelwood (L-H) mechanism, the Eley–Rideal (E—R) mechanism, and the Mars–Van Krevelen (MvK) mechanism. The L-H mechanism involves the adsorption of two reaction gas molecules on the catalyst's surface, where the reaction proceeds through the interactions between the adsorbed species [50–52]. In contrast, the E-R mechanism entails the adsorption of a reactant gas on the catalyst's surface, with the reaction occurring through molecular interactions between the adsorbed species and the gas-phase molecules [53–55]. The MvK mechanism primarily pertains to reactions that involve active lattice oxygen [50,56,57]. For Mn-based catalysts, many studies have identified the prevalent mechanisms governing the CO-SCR reaction as the L-H mechanism [31,33,38,58] and the E-R mechanism [23–25,49]. To date, there is no literature reporting the involvement of the MvK mechanism in the CO-SCR reaction, where N_2O acts as the principal reaction intermediate in this process.

The E-R mechanism is likely to occur when Mn predominantly exists in a low valence state in Mn-based catalysts. For the E-R mechanism, the formation of N_2O can be described as follows: NO initially adsorbs on the catalyst to produce NO* (Equation (5)). Subsequently, NO* dissociates into N* and O* (Equation (6)), and N* reacts with another adsorbed NO to form N_2O^* (Equation (7)). Finally, the adsorbed N_2O is reduced to N_2 and CO_2 by gaseous CO (Equation (8), where * denotes the adsorbed state).

$$NO \rightarrow NO^* \tag{5}$$

$$NO^* \rightarrow N^* + O^* \tag{6}$$

$$N^* + NO^* \rightarrow N_2O^* \tag{7}$$

$$N_2O^* + CO \rightarrow N_2 + CO_2 \tag{8}$$

The E-R mechanism is more likely to occur at low temperatures. Shan et al. [24] synthesized α-MnO_2 nanorods using a colloidal method and studied their catalytic behavior in the reduction of NO by CO. They found that the reduction of NO by CO on α-MnO_2 proceeds by initially generating N_2O as an intermediate, which then dissociates into N_2 (Figure 4a). During the catalytic process, Mn^{2+} is easily oxidized to Mn^{3+}, which can subsequently adsorb and dissociate NO, thus forming the intermediate N_2O. In situ investigations revealed the reconstruction of the α-MnO_2 surface, leading to the formation of a new active surface phase consisting of Mn^{3+} and OVs serving as active sites. Among the various TiO_2-supported transition metal-based catalysts, MnO_x/TiO_2 shows the most promising CO-SCR activity at low temperatures. As illustrated in Figure 4b, the presence of NO on MnO_x/TiO_2 inhibits the oxidation of CO to CO_2. Instead, in situ Fourier transform infrared (FT-IR) studies found that the dissociation of NO takes place at the reduced Mn site, resulting in the generation of N_2O. N_2O acts as an intermediate and then reacts with CO to form N_2 [25].

When Mn in the Mn-based catalysts is present in a high valence state with numerous OVs, it leads to an increased adsorption capacity for NO and CO. This condition also facilitates the occurrence of the L-H mechanism. In the L-H mechanism, the formation process of N_2O occurs as follows: NO and CO are adsorbed on the catalyst to form NO* and CO* (Equations (9) and (10)); then, NO* dissociated into N* and O* (Equation (11)). Subsequently, N* combines with another adsorbed NO to produce N_2O^* (Equation (12)), which is further reduced by adsorbed CO to yield N_2 and CO_2 (Equation (13)):

$$NO \rightarrow NO^* \quad (9)$$

$$CO \rightarrow CO^* \quad (10)$$

$$NO^* \rightarrow N^* + O^* \quad (11)$$

$$N^* + NO^* \rightarrow N_2O^* \quad (12)$$

$$N_2O^* + CO^* \rightarrow N_2 + CO_2 \quad (13)$$

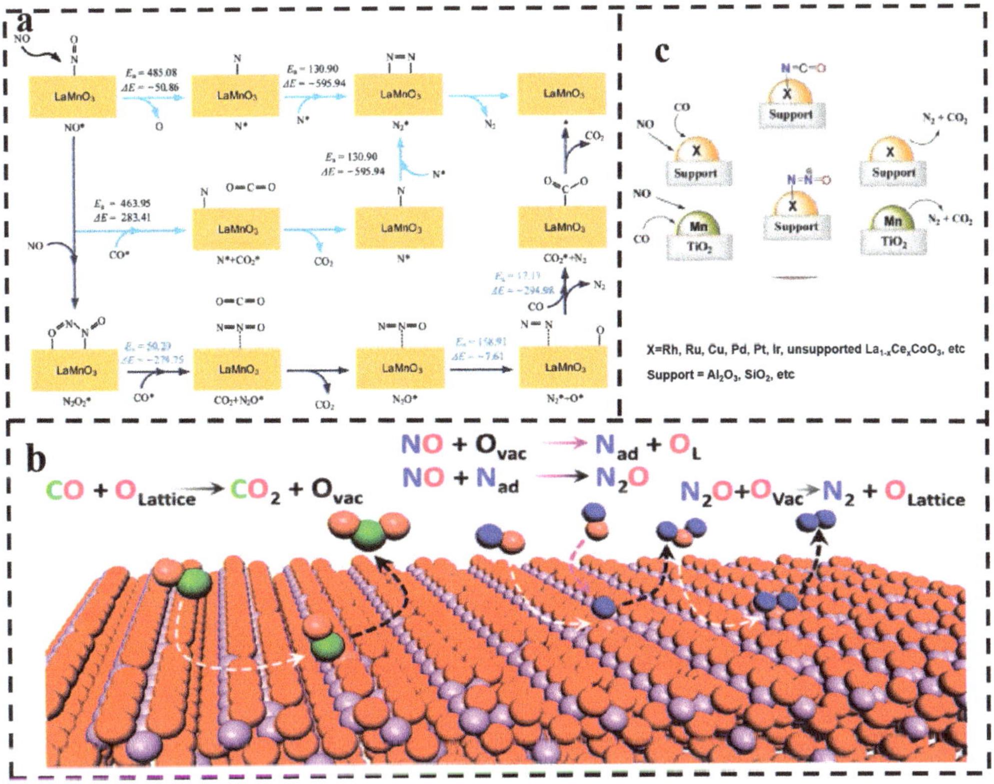

Figure 4. (**a**) Reaction mechanism on α-MnO_2 nanorod catalyst [24]; (**b**) reaction mechanism on MnO_x/TiO_2 catalyst [25] (**c**) reaction mechanism on $LaMnO_3$ catalyst [58].

For the $LaMnO_3$ catalyst, the Mn site on its surface serves as the primary active site for NO reduction. In this case, $N_2O_2^*$ generated through the NO coupling reaction acts as a crucial intermediate in the CO-SCR process. $N_2O_2^*$ subsequently reacts with chemisorbed CO to produce CO_2 and N_2O^* intermediates. The reaction path of CO-SCR on the $LaMnO_3$ surface consists of three sequential steps: $2NO^* \rightarrow N_2O_2^*$, $N_2O_2^* + CO^* \rightarrow N_2O^* + CO_2$, and $N_2O^* + CO^* \rightarrow N_2 + CO_2^*$ (Figure 4c). The predominance of the bimolecular coupling mechanism in the CO-SCR reaction on the $LaMnO_3$ catalyst is attributed to the higher activation energy barrier for NO decomposition to N_2O on the $LaMnO_3$ surface compared to that of the $N_2O_2^*$ formation [58]. In the case of the NiMn-MOF-74 catalyst, the in situ FT-IR results showed that NO molecules preferentially adsorbed on the catalyst surface and dissociated to N* and O* under the action of surface SSOVs. This process weakens the competitive adsorption of NO and CO on the catalyst. The dissociated NO undergoes subsequent reactions with CO to form N_2O intermediates, which are ultimately transformed into N_2 [31].

The presence of abundant variable valence states plays a crucial role in the catalytic activity and selectivity of Mn-based catalysts in the presence of oxygen. Mn^{3+} ions exhibit efficient adsorption and dissociation of NO. Lewis acid sites are very important for the CO-SCR reaction of MnO_x-based catalysts by enhancing the adsorption capability of CO molecules [41]. For the V_O-OMS-2 catalyst, the reaction mechanism can be described as follows:

$$Mn^{3+}\ (s) + NO\ (g) \rightarrow N_2\ (g) + Mn^{4+}\ (s) + O_2\ (g) \quad (14)$$

$$Mn^{4+}\ (s) + O_2\ (g) + CO\ (g) \rightarrow CO_2\ (g) + Mn^{3+}\ (s) \quad (15)$$

In summary, different Mn-based catalysts show different reaction mechanisms. When an N_2O intermediate is generated, the selectivity of the CO-SCR reaction may be compromised if N_2O, functioning as an intermediate, is unable to undergo further reduction by CO to form N_2 and CO_2. Therefore, it is necessary to modify the composition, structure, oxidation state, and acid site of the Mn-based catalysts to facilitate the conversion of N_2O into N_2. The diverse valence states of Mn play a crucial role in determining the catalytic activity and selectivity of the catalyst during the CO-SCR reaction. By adjusting the catalyst to achieve an appropriate oxidation state, it demonstrates improved resistance to oxygen.

5. Challenges

The concentrations of NO and CO in various industrial flue gases are different. Typically, the concentration of NO ranges between 500 ppm and 1600 ppm, whereas the CO concentration falls within the range of 1 vol% to 1.25 vol%. Besides CO and NO gases, there are also significant amounts of O_2 ($\geq$5 vol%) and H_2O ($\geq$10 vol%), as well as a minor quantity of SO_2 ($\leq$50 ppm), which have a discernible impact on the catalytic activity of the catalyst in the CO-SCR reaction. For instance, an excess of oxygen may lead to favoring of the oxidation of CO and NO. SO_2 tends to form sulfates on the catalyst's surface, which can obstruct the surface active sites. H_2O can potentially compete with CO or NO for adsorption sites. On the other hand, the high catalytic activity required at low temperatures poses another challenge for the reaction. While the flue gas emission temperature in the power industry typically exceeds 350 °C, most industrial furnaces operate at lower temperatures, with flue gas emissions ranging between 200 and 230 °C. The flue gas temperature produced by industrial processes like steel sintering is typically below 150 °C. Hence, the catalytic efficiency of the catalyst in the CO-SCR reaction at low temperatures is of significant importance. Although the precious metal catalysts described in the literature exhibit commendable catalytic activity under oxygen-rich conditions, their performance notably diminishes at lower temperatures. In contrast, low-cost Mn-based catalysts demonstrate significant potential in catalyzing the CO-SCR reaction at low temperatures (Table 2). Further investigation in this field is anticipated to result in the creation of catalysts capable of functioning at reduced reaction temperatures in oxygen-rich environments.

Table 2. Comparison of catalytic properties of Mn-based catalysts and other transition metal catalysts.

Catalyst	Gas Composition (%)					GHSV or WHSV	Temperature (°C)	NO Conversion (%)	CO Conversion (%)	N_2 Selectivity (%)	Ref.
	NO	CO	O_2	SO_2	H_2O						
MnO_x/TiO_2	0.4	0.4	2.0	---	---	50,000 h^{-1}	175	100	---	100	[29]
NiMn-MOF-74	0.5	1.0	---	0.1	5.0	30,000 h^{-1}	200	100	---	---	[25]
Cu-Ce-Fe-Mn/TiO_2	0.02	0.02	1.0	---	---	10,000 h^{-1}	200	100	82	---	[28]
Mn-CeO_2@Co_3O_4	1.0	2.0	5.0	0.05	10.0	24,000 h^{-1}	200	82	100	78	[38]
$Mn_xCo_{3-x}O4$	1.0	2.0	5.0	0.05	5.0	20,000 h^{-1}	180	100	---	100	[37]
$Cu_{1.2}FeMg_{2.8}$-LDO	0.03	0.12	1.0	0.05	5.0	60,000 mL $g^{-1}h^{-1}$	225	100	---	100	[59]
Ce-Cu-BTC	0.1	0.1	5	---	---	---	300	100	70	100	[60]
CuO/TiO_2	1.0	1.0	0.5	---	---	12,000 h^{-1}	250	100	---	90	[61]
NiFe/CeO_2	0.05	0.5	0.5	---	---	22,800 h^{-1}	250	100	---	100	[62]
Fe-Co/ASC	0.1	0.5	0.1	---	---	20,000 h^{-1}	300	100	100	100	[63]
Fe@CeO_2-ZIF-8	0.5	0.1	---	---	5.0	22,800 h^{-1}	300	100	50	99	[64]
Co_3O_4-CeO_2-IOV	0.1	0.2	5.0	0.05	10.0	20,000 h^{-1}	200	100	100	100	[65]
Co SA + CoO_x NC/CZO	0.1	0.2	5.0	0.05	5.0	20,000 h^{-1}	250	100	100	100	[12]

5.1. O_2

Due to the abundance of O_2, CO and NO may exhibit a preference for reacting with O_2. This can lead to enhanced CO consumption while preventing the reduction of NO to N_2. In particular, the hindrance of the CO-SCR reaction by an excess of O_2 (typically exceeding 5 vol.%) poses a significant challenge for the practical application of CO-SCR. On a Ni-Mn_2 catalyst, the presence of O_2 could inhibit the deNO_x process. Under the condition of a high O_2 concentration, part of NO could be converted to NO_2. Especially at high temperatures, the amount of NO_2 produced was greater than that of N_2, leading to the poor N_2 selectivity of the Ni-Mn_2 catalyst [66]. In the presence of 6% O_2, the CO conversion of the Mn-Ce-Fe-Co/TiO_2 catalyst reached 92%. Nevertheless, the NO conversion by Mn-Ce-Fe-Co/TiO_2 decreased as the O_2 content increased. This could be elucidated by the observation that in the presence of O_2, a portion of NO was oxidized to NO_2 instead of being converted to N_2, consequently leading to reduced NO_x conversion [34]. At an O_2 concentration of 8%, the NO conversion by a 5Fe-10Mn/AC catalyst was high in the reaction's early stages. However, its stability was notably poor, decreasing to 42% at 230 °C after only 7 min [67]. There are abundant variable valence states in Mn-based catalysts, and Mn^{3+} has been shown to be able to effectively adsorb and activate NO and strengthen N-O bond cleavage. Thus, regulating the valence state of Mn in Mn-based catalysts may be a solution to their poor catalytic activity and selectivity under oxygen-rich conditions.

Currently, it has been established that the dispersion of nanoparticles into single atoms (SAs) and the manipulation of the charge state of SAs can enhance the resistance of catalysts to O_2 in the CO-SCR reaction to a certain degree [9,10,68]. The underlying principle is that SACs possess an adaptable coordination structure, offering a favorable opportunity to modulate their catalytic efficacy [69]. Theoretically, Mn displays a wide range of variable valence states, and by adjusting the coordination environment of Mn SAs, distinct valence states of Mn can be achieved. The distinctive valence of Mn affects its adsorption capacity for reaction gases differently than that of O_2, potentially improving the O_2 resistance of Mn-based catalysts. While the utilization of Mn SAs in CO-SCR reactions has not been documented in the literature, studies have shown that Mn^{4+} in MnO_2 exhibits a higher affinity for NO compared to O_2, leading to a decrease in O_2 absorption [70]. Theoretically, the dispersion of Mn species in Mn-based catalysts into single atoms and the exposure of more Mn^{4+} species may enhance their O_2 resistance.

5.2. H_2O

In actual flue gas, a significant quantity of water is typically present. The presence of water will compete with NO for adsorption sites, leading to a reduction in NO conversion and N_2 selectivity [71]. However, the impact of water on the catalyst is reversible, and the catalyst can regain certain performance levels after the water is removed. For a $CuMnO_2$ nanosheet catalyst with the synergistic effect, it could achieve 100% NO conversion to N_2 at 310 °C. However, when 5% water vapor was introduced, the N_2 selectivity suddenly dropped from 90.2% to 10.1%, although the NO conversion increased from 89.2% to 100% (Figure 5a) [72]. When MnO_x was doped into CuO/CeO_2, the Cu/CeMn-10:1 catalyst (the molar ratio of CuO/CeO_2 to MnO_x was 10:1) showed the best catalytic activity in the NO + CO reaction, with 100% NO conversion and 83% N_2 selectivity at 220 °C in the absence of oxygen. After the introduction of 10 vol% H_2O into the reaction system, both the NO conversion and N_2 selectivity decreased, especially the N_2 selectivity [73]. The Mn-CeO_2@Co_3O_4 catalyst achieved 82% NO conversion and 78% N_2 selectivity at 200 °C with 5% O_2 present. When 10 vol.% H_2O was introduced, the NO conversion decreased to 65% (Figure 5b). The main reason for this decline was identified as the competitive adsorption of NO and H_2O [38].

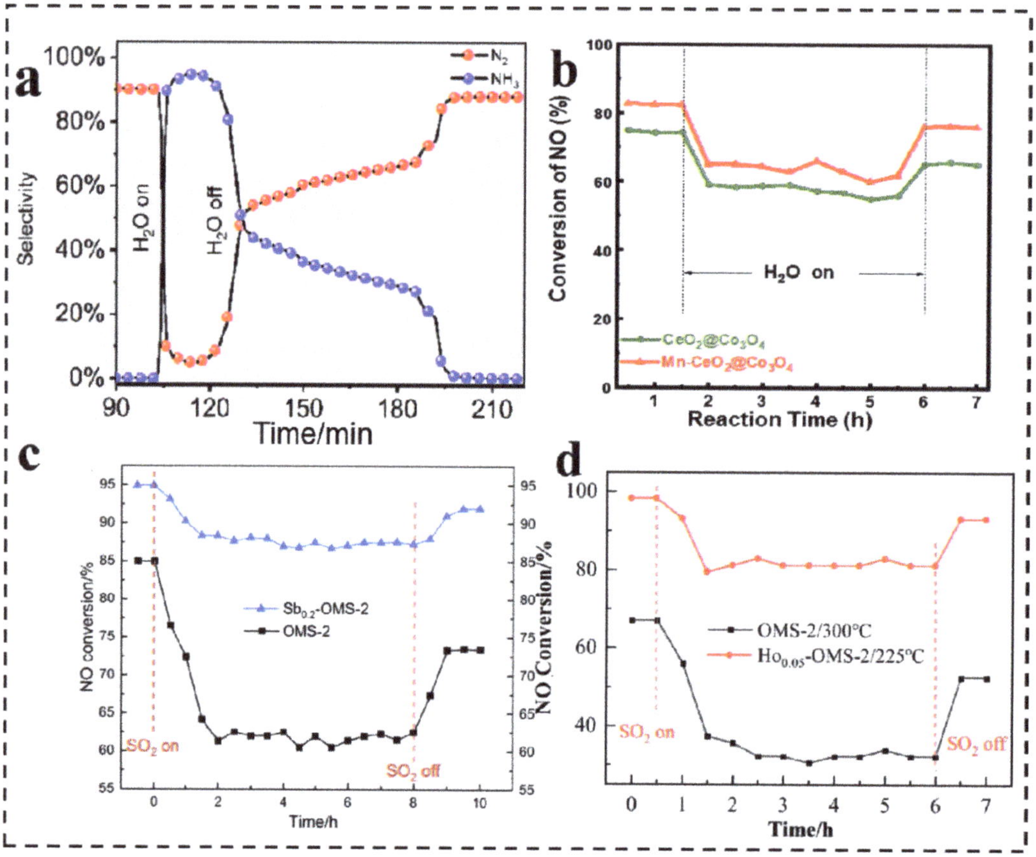

Figure 5. (**a**) N_2 and NH_3 selectivity of $CuMnO_2$/SSM sample in the CO-SCR test under the conditions of 5% water, 1000 ppm NO, and 2000 ppm CO [72]; (**b**) effect of H_2O on NO conversion on Mn-CeO_2@Co_3O_4 catalyst activity at 200 °C [38]; (**c**) sulfur resistance stability test of OMS-2 and $Sb_{0.2}$-OMS-2 [42]; (**d**) influence of SO_2 on $Ho_{0.05}$-OMS-2 catalyst activity. Reaction conditions: NO = 0.05%, CO = 0.05%, O_2 = 2%, SO_2 = 0.02%, N_2 = balance, and GHSV = 15,000 h^{-1} [43].

5.3. SO_2

Small quantities of SO_2 are present in flue gas. In the case of Mn-based catalysts, their ability to tolerate sulfur is comparatively low. This is due to the formation of sulfides by SO_2 on the catalyst surface. In comparison to undoped OMS-2, an Sb-doped $Sb_{0.2}$-OMS-2 catalyst exhibited an increased proportion of OVs on the catalyst surface and enriched Lewis acid sites, thus facilitating the conversion of NO. The NO conversion improved from 60% in OMS-2 to 83% in Sb-doped $Sb_{0.2}$-OMS-2 at 150 °C and from 80% to 95% at 300 °C. When SO_2 was introduced, the NO conversion of OMS-2 plummeted. While the introduction of Sb led to some improvement in sulfur resistance, the conversion of NO also showed a continuous decline, and the sulfur resistance was still limited (Figure 5c), which required further improvement [42]. Moreover, incorporating the metal Ho into OMS-2 also resulted in an enhancement in catalytic activity. Nonetheless, its resistance to sulfur was comparable to that of Sb0.2-OMS-2 (Figure 5d) [43]. Likewise, the presence of SO_2 could have a substantial impact on the catalytic performance of Mn-Ce-Fe-Co/TiO_2 catalysts. When 50 ppm of SO_2 was introduced, the NO conversion of Mn-Ce-Fe-Co/TiO_2 experienced a notable decrease, with the lowest NO conversion being only 20% [34]. Indeed, as mentioned above, the addition of metal can enhance sulfur resistance to some degree; however, there remains significant potential for further enhancement.

By coordinating with a low-electronegativity atom, the central atom acquires a negative charge, thereby enhancing the resistance to O_2 and SO_2 of the catalyst to some degree. This is based on the fact that the negatively charged single atom can boost the adsorption of NO while reducing the adsorption of O_2. Additionally, the adsorption capability of SO_2 is also diminished [9,69].

6. Conclusions and Perspectives

In recent years, CO-SCR technology has emerged as a prominent research focus in the denitrification field. Currently, the crucial aspect for the industrial implementation of CO-SCR technology lies in the development of cost-effective and high-performance catalysts. Mn-based catalysts offer advantages such as affordability and high activity at low temperatures. However, their development is hindered primarily by poor N_2 selectivity and stability in the presence of oxygen, as well as their limited water and sulfur resistance. In order to promote their industrialized application, it is crucial to tackle these issues. Subsequent research endeavors may be pursued with a focus on the following aspects:

(i) Presently, numerous studies have been published on SACs of precious metals, such as those based on Ir, Pd, and Pt. The dispersion of transition metals into single atoms to form effective catalysts has also been documented. For instance, a single-atom Mn catalyst has been utilized in various reactions such as oxygen reduction reactions [74] and the photocatalytic removal of water-based organics [75]. These reports provide support for the feasibility of synthesizing Mn SACs. However, currently, there is a lack of literature regarding the use of Mn SACs for the CO-SCR reaction. Previously, under the guidance of DFT, our research group successfully developed a catalyst consisting of Co SAs and CoO_x nanoclusters (NCs) with dual active centers co-anchored on the CZO support, which achieved superior catalytic performance at a broad temperature range between 250 and 400 °C and under 5 vol% O_2 conditions [12]. Therefore, by leveraging the benefits of the high low-temperature activity of Mn-based catalysts, there is the potential to obtain low-temperature performance. Furthermore, combining the Mn SAs with a support that possesses oxygen resistance may lead to outstanding low-temperature performance under oxygen-rich conditions.

(ii) Our group discovered that modifying the type and quantity of coordination atoms can alter their valence state in SACs. The adsorption behaviors of NO, CO, O_2, and SO_2 on negatively charged SACs significantly differ from those on positively charged SACs or nanoparticle catalysts [9,10]. For Mn-based catalysts, upon dispersing them into single atoms, the coordination type of the Mn atoms is adjusted to induce a negative charge, with the extent of negativity regulated by the quantity of coordinating atoms. By synthesizing negatively charged Mn SACs, it is possible to enhance their antioxidant properties and resistance to sulfur dioxide.

(iii) Various in situ characterization methods, such as in situ electron paramagnetic resonance (EPR), in situ transmission electron microscope (TEM), and in situ X-ray photoelectron spectroscopy (XPS), are frequently employed for defect identification of reaction intermediates and the changes in the structure and oxidation states of catalysts. Mn-based catalysts usually exhibit a significant number of defects, underscoring the critical necessity of accurately determining the defect type and density to gain insights into the underlying reaction mechanisms. Utilizing in situ techniques for observing the changes throughout the reaction process and integrating a variety of in situ characterization methods to thoroughly investigate the CO-SCR reaction process and elucidate the dynamic structure–activity relationship are crucial aspects. Hence, various in situ characterization techniques can be employed to observe reaction intermediates and clarify the reaction pathways for Mn-based catalysts, which is significant for designing more efficient Mn-based catalysts.

(iv) Our group employed solid-state ball milling technology to synthesize Co-based catalysts with excellent oxygen resistance and relatively high activity levels at low temperatures [65]. The application of catalysts in industry necessitates simple preparation methods, superior catalytic performance, reproducibility, cost efficiency, and scalability for mass production. Hence, the development of straightforward, replicable, scalable, and promising preparation methodologies for Mn-based catalyst are imperative.

Author Contributions: D.L.: Writing—original draft, Visualization. M.C. and H.W.: Date curation. C.L. and B.L.: Resources. G.D. and S.H.: Software. Y.L.: Supervision. Y.J.: Writing—review and editing, Supervision, Visualization, Funding acquisition. All authors have read and agreed to the published version of the manuscript.

Funding: This research was funded by the National Natural Science Foundation of China grant number 21978299 from Y.J.; the Research Foundation for Advanced Talents of Beijing Technology and Business University grant number 19008020159 from Y.J.; the Graduate Research Ability Enhancement Program of Beijing Technology and Business University grant number 19008024042 from D.L.; and the Shccig-Qinling Program grant number SMYJY202400134C from Y.J.

Conflicts of Interest: The authors declare no competing interests.

References

1. Skalska, K.; Miller, J.S.; Ledakowicz, S. Trends in NO_x abatement: A review. *Sci. Total Environ.* **2010**, *408*, 3976–3989. [CrossRef] [PubMed]
2. Yun, L.; Li, Y.; Zhou, C.; Lan, L.; Zeng, M.; Mao, M.; Liu, H.; Zhao, X. The formation of CuO/OMS-2 nanocomposite leads to a significant improvement in catalytic performance for NO reduction by CO. *Appl. Catal. A Gen.* **2017**, *530*, 1–11.
3. Wang, C.; Wang, W.; Sardans, J.; An, W.; Zeng, C.; Abid, A.A.; Penuelas, J. Effect of simulated acid rain on CO_2, CH_4 and N_2O fluxes and rice productivity in a subtropical Chinese paddy field. *Environ. Pollut.* **2018**, *243*, 1196–1205. [CrossRef] [PubMed]
4. Andana, T.; Rappé, K.G.; Gao, F.; Szanyi, J.; Pereira-Hernandez, X.; Wang, Y. Recent advances in hybrid metal oxide-zeolite catalysts for low-temperature selective catalytic reduction of NO_x by ammonia. *Appl. Catal. B Environ.* **2021**, *291*, 120054. [CrossRef]
5. Inomata, Y.; Kubota, H.; Hata, S.; Kiyonaga, E.; Morita, K.; Yoshida, K.; Sakaguchi, N.; Toyao, T.; Shimizu, K.I.; Ishikawa, S.; et al. Tungsten-substituted vanadium oxide for low-temperature NO_x removal in the presence of water. *Nat. Commun.* **2021**, *12*, 557. [CrossRef] [PubMed]
6. Ren, D.; Gui, K.; Gu, S.; Wei, Y. Study of the nitric oxide reduction of SCR-NH_3 on γ-Fe_2O_3 catalyst surface with quantum chemistry. *Appl. Surf. Sci.* **2020**, *509*, 144659. [CrossRef]
7. Hu, Y.; Griffiths, K.; Norton, P.R. Surface science studies of selective catalytic reduction of NO: Progress in the last ten years. *Surf. Sci.* **2009**, *603*, 1740–1750. [CrossRef]
8. Jeon, J.; Ham, H.; Xing, F.; Nakaya, Y.; Shimizu, K.-i.; Furukawa, S. PdIn-Based Pseudo-Binary Alloy as a Catalyst for NO_x Removal under Lean Conditions. *ACS Catal.* **2020**, *10*, 11380–11384. [CrossRef]
9. Ji, Y.; Liu, S.; Song, S.; Xu, W.; Li, L.; Zhang, Y.; Chen, W.; Li, H.; Jiang, J.; Zhu, T.; et al. Negatively Charged Single-Atom Pt Catalyst Shows Superior SO_2 Tolerance in NO_x Reduction by CO. *ACS Catal.* **2023**, *13*, 224–236. [CrossRef]
10. Ji, Y.; Liu, S.; Zhu, H.; Xu, W.; Jiang, R.; Zhang, Y.; Yu, J.; Chen, W.; Jia, L.; Jiang, J.; et al. Isolating Contiguous Ir Atoms and Forming Ir-W Intermetallics with Negatively Charged Ir for Efficient NO Reduction by CO. *Adv. Mater.* **2022**, *34*, 2205703. [CrossRef]
11. Hungría, A.B.; Fernández-García, M.; Anderson, J.A.; Martínez-Arias, A. The effect of Ni in Pd-Ni/(Ce, Zr)O_x/Al_2O_3 catalysts used for stoichiometric CO and NO elimination. Part 2: Catalytic activity and in situ spectroscopic studies. *J. Catal.* **2005**, *235*, 262–271. [CrossRef]
12. Liu, S.; Ji, Y.; Liu, B.; Xu, W.; Chen, W.; Yu, J.; Zhong, Z.; Xu, G.; Zhu, T.; Su, F. Co Single Atoms and CoO_x Nanoclusters Anchored on $Ce_{0.75}Zr_{0.25}O_2$ Synergistically Boosts the NO Reduction by CO. *Adv. Funct. Mater.* **2023**, *33*, 2303297. [CrossRef]
13. He, J.; Kang, R.; Wei, X.; Huang, J.; Feng, B.; Nam Hui, K.; San Hui, K.; Wu, D. Comparative study of $MCe_{0.75}Zr_{0.25}O_y$ (M = Cu, Mn, Fe) catalysts for selective reduction of NO by CO: Activity and reaction pathways. *Carbon Resour. Convers.* **2021**, *4*, 205–213. [CrossRef]
14. Wang, X.; Li, X.; Mu, J.; Fan, S.; Wang, L.; Gan, G.; Qin, M.; Li, J.; Li, Z.; Zhang, D. Facile Design of Highly Effective $CuCe_xCo_{1-x}O_y$ Catalysts with Diverse Surface/Interface Structures toward NO Reduction by CO at Low Temperatures. *Ind. Eng. Chem. Res.* **2019**, *58*, 15459–15469. [CrossRef]
15. Hamada, H.; Haneda, M. A review of selective catalytic reduction of nitrogen oxides with hydrogen and carbon monoxide. *Appl. Catal. A Gen.* **2012**, *421*, 1–13. [CrossRef]
16. Wang, J.; Gao, F.; Dang, P.; Tang, X.; Lu, M.; Du, Y.; Zhou, Y.; Yi, H.; Duan, E. Recent advances in NO reduction with CO over copper-based catalysts: Reaction mechanisms, optimization strategies, and anti-inactivation measures. *Chem. Eng. J.* **2022**, *450*, 137374. [CrossRef]
17. Chen, X.; Liu, Y.; Liu, Y.; Lian, D.; Chen, M.; Ji, Y.; Xing, L.; Wu, K.; Liu, S. Recent Advances of Cu-Based Catalysts for NO Reduction by CO under O_2-Containing Conditions. *Catalysts* **2022**, *12*, 1402. [CrossRef]

18. Liu, S.; Gao, J.; Xu, W.; Ji, Y.; Zhu, T.; Xu, G.; Zhong, Z.; Su, F. A review of the catalysts used in the reduction of NO by CO for gas purification. *Chem. Eng. J.* **2024**, *486*, 150285. [CrossRef]
19. Wang, Y.; Xu, W.; Liu, H.; Chen, W.; Zhu, T. Catalytic removal of gaseous pollutant NO using CO: Catalyst structure and reaction mechanism. *Environ. Res.* **2024**, *246*, 150285. [CrossRef]
20. Gholami, Z.; Luo, G.; Gholami, F.; Yang, F. Recent advances in selective catalytic reduction of NO_x by carbon monoxide for flue gas cleaning process: A review. *Environ. Sci. Pollut. Res.* **2021**, *63*, 68–119. [CrossRef]
21. Li, S.; Wang, F.; Ng, D.; Shen, B.; Xie, Z. Stability challenges and prospects for the industrial application of non-noble catalysts for selective catalytic reduction of NO_x by CO (CO-SCR). *ChemCatChem* **2024**, *16*, e202301246. [CrossRef]
22. Xie, J.; Fang, D.; He, F.; Chen, J.; Fu, Z.; Chen, X. Performance and mechanism about MnO_x species included in MnO_x/TiO_2 catalysts for SCR at low temperature. *Catal. Commun.* **2012**, *28*, 77–81. [CrossRef]
23. Jin, Q.; Han, L.; Li, N.; Zhang, T.; Gao, E.; Yao, M.; Yao, S.; Wu, Z.; Li, J.; Zhu, J.; et al. Exploring the influence of chemical state of Cu species on CO-SCR performance in spinel-type CuM_2O_4 (M = Co, Mn, Fe, Ni, and Cr): The synergy between Cu^{2+} and surface oxygen vacancy. *Fuel* **2024**, *360*, 130553. [CrossRef]
24. Shan, J.; Zhu, Y.; Zhang, S.; Zhu, T.; Rouvimov, S.; Tao, F. Catalytic Performance and in Situ Surface Chemistry of Pure α-MnO_2 Nanorods in Selective Reduction of NO and N_2O with CO. *J. Phy. Chem. C* **2013**, *117*, 8329–8335. [CrossRef]
25. Boningari, T.; Pavani, S.M.; Ettireddy, P.R.; Chuang, S.S.C.; Smirniotis, P.G. Mechanistic investigations on NO reduction with CO over Mn/TiO_2 catalyst at low temperatures. *Mol. Catal.* **2018**, *451*, 33–42. [CrossRef]
26. Zhang, L.; Li, B.; Liu, C.; Tian, H.; Hong, M.; Yin, X.; Feng, X. NO reduction with CO over a highly dispersed Mn/TiO_2 catalyst at low temperature: A combined experimental and theoretical study. *Nanotechnology* **2021**, *32*, 505717. [CrossRef] [PubMed]
27. Sreekanth, P.M.; Smirniotis, P.G. Selective reduction of NO with CO over titania supported transition metal oxide catalysts. *Catal. Lett.* **2008**, *122*, 37–42. [CrossRef]
28. Zhang, Y.; Zhao, L.; Chen, Z.; Li, X. Promotional effect for SCR of NO with CO over MnO_x-doped Fe_3O_4 nanoparticles derived from metal-organic frameworks. *Chin. J. Chem. Eng.* **2022**, *46*, 113–125. [CrossRef]
29. Liu, T.; Yao, Y.; Wei, L.; Shi, Z.; Han, L.; Yuan, H.; Li, B.; Dong, L.; Wang, F.; Sun, C. Preparation and Evaluation of Copper Manganese Oxide as a High-Efficiency Catalyst for CO Oxidation and NO Reduction by CO. *J. Phy. Chem. C* **2017**, *121*, 12757–12770. [CrossRef]
30. Li, D.; Yu, Q.; Li, S.-S.; Wan, H.-Q.; Liu, L.-J.; Qi, L.; Liu, B.; Gao, F.; Dong, L.; Chen, Y. The Remarkable Enhancement of CO-Pretreated CuO-Mn_2O_3/γ-Al_2O_3 Supported Catalyst for the Reduction of NO with CO: The Formation of Surface Synergetic Oxygen Vacancy. *Chem. Eur. J.* **2011**, *17*, 5668–5679. [CrossRef]
31. Shi, Y.; Chu, Q.; Xiong, W.; Gao, J.; Huang, L.; Zhang, Y.; Ding, Y. A new type bimetallic NiMn-MOF-74 as an efficient low-temperatures catalyst for selective catalytic reduction of NO by CO. *Chem. Eng. Process.* **2021**, *159*, 108232. [CrossRef]
32. Li, L.; Wang, Y.; Zhang, L.; Yu, Y.; He, H. Low-Temperature Selective Catalytic Reduction of NO_x on MnO_2 Octahedral Molecular Sieves (OMS-2) Doped with Co. *Catalysts* **2020**, *10*, 396. [CrossRef]
33. Fan, F.; Wang, L.; Wang, L.; Liu, J.; Wang, M. Low-Temperature Selective NO Reduction by CO over Copper-Manganese Oxide Spinels. *Catalysts* **2022**, *12*, 591. [CrossRef]
34. Pan, K.L.; Young, C.W.; Pan, G.T.; Chang, M.B. Catalytic reduction of NO by CO with Cu-based and Mn-based catalysts. *Catal. Today* **2020**, *348*, 15–25. [CrossRef]
35. Shi, X.; Chu, B.; Wang, F.; Wei, X.; Teng, L.; Fan, M.; Li, B.; Dong, L.; Dong, L. Mn-Modified CuO, $CuFe_2O_4$, and γ-Fe_2O_3 Three-Phase Strong Synergistic Coexistence Catalyst System for NO Reduction by CO with a Wider Active Window. *ACS Appl. Mater. Int.* **2018**, *10*, 40509–40522. [CrossRef] [PubMed]
36. Wang, J.; Xing, Y.; Su, W.; Li, K.; Zhang, W. Bifunctional Mn-Cu-CeO_x/γ-Al_2O_3 catalysts for low-temperature simultaneous removal of NO_x and CO. *Fuel* **2022**, *321*, 124050. [CrossRef]
37. Gholami, F.; Gholami, Z.; Tomas, M.; Vavrunkova, V.; Mirzaei, S.; Vakili, M. Promotional Effect of Manganese on Selective Catalytic Reduction of NO by CO in the Presence of Excess O_2 over M@La-Fe/AC (M = Mn, Ce) Catalyst. *Catalysts* **2020**, *10*, 1322. [CrossRef]
38. Meng, Y.; Liu, S.; Wang, Y.; Xu, W.; Gao, J.; Yu, S.; Su, F.; Zhu, T. Hollow Mn-doped CeO_2@Co_3O_4 catalyst for NO reduction by CO. *J. Catal.* **2024**, *430*, 115311. [CrossRef]
39. Li, X.; Ren, S.; Chen, Z.; Wang, M.; Chen, L.; Chen, H.; Yin, X. A Review of Mn-Based Catalysts for Abating NO_x and CO in Low-Temperature Flue Gas: Performance and Mechanisms. *Molecules* **2023**, *28*, 6885. [CrossRef]
40. Wei, L.; Liu, T.; Wu, Y.; Liu, H.; Dong, L.; Li, B. Design of co-symbiotic Fe_xMn_yO catalysts for NO reduction by CO. *Catal. Sci. Technol.* **2020**, *10*, 7894–7903. [CrossRef]
41. He, H.; Wang, Y.; Fu, W.; Zhang, L.; Zeng, J.; Zhu, G. Study on the CO-SCR anti-sulfur and denitration performance of V-doped OMS-2 catalysts. *Ceram. Inter.* **2021**, *47*, 33120–33126. [CrossRef]
42. Wang, Y.; Zhang, C.; Zhang, L.; Zeng, J.; He, H. Anti-sulfur selective catalytic reduction of NO_x on Sb-doped OMS-2. *Appl. Catal. A Gen.* **2022**, *641*, 118684. [CrossRef]
43. Luo, B.; Wang, Z.; Huang, J.; Ning, S.; Deng, W.; Zhao, B.; Su, Y. Study on CO-SCR denitrification performance of Ho-modified OMS-2 catalyst. *J. Environ. Chem. Eng.* **2023**, *11*, 119050. [CrossRef]
44. Yao, X.; Xiong, Y.; Sun, J.; Gao, F.; Deng, Y.; Tang, C.; Dong, L. Influence of MnO_2 modification methods on the catalytic performance of CuO/CeO_2 for NO reduction by CO. *J. Rare Earths* **2014**, *32*, 131–138. [CrossRef]

45. Deng, C.S.; Qian, J.N.; Yu, C.X.; Yi, Y.N.; Zhang, P.; Li, W.; Dong, L.H.; Li, B.; Fan, M.G. Influences of doping and thermal stability on the catalytic performance of $CuO/Ce_{20}M_1O_x$(M = Zr, Cr, Mn, Fe, Co, Sn) catalysts for NO reduction by CO. *Rsc. Adv.* **2016**, *6*, 113630–113647. [CrossRef]
46. Liu, S.; Ji, Y.; Xu, W.; Zhang, J.; Jiang, R.; Li, L.; Jia, L.; Zhong, Z.; Xu, G.; Zhu, T.; et al. Hierarchically interconnected porous $Mn_xCo_{3-x}O_4$ spinels for Low-temperature catalytic reduction of NO by CO. *J. Catal.* **2022**, *406*, 72–86. [CrossRef]
47. Wan, H.; Li, D.; Dai, Y.; Hu, Y.; Liu, B.; Dong, L. Catalytic behaviors of CuO supported on Mn_2O_3 modified γ-Al_2O_3 for NO reduction by CO. *J. Mol. Catal. A Chem.* **2010**, *332*, 32–44. [CrossRef]
48. Li, C.; Shi, Y.; Zhao, Q.D.; Xiong, W.; Ding, Y.; Sun, J.H.; Huang, Y.L.; Zhao, Z.F. A stable spherical MOF-derived Mn_x-Fe_2O_3/C catalysts for low-temperature CO-SCR. *Chem. Eng. J.* **2023**, *475*, 146388. [CrossRef]
49. Liu, T.; Wei, L.; Yao, Y.; Dong, L.; Li, B. La promoted CuO-MnO_x catalysts for optimizing SCR performance of NO with CO. *Appl. Surf. Sci.* **2021**, *546*, 148971. [CrossRef]
50. Wang, L.; Cheng, X.; Wang, Z.; Ma, C.; Qin, Y. Investigation on Fe-Co binary metal oxides supported on activated semi-coke for NO reduction by CO. *Appl. Catal. B Environ.* **2017**, *201*, 636–651. [CrossRef]
51. Huang, B.-f.; Wang, D.-f.; Li, J.-l.; Shi, Z. Cu-Fe/activated carbon catalyst for low-temperature CO-selective catalytic: Modification and denitration mechanism. *Environ. Prog. Sustain. Energy* **2022**, *41*, e13827. [CrossRef]
52. Li, J.; Zhu, J.; Fu, S.; Tao, L.; Chu, B.; Qin, Q.; Wang, J.; Li, B.; Dong, L. Insight into copper-cerium catalysts with different Cu valence states for CO-SCR and in-situ DRIFTS study on reaction mechanism. *Fuel* **2023**, *339*, 126962. [CrossRef]
53. Sun, Y.; Wu, Y.; Bai, Y.; Wu, X.; Wang, H.; Wu, Z. High performance iridium loaded on natural halloysite nanotubes for CO-SCR reaction. *Fuel* **2024**, *357*, 129938. [CrossRef]
54. Tao, L.; Wang, J.; Qin, Q.; Chu, B.; Gao, P.; Qiu, J.; Li, Q.; Du, X.; Dong, L.; Li, B. Simple anion-modified layered double oxides use for controlling Cu valence states for low-temperature CO-SCR. *Surf. Interfaces* **2024**, *44*, 103654. [CrossRef]
55. Li, W.; Liu, Z.; Yu, F.; Pan, K.; Zhao, H.; Gao, F.; Zhou, M.; Dai, B.; Dan, J. $CuCeO_x$/VMT powder and monolithic catalyst for CO-selective catalytic reduction of NO with CO. *New J. Chem.* **2022**, *46*, 10422–10432. [CrossRef]
56. Du, Y.; Gao, F.; Tang, X.; Yi, H.; Zhou, Y.; Zhao, S.; Duan, E.; Wang, J.; Qi, Z. Mechanistic insight into the enhanced NO reduction by CO over a pre-reduced Cu_xO_y-CeO_2 multiphase catalyst. *J. Environ. Chem. Eng.* **2023**, *11*, 110386. [CrossRef]
57. Liu, T.; Qian, J.; Yao, Y.; Shi, Z.; Han, L.; Liang, C.; Li, B.; Dong, L.; Fan, M.; Zhang, L. Research on SCR of NO with CO over the $Cu_{0.1}La_{0.1}Ce_{0.8}O$ mixed-oxide catalysts: Effect of the grinding. *Mol. Catal.* **2017**, *430*, 43–53. [CrossRef]
58. Yan, X.; Liu, J.; Yang, Y.; Wang, Z.; Zheng, Y. A catalytic reaction scheme for NO reduction by CO over Mn-terminated $LaMnO_3$ perovskite: A DFT study. *Fuel Process. Technol.* **2021**, *216*, 106798. [CrossRef]
59. Liu, J.; Zang, P.; Liu, X.; Mi, J.; Wang, Y.; Zhang, G.; Chen, J.; Zhang, Y.; Li, J. A novel highly active catalyst form CuFeMg layered double oxides for the selective catalytic reduction of NO by CO. *Fuel* **2022**, *317*, 123469. [CrossRef]
60. Zhang, Y.; Zhao, L.; Duan, J.; Bi, S. Insights into $deNO_x$ processing over Ce-modified Cu-BTC catalysts for the CO-SCR reaction at low temperature by in situ DRIFTS. *Sep. Purif. Technol.* **2020**, *234*, 116081. [CrossRef]
61. Amano, F.; Suzuki, S.; Yamamoto, T.; Tanaka, T. One-electron reducibility of isolated copper oxide on alumina for selective NO-CO reaction. *Appl. Catal. B Environ.* **2006**, *64*, 282–289. [CrossRef]
62. Li, S.; Chen, X.; Wang, F.; Xie, Z.; Hao, Z.; Liu, L.; Shen, B. Promotion effect of Ni doping on the oxygen resistance property of Fe/CeO_2 catalyst for CO-SCR reaction: Activity test and mechanism investigation. *J. Hazard. Mater.* **2022**, *431*, 128622. [CrossRef] [PubMed]
63. Cheng, X.; Wang, L.; Wang, Z.; Zhang, M.; Ma, C. Catalytic Performance of NO Reduction by CO over Activated Semicoke Supported Fe/Co Catalysts. *Ind. Eng. Chem. Res.* **2016**, *55*, 12710–12722. [CrossRef]
64. Li, S.; Wang, F.; Xie, Z.; Ng, D.; Shen, B. A novel core-shell structured Fe@CeO_2-ZIF-8 catalyst for the reduction of NO by CO. *J. Catal.* **2023**, *421*, 240–251. [CrossRef]
65. Liu, S.; Xue, W.; Ji, Y.; Xu, W.; Chen, W.; Jia, L.; Zhu, T.; Zhong, Z.; Xu, G.; Mei, D.; et al. Interfacial oxygen vacancies at Co_3O_4-CeO_2 heterointerfaces boost the catalytic reduction of NO by CO in the presence of O_2. *Appl. Catal. B Environ.* **2023**, *323*, 122151. [CrossRef]
66. Sun, P.; Li, X.; Cheng, X.; Wang, Z.; Wang, P. Transition metal modified Mn-based catalysts for CO-SCR in the presence of excess oxygen. *Process Saf. Environ. Protect.* **2023**, *176*, 389–401. [CrossRef]
67. Huang, B.; Shi, Z.; Yang, Z.; Dai, M.; Wen, Z.; Li, W.; Zi, G.; Luo, L. Mechanism of CO selective catalytic reduction denitration on Fe-Mn/AC catalysts at medium and low temperatures under oxygen atmosphere. *Chem. Eng. J.* **2022**, *446*, 137412. [CrossRef]
68. Ji, Y.; Chen, X.; Liu, S.; Song, S.; Xu, W.; Jiang, R.; Chen, W.; Li, H.; Zhu, T.; Li, Z.; et al. Tailoring the Electronic Structure of Single Ag Atoms in Ag/WO_3 for Efficient NO Reduction by CO in the Presence of O_2. *ACS Catal.* **2023**, *13*, 1230–1239. [CrossRef]
69. Zhang, X.; Gao, D.; Zhu, B.; Cheng, B.; Yu, J.; Yu, H. Enhancing photocatalytic H_2O_2 production with Au co-catalysts through electronic structure modification. *Nat. Commun.* **2024**, *15*, 3212. [CrossRef]
70. Li, H.; Wang, D.; Hui, S. Adsorption of NO and O_2 on MnO_2 and $(MnO_2)_3/Al_2O_3$. *Appl. Surf. Sci.* **2021**, *569*, 150994. [CrossRef]
71. Xia, Z.; Zhang, R.; Duan, J.; Liu, Y.; Li, Z.; Gou, X. Performance of Mn-Ce-Fe/FA Catalysts on Selective Catalytic Reduction of NO_x with CO under Different Atmospheres. *Energies* **2023**, *16*, 3859. [CrossRef]
72. Yang, Y.; Yang, L.; Cao, G.; Cui, Y.; Feng, L.; Jia, L.; Zhang, X.; Li, J.; Liu, B. Monolithic $CuMnO_2$-Nanosheet-Based Catalysts In Situ Grown on Stainless Steel Mesh for Selective Catalytic Reduction of NO with CO. *ACS Appl. Nano Mater.* **2023**, *6*, 4803–4811. [CrossRef]

73. Deng, C.; Huang, Q.; Zhu, X.; Hu, Q.; Su, W.; Qian, J.; Dong, L.; Li, B.; Fan, M.; Liang, C. The influence of Mn-doped CeO_2 on the activity of CuO/CeO_2 in CO oxidation and NO plus CO model reaction. *Appl. Surf. Sci.* **2016**, *389*, 1033–1049. [CrossRef]
74. Li, Y.; Li, Z.; Shi, K.; Luo, L.; Jiang, H.; He, Y.; Zhao, Y.; He, J.; Lin, L.; Sun, Z.; et al. Single-Atom Mn Catalysts via Integration with Mn Sub Nano-Clusters Synergistically Enhance Oxygen Reduction Reaction. *Small* **2024**, *20*, 2309727. [CrossRef]
75. Wang, J.; Xu, Z.; Xu, Y.; Liu, S.; Qiu, J.; Wang, Z.; Wang, Y.; Weng, Y.; Cao, H.; Wang, S.; et al. Mn single atoms for one-electron photoozonation of aqueous organics. *Appl. Catal. B Environ. Energy* **2024**, *349*, 123900. [CrossRef]

catalysts

MDPI

Article

Efficient Green Synthesis of Hydrazide Derivatives Using L-Proline: Structural Characterization, Anticancer Activity, and Molecular Docking Studies

Sobhi M. Gomha [1,*], Tariq Z. Abolibda [1], Awatif H. Alruwaili [2], Basant Farag [3], Waleed E. Boraie [4], Sami A. Al-Hussain [5], Magdi E. A. Zaki [5,*] and Ahmed M. Hussein [6,7]

1 Chemistry Department, Faculty of Science, Islamic University of Madinah, Madinah 42351, Saudi Arabia; t.z.a@iu.edu.sa
2 Department of Chemistry, Faculty of Science, Northern Border University, Arar 73222, Saudi Arabia; awatef.alrowili@nbu.edu.sa
3 Department of Chemistry, Faculty of Science, Zagazig University, Zagazig 44519, Egypt; basantfarag@zu.edu.eg
4 Department of Chemistry, College of Science, King Faisal University, Hofuf 31982, Saudi Arabia; wboraie@kfu.edu.sa
5 Department of Chemistry, Faculty of Science, Imam Mohammad Ibn Saud Islamic University (IMSIU), Riyadh 11623, Saudi Arabia; sahussain@imamu.edu
6 Chemistry Department, Faculty of Science, Beni-Suef University, Beni-Suef 62511, Egypt; ahussein@su.edu.sa
7 Chemistry Department, College of Science and Humanities—Al Quwaiiyah, Shaqra University, Al-Dawadmi 11911, Saudi Arabia
* Correspondence: smgomha@iu.edu.sa (S.M.G.); mezaki@imamu.edu.sa (M.E.A.Z.)

Abstract: Green synthesis using L-proline as an organocatalyst is crucial due to its reusability, mild conditions, clean reactions, easy workup, high purity, short reaction times, and high yields. However, existing methods often involve harsh conditions and longer reaction times. In this study, 2-cyano-*N′*-(2-cyanoacetyl)acetohydrazide (**3**) was prepared and condensed with various benzaldehyde derivatives to yield 2-cyano-*N′*-(2-cyano-3-phenylacryloyl)-3-phenylacrylohydrazide derivatives (**5a–e**, **7a,b**) using a grinding technique with moist L-proline. Additionally, three 2-cyano-*N′*-(2-cyano-3-heterylbut-2-enoyl)-3-heterylbut-2-enehydrazides (**9**, **11**, **13**) were synthesized by condensing compound 3 with respective (heteraryl)ketones (**8**, **10**, **12**) following the same method. The synthesized compounds were characterized using IR, NMR, and MS spectroscopy. L-proline's reusability was confirmed for up to four cycles without significant yield loss, showcasing the protocol's efficiency and sustainability. The new compounds were screened for anticancer activities against the HCT-116 colon carcinoma cell line using the MTT assay. Molecular docking studies revealed the binding conformations of the most potent compounds to the target protein (PDB ID 6MTU), correlating well with in vitro results. In silico ADMET analysis indicated favorable pharmacokinetic properties, highlighting these novel compounds as promising targeted anti-colon cancer agents.

Keywords: 2-cyanoacetohydrazide; condensation; grinding; hydrazides; anticancer activity; molecular docking study; in silico study

1. Introduction

Green chemistry aims to design products and processes that reduce or eliminate hazardous substances [1]. This field emphasizes the importance of developing safer chemicals, reducing waste, and improving energy efficiency. An overview of the twelve principles of green chemistry highlights key aspects such as the design of safer chemicals, waste reduction, and energy efficiency improvements. The focus is on creating environmentally sustainable chemical procedures that are safe for biological activity and industry-leading molecules [1–4]. It emphasizes alternatives such as reduced solvent use, high atom economy and selectivity, elimination of hazardous wastes, simple product separation and purification methods, and alternative energy sources for organic transformations [5–10]. Mechanosynthesis, using "ball milling" or simple grinding methods [11,12], fits this sustainable and "greener" definition. Manual grinding with a mortar and pestle is especially useful at the laboratory scale, offering an energy-efficient, economical, and ecologically advantageous process without external heating [13]. Toda and colleagues demonstrated that many organic exothermic reactions could be performed using mortar and pestle grinding with good yields [14], including condensation reactions like Schiff base and oxime formation [15]. Recently, this method has also proven efficient for constructing biologically important heteroaromatic compounds [16–19].

Organocatalysis has gained significant interest due to its ease of use, stability, non-toxicity, and cost-effectiveness. It does not require an inert atmosphere or anhydrous conditions, making it suitable for mild conditions [20,21]. The field has grown exponentially, with new catalysts and reactions producing asymmetric products in classical reactions such as Diels–Alder, Mannich, and Michael reactions [22,23]. L-proline, a naturally occurring amino acid, has gained significant attention as a catalyst in green chemistry due to its effectiveness and environmental friendliness [24–27]. As an organocatalyst, L-proline is used in various organic transformations, such as aldol reactions, Michael additions, and Mannich reactions. It offers advantages, including biodegradability, non-toxicity, high selectivity, and mild reaction conditions, making it a valuable tool for synthesizing complex molecules with high enantioselectivity and yield [24–27]. Proline's biological relevance extends to both natural and synthetic compounds. In nature, proline is a significant component of proteins, particularly collagen, contributing to structural stability. It also plays a role in neurotransmission and enzyme catalysis. In synthetic compounds, proline and its derivatives are used to design pharmaceuticals with enhanced stability, bioavailability, and therapeutic efficacy, including antiviral, anticancer, and anti-inflammatory agents [28–30].

Cancer continues to be a significant global health challenge, with its incidence rising in both developed and developing countries. Colorectal cancer (CRC) is the third most commonly diagnosed cancer and the fourth leading cause of cancer-related deaths worldwide [31]. This disease affects both men and women similarly, being the third most common cancer in men and the second most common in women [32]. Despite its prevalence, CRC is considered one of the most preventable cancers through lifestyle modifications [17,33,34]. Current CRC chemotherapy mainly involves 5-fluorouracil, often combined with oxaliplatin, irinotecan, leucovorin, or folic acid in various treatment protocols. While these treatments are effective, they come with significant side effects, including neurological and gastrointestinal disorders, myelosuppression, anemia, and neutropenia [35–38]. Given the drawbacks of conventional chemotherapy, there is an urgent need to develop new treatment strategies for colorectal cancer. Chemoprevention, which involves using natural, synthetic, or biological agents to reverse, suppress, or prevent carcinogenesis at any stage (initiation, promotion, or progression), has emerged as a promising approach [39,40].

The pharmacological activities of hydrazides (–CO–NH–NH–) and hydrazones (–CO–NH–N=) have been extensively evaluated, making them a significant class of compounds [41–43]. Hydrazones, important compounds with diverse pharmacological activities, are currently in clinical trials, such as pyridoxal isonicotinoyl hydrazone (**I**) [44]. For instance, compounds (**II**) and (**III**) demonstrated antiproliferative activity with IC_{50} values of 0.29 and 3.1 μM against HCT 116 cells, outperforming the drug 5-fluorouracil (IC_{50} = 5 μM) [45]. Hydrazone (**IV**) induced a cytostatic effect in p53-competent HCT116 cells, leading to cell cycle arrest at the S/G2 phase [46]. Acylhydrazone (V) exhibited toxicity in colon cancer cells without harming nonmalignant cells, triggering apoptosis and cell cycle arrest in the G2/M phase [47]. Compound (**VII**) showed greater antiproliferative activity than compound (I) due to its high chelation efficacy [48], while compound (**VI**) displayed strong antiproliferative activity on esophageal carcinoma cells with IC_{50} values of 1.09 and 2.79 μM, inducing apoptosis and cell cycle arrest at the G0/G1 phase [49] (Figure 1). This makes hydrazides and their derivatives promising candidates for developing effective cancer therapies.

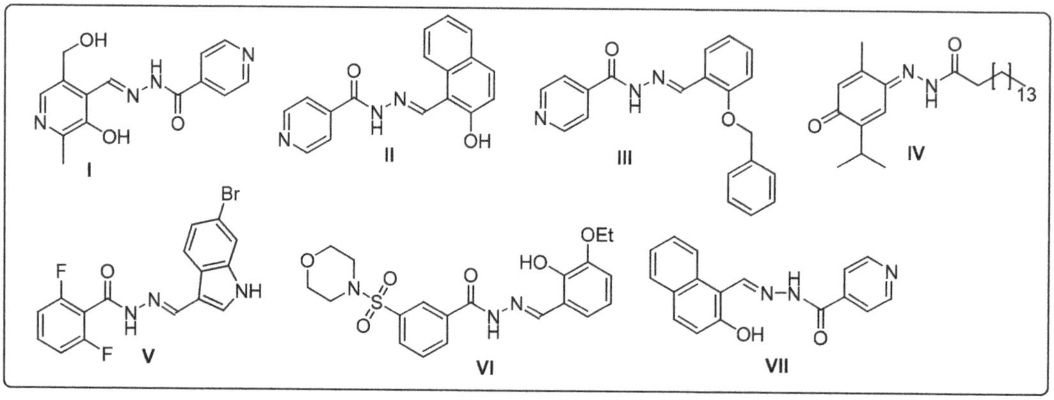

Figure 1. Reported hydrazones with anticancer activity.

Furthermore, the biological profile of the new hydrazides was evaluated, and the results were followed by an in silico docking study. Recently, in silico computational methods have become important in the drug discovery area [50]. Using molecular docking, an important computational technique, to understand the theoretical interactions of compounds with a macromolecular structure and provide assistance to the results of in vitro assays. In this research, we targeted this protein with PDB ID (6MTU) and docked it with novel synthesized compounds to present its effect as anti-colon cancer. For the majority of therapeutic drugs, one of these investigations combined the methods of Absorption, Distribution, Metabolism, and Toxicity (ADMET). The procedure by which a medication enters the systemic circulation after administration is known as absorption. Distribution is the movement of medication from systemic circulation to extravascular sites. Metabolism involves the enzymatic breakdown of a drug into metabolites that are cleared from the body. Excretion is the passive or active transport of intact drug molecules into the urine or bile [51]. Toxicology is described in this article as the negative effects that drug substances have on humanity. Early-stage medication development is drawing attention to ADMET characteristics due to their increased significance [52,53].

Building on our research into bioactive heterocyclic frameworks [5,9,17,19,54–60], we synthesized novel hydrazide derivatives via solvent-free mechanical grinding of 2-cyano-*N′*-(2-cyanoacetyl)acetohydrazide with various aldehydes and ketones, using L-proline as an efficient organocatalyst. This method achieved high yields, short reaction times, and low costs. The synthesis was conducted using both thermal and grinding techniques. We then assessed the in vitro and in silico cytotoxicity of these derivatives against the HCT-116 colon carcinoma cell line. This comprehensive approach, integrating synthetic chemistry, biological evaluation, and computational modeling, highlights the potential of hydrazide derivatives as anticancer agents. The promising results from both experimental and in silico analyses support further development and optimization of these compounds for cancer therapy.

2. Results and Discussion

The main precursor for synthesizing new compounds, 2-cyano-*N′*-(2-cyanoacetyl) acetohydrazide (**3**), is formed by reacting 3-(3,5-dimethyl-1*H*-pyrazol-1-yl)-3-oxopropanenitrile with 2-cyanoacetohydrazide under reflux in dioxane [61]. Carbonyl compounds, including aldehydes and ketones, can react with compound **3** to produce hydrazide derivatives through either conventional reflux or green grinding processes, using moist L-proline as the basic organocatalyst. Specifically, reacting compound **3** with two equivalents of various benzaldehyde derivatives (**4a–e**) yields the corresponding hydrazide derivatives (**5a–e**) in high to excellent yields, irrespective of the substituents being electron-withdrawing or electron-donating, as shown in Scheme 1.

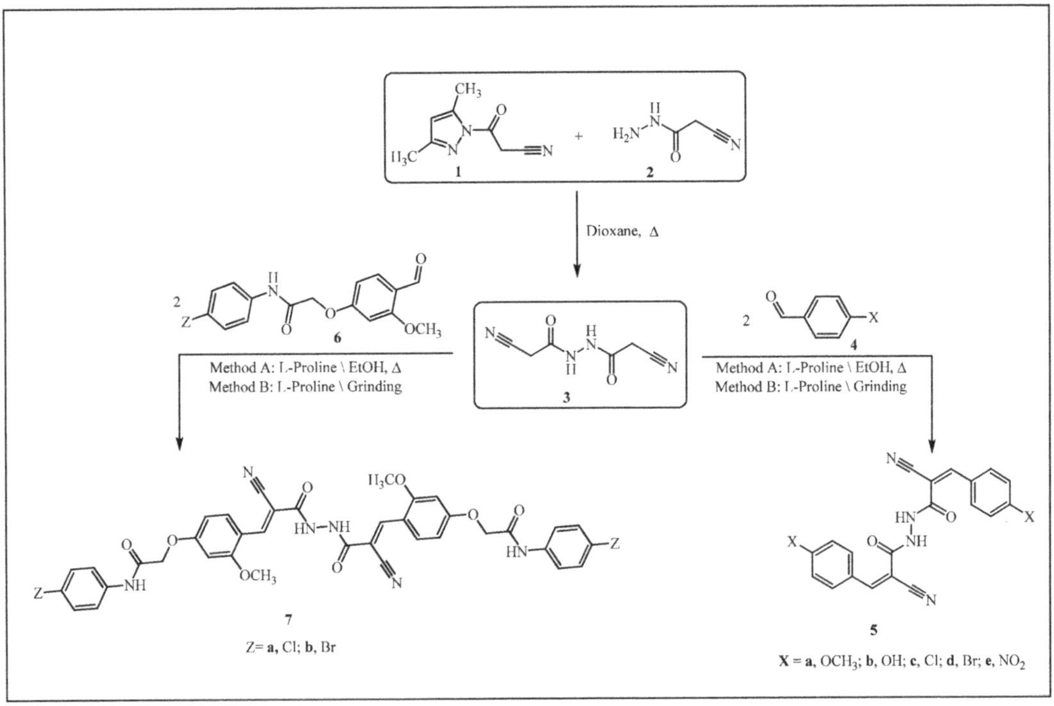

Scheme 1. Synthesis of compounds **5a–e** and **7a,b**.

The identification of the structures of the isolated products **5a–e** was established through spectral data and elemental analyses. The ^{1}H-NMR spectra of the condensation products **5a–e** revealed the absence of signals associated with CH_2 groups, indicating the formation of enone. Additionally, the methine proton –CH= was distinctly observed at approximately δ 8.14 ppm, along with the anticipated aromatic protons. The IR spectrum of compound **5a** showed absorption bands at 3277 cm^{-1} (NH), 2216 cm^{-1} (CN), and 1661 cm^{-1} (C=O). The 3277 cm^{-1} band indicates NH groups, typical of amines. The 2216 cm^{-1} band confirms the cyano (C≡N) group. The 1661 cm^{-1} band suggests a carbonyl (C=O) group. The mass spectra of products **5a–e** displayed molecular ion peaks matching their molecular weights.

Similarly, in the presence of moist L-proline organocatalyst, 2-cyano-*N*′-(2-cyanoacetyl) acetohydrazide (**3**) was reacted with two equivalents of 2-(4-formyl-3-methoxyaryloxy)-*N*-phenylacetamides (**6a,b**) to form the respective 2,2′-(((hydrazine-1,2-diylbis(2-cyano-3-oxoprop-1-ene-3,1-diyl))bis(3-methoxy-4,1-phenylene))bis(oxy))bis(*N*-arylacetamide) derivatives (**7a,b**) as depicted in Scheme 1.

We extended our synthesis efforts to produce ketonic hydrazides by reacting compound **3** with three different acetyl compounds: 1-(4-bromophenyl)ethan-1-one (**8**), 1-(1*H*-indol-3-yl)ethan-1-one (**10**), and 1-(pyridin-3-yl)ethan-1-one (**12**) (two equivalents). These reactions were carried out using both thermal and grinding methods, resulting in the formation of the corresponding hydrazides (**9**, **11**, and **13**) in satisfactory yields, as shown in Scheme 2. The structures of these compounds were confirmed through spectral analysis (IR, ^{1}H-NMR, and MS) and elemental analyses. For instance, the IR spectrum of compound **11** shows absorption bands at 3403 cm^{-1} and 3233 cm^{-1} (NH groups), 2216 cm^{-1} (CN group), and 1670 cm^{-1} (C=O group). Its ^{1}H-NMR spectrum revealed three singlet signals at δ 2.41, 10.46, and 11.88 ppm, corresponding to the methyl group and two NH groups. In addition, the ten aromatic protons resonated as multiplet signals at δ 7.13–8.27 ppm. Mass analysis further identified the molecular ion peak at $m/z = 448$ (73%), consistent with the molecular formula ($C_{26}H_{20}N_6O_2$).

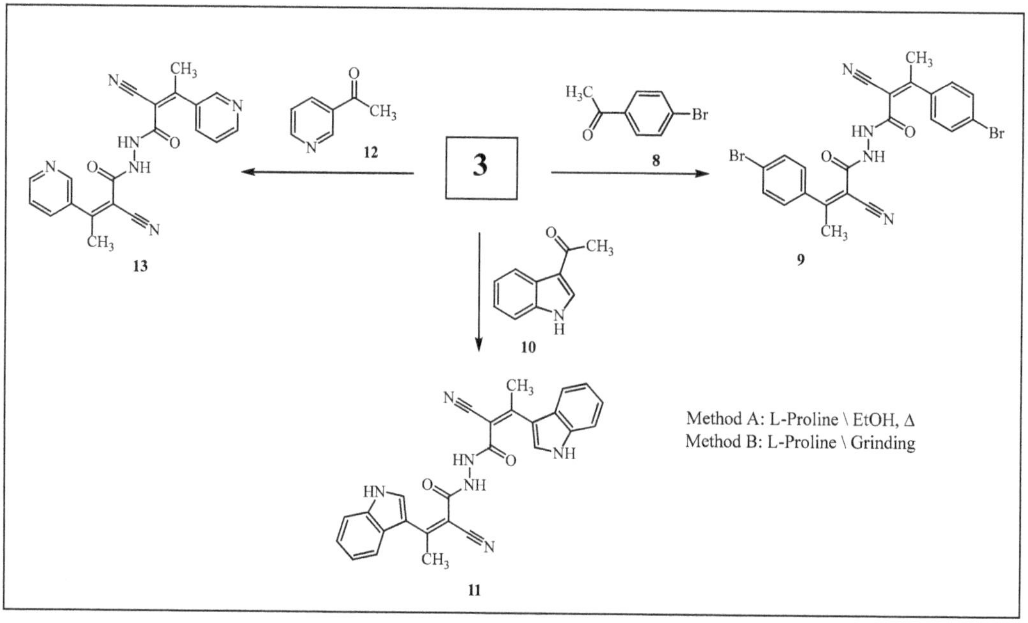

Scheme 2. Synthesis of compounds **9**, **11**, and **13**.

Additionally, a comparative study of the conventional (thermal) method and the green (grinding) method using moist L-proline as the basic catalyst is presented in Table 1.

Table 1. Comparison of the synthesis of products **5a–e**, **7a,b**, **9**, **11**, and **13** using thermal and green methods with L-proline organocatalyst.

Compounds	Thermal Method		Grinding Method	
	Time (h)	(%) Yield	Time (min)	(%) Yield
5a	5	75	25	90
5b	4	74	27	89
5c	3	79	21	92
5d	4	73	26	91
5e	3	78	20	89
7a	5	72	29	84
7b	5	70	30	85
9	4	78	27	91
11	5	77	23	90
13	4	79	28	90

Table 1 compares the synthesis of products **5a–f**, **7a,b**, **9**, **11**, and **13** using both traditional reflux and the eco-friendly grinding method with moist L-proline as the basic catalyst. The green method demonstrated significantly shorter reaction times and generally higher yields than the conventional reflux method. For example, compound **5a** achieved a yield of 75% in 5 h under reflux, whereas the grinding method yielded 90% in just 25 min. This trend was consistent for other compounds, highlighting that the grinding method is not only more time-efficient but also produces comparable or improved yields, making it a promising greener alternative for synthesizing these compounds.

Additionally, Table 2 presents the optimization of catalyst loading for the synthesis of compound **5a** using the grinding method with moist L-proline as the basic catalyst.

Table 2. Optimization of the catalyst loading for the synthesis of compound **5a** using the grinding method.

Entry	Catalyst (mol%)	Time (min)	Temperature (°C)	Yield (%)
1	1	25	25	71
2	3	25	25	83
3 [a]	5	25	25	90
4	10	25	25	90

[a] The best reaction condition for the synthesis of compound **5a**.

Table 2 presents the optimization process for catalyst loading in the synthesis of compound **5a**. Each entry reflects a different catalyst percentage, along with corresponding reaction times, temperatures, and yields. Notably, entry 3, with a catalyst loading of 5 mol%, achieved the highest yield of 90%, indicating the optimal conditions for synthesizing compound **5a** in this experimental setup.

Additionally, we explored the catalyst's regeneration and reusability. The reaction between compound **3** and *p*-anisaldehyde **4a** was conducted under optimized conditions. Upon completion of the reaction, the mixture was diluted with 10 mL of water, and the crude solid product was filtered and washed with 2–5 mL of water. The L-proline catalyst was removed by washing with water, as it is water-soluble. The catalyst was then recovered by evaporating the water, washed with diethyl ether, and reused for subsequent catalytic runs, as detailed in Table 3.

Table 3. Recyclability of L-proline (5 mol%).

State of Catalyst	New Catalyst	Recycled Catalyst				
		(1)	(2)	(3)	(4)	(5)
Yield of product **5a**	90	87	81	78	74	49

Table 3 showcases the recyclability data for the L-proline organocatalyst, comparing its performance as a fresh catalyst (1) and after multiple recycling cycles (2 to 5). The yield of compound **5a** serves as a benchmark for assessing the catalyst's efficiency. Initially, the new catalyst achieves a 90% yield. However, with each recycling cycle, a gradual decrease in yield is observed: 87% after the first cycle, 81% after the second, 78% after the third, and 74% after the fourth cycle. A notable decline in performance is evident after the third cycle, with the yield dropping to 49% by the fifth cycle. This trend indicates a reduction in the catalyst's recyclability over successive uses, although it remains effective for up to four cycles with minimal loss in activity. The loss in yield after the fifth cycle of recycling L-proline as a catalyst is primarily due to cumulative structural degradation, loss of catalyst quantity, contamination, and changes in physical properties affecting its catalytic performance. These cumulative effects lead to a significant reduction in yield, dropping from 90% with a new catalyst to 49% after the fifth recycle.

2.1. Antitumor Study

The in vitro cytotoxic activity of ten newly synthesized target compounds was evaluated against the HCT-116 colon carcinoma cell line. The bioactivity of these compounds was assessed using the MTT assay, which measures cell viability and proliferation. Cisplatin, a well-known chemotherapeutic agent, was used as a reference standard to compare the calculated IC_{50} values of the new compounds. The results of this comparative study are presented in Table 4, providing insights into the potential efficacy of the synthesized compounds relative to the established drug.

Table 4. In vitro cytotoxic effect of synthesized compounds against the HCT-116 cell line.

Tested Compounds	IC_{50} Values (μM)	Tested Compounds	IC_{50} Values (μM)
5a	3.8 ± 0.7	**7b**	37.2 ± 4.2
5b	3.2 ± 1.1	**9**	9.3 ± 1.7
5c	9.3 ± 1.4	**11**	2.5 ± 0.81
5d	8.5 ± 1.3	**13**	3.7 ± 1.0
5e	27.0 ± 2.5	**Cisplatin***	2.43 ± 1.1
7a	35.3 ± 3.1		

* Reference drug.

The outline data in Table 4 showed that the in vitro cytotoxic effect of synthesized compounds against the HCT-116 cell line reveals a range of efficacies, as indicated by their IC_{50} values (μM). Among the synthesized compounds, compound **11** (IC_{50} = 2.5 ± 0.81 μM) exhibits the highest cytotoxicity, closely followed by compound **5b** (IC_{50} = 3.2 ± 1.1 μM) and compound **13** (IC_{50} = 3.7 ± 1.0 μM). These compounds show comparable efficacy to Cisplatin, a known chemotherapeutic agent, which has an IC_{50} value of 2.43 ± 1.1 μM. Compounds **5a** (IC_{50} = 3.8 ± 0.7 μM) and **9** (IC_{50} = 9.3 ± 1.7 μM) demonstrate moderate cytotoxicity, while compounds **5c** (IC_{50} = 9.3 ± 1.4 μM) and **5d** (IC_{50} = 8.5 ± 1.3 μM) exhibit higher IC_{50} values, indicating lower cytotoxicity. Compounds **7a** (IC_{50} = 35.3 ± 3.1 μM) and **7b** (IC_{50} = 37.2 ± 4.2 μM) are the least effective, with the highest IC_{50} values among the tested compounds. Compound **5e** (IC_{50} = 27.0 ± 2.5 μM) also shows relatively lower efficacy. These findings highlight compounds **11**, **5b**, and **13** as potential candidates for further development as anticancer agents, given their promising cytotoxic effects against the HCT-116 cell line.

The structure–activity relationship (SAR) analysis examines how the chemical structure of the synthesized compounds affects their cytotoxicity against the HCT-116 cell line. Compounds with aromatic substituents exhibit varying cytotoxicity, with specific patterns enhancing activity, as seen with compound **5b** (IC_{50} = 3.2 ± 1.1 μM) compared to **5a** (IC_{50} = 3.8 ± 0.7 μM) and **5c** (IC_{50} = 9.3 ± 1.4 μM). Chalcone derivatives, such as **7a** (IC_{50} = 35.3 ± 3.1 μM) and **7b** (IC_{50} = 37.2 ± 4.2 μM), show higher IC_{50} values, indicating lower cytotoxicity. Conversely, heterocyclic compounds like **11** (IC_{50} = 2.5 ± 0.81 μM) demonstrate the lowest IC_{50} values, suggesting enhanced effectiveness. Larger and more flexible compounds, such as **5e** (IC_{50} = 27.0 ± 2.5 μM), have reduced activity, highlighting the importance of balancing rigidity and flexibility for effective interaction with the target protein. The comparable efficacy of compounds **11** and **5b** to Cisplatin (IC_{50} = 2.43 ± 1.1 μM) underscores their potential as anticancer agents. These findings indicate that the nature and position of substituents, the core structure, and the overall size and flexibility of the compounds play significant roles in their cytotoxicity, providing a basis for optimizing structures to improve anticancer properties.

2.2. Molecular Docking Study

After identifying the potent compounds against the HCT-116 colon carcinoma cell line, docking studies were performed to understand their binding interactions with the prepared protein (PDB ID: 6MTU). Molecular docking was used to examine how the ligands interact with the protein's active site [62]. The results, compared to the co-crystallized ligand SEP, revealed significant interactions with the protein residues, as shown in Figure 2. Key findings include (i) more negative binding energies for our ligands than the CCL (co-crystallized ligand), indicating greater stability; (ii) various interactions, including hydrogen donors, hydrogen acceptors, π–H, and π–π interactions; (iii) key binding residues were glycine, isoleucine, arginine, and leucine; (iv) primary interaction sites were oxygen, nitrogen, and six-membered rings; and (v) compounds **5a** and **5b** are predicted to exhibit the strongest inhibition against the 6MTU protein in cancer. As shown in Table 5, the test compound 5a formed four H-bond acceptors: two between oxygen atoms of the carbonyl moiety with the active site residues Gly739 and Ile740, and another two H-bond acceptors between nitrogen atoms of the cyano group with Arg733 and Gly737. Additionally, in test compound **5b**, four H-bond acceptors were formed: three between the oxygen atoms of carbonyl groups with the active site residues Ile740, Gly737, and Gly739, and another one between the nitrogen atoms of the cyano moiety with Gly737. The 3D model of compound **5c** showed one H-bond donor between its *N* and Ile740. On the contrary, the docking of **5d** gave one H-bond acceptor between the oxygen atom and the residue of Gly739. Finally, the docking of **11** gave one H-bond donor between the *N* of the hydrazinylidene moiety and the residue of Ile740; additionally, there were three H-bond acceptors: two between the nitrogen atom of the cyano group with the Arg733 and Arg801, and another one between the oxygen atom of the carbonyl group with the Leu738.

Table 5. Docking score (kcal/mol), No. of H-bonding, and No. of arene interaction of potent synthesized compounds **5a–d** and **11** with the 6MTU receptor relative to the CCL.

Compounds	Docking Score (kcal/mol)	No. of Hydrogen Bonding	No. of Arene Interaction	Donor Atom	Acceptor Atom
5a	−5.75	1 (Gly739) 1 (Ile740) 1 (Arg733) 1 (Gly737)	1 (π-*H*) [Gly739]	-	*O* *N*
5b	−5.78	1 (Ile740) 2 (Gly737) 1 (Gly739)	-	-	*O* *O/N* *O*
5c	−5.61	1 (Ile740)	1 (π-*H*) [Leu738] 1 (π-Cation) [Arg801]	*N*	-
5d	−5.72	1(Gly739)	1 (π-*H*) [Gly739]	-	*O*
11	−6.40	1 (Ile740) 1 (Leu738) 1 (Arg733) 1 (Arg801)	-	*N* - - -	- *O* *N* *N*
CCL	−4.60	2 (Ile740) 1 (Gly739) 1 (Leu738) 1 (Arg733)	-	*N*/- - - -	-/*O* *O* *O* *O*

Compounds	Three-Dimensional Ligand Interaction	Two-Dimensional Ligand Interaction
5a	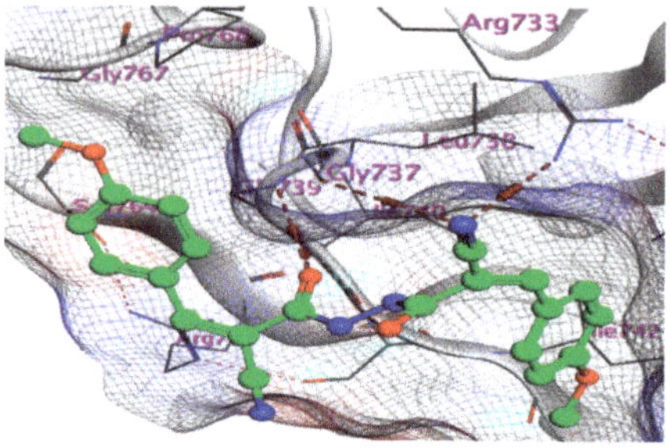	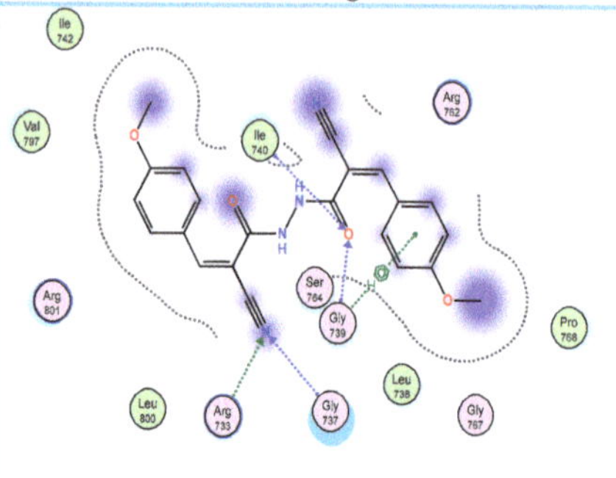
5b	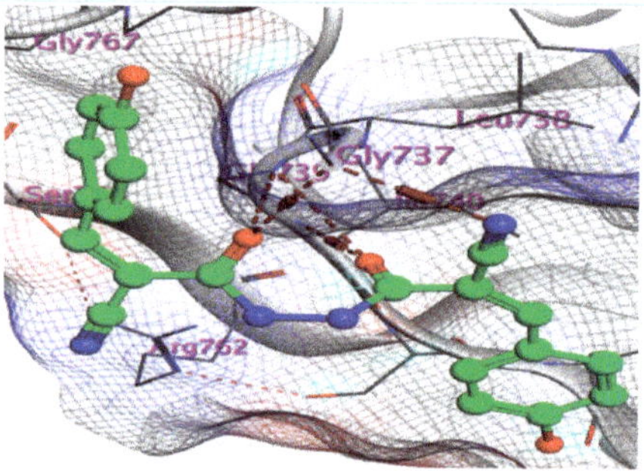	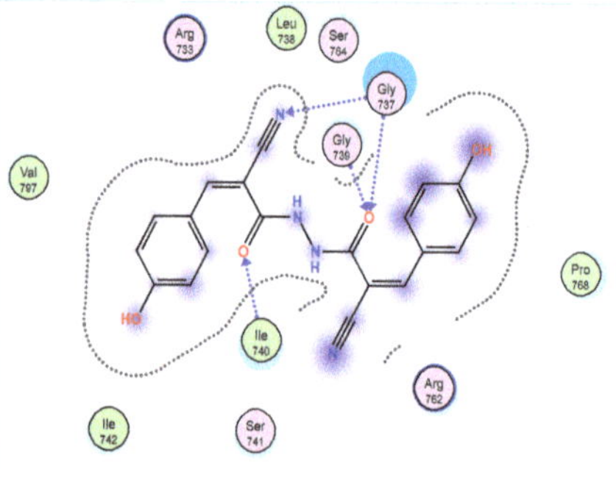
5c	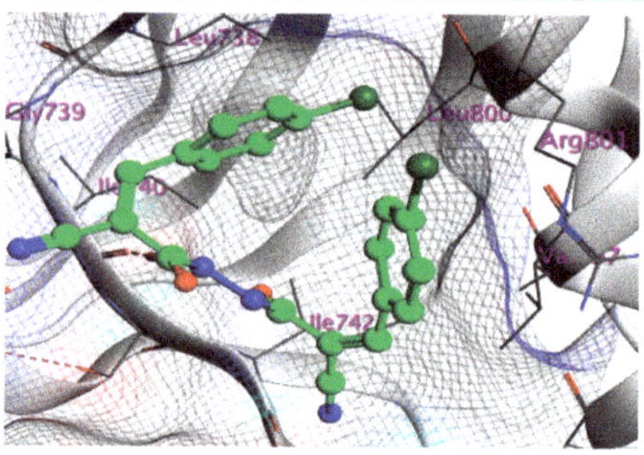	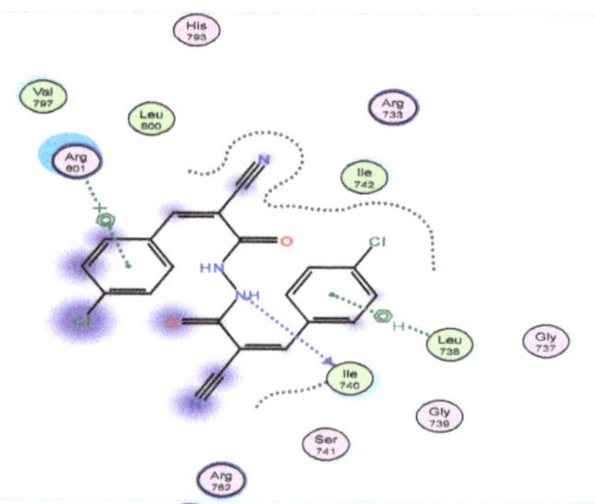
5d	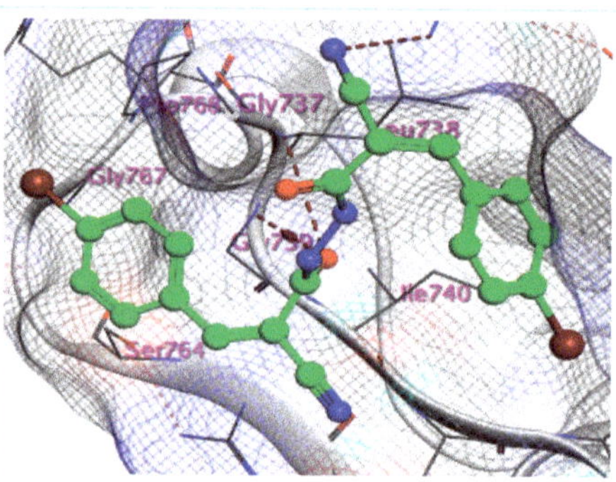	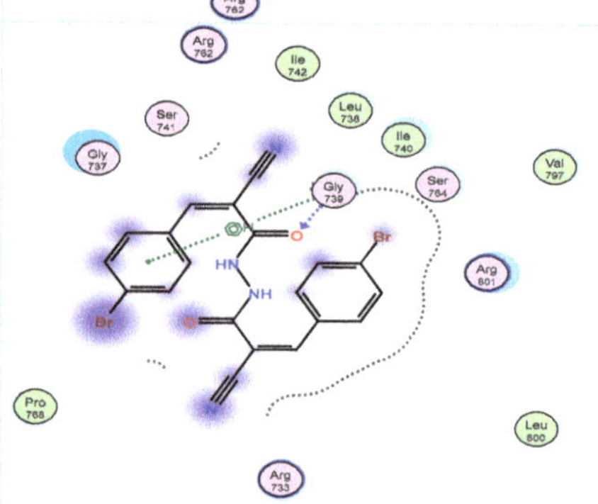

Figure 2. *Cont.*

11

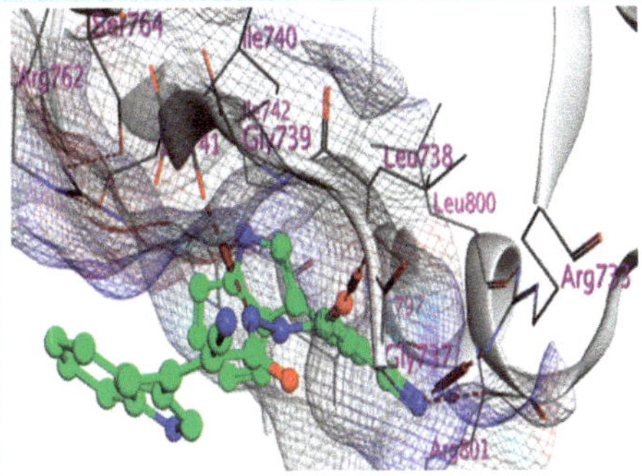

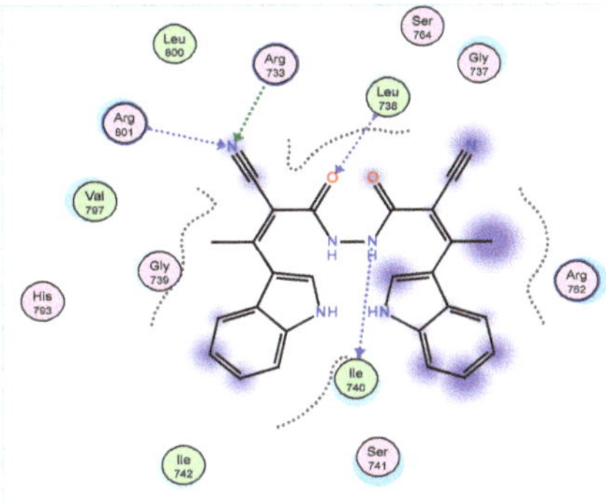

CCL

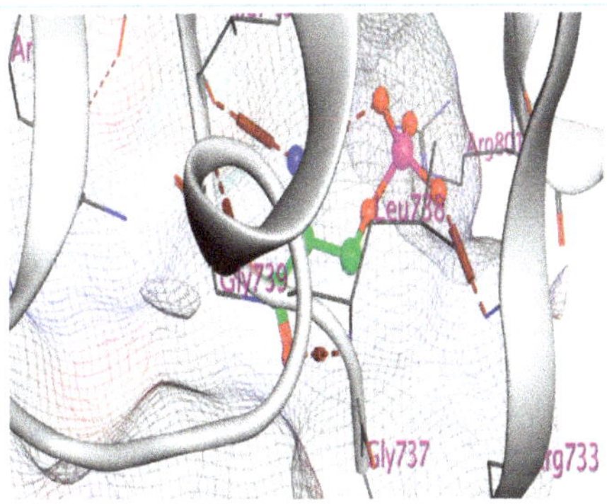

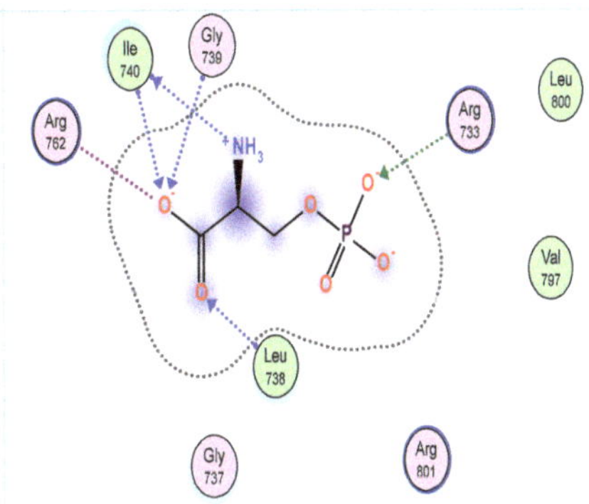

Figure 2. Three-dimensional and two-dimensional ligand interactions within the binding site of 6MTU for potent synthesized compounds **5a–d**, **11**, and CCL.

2.3. *In Silico Pharmacokinetic Profile (ADMET)*

The tested compounds showed intestinal absorption rates between 65.21% and 92.60%, with a volume of distribution (log L/kg) ranging from −0.51 to 0.30. The derivative **5a** inhibited CYP3A4, while **5b** and **5c** were metabolized by CYP1A2. Additionally, derivatives **5c** and **5d** inhibited CYP3A4, CYP2C19, and CYP2C9. The compound **11** was metabolized by all of the cytochrome substrates and inhibitors. The evaluated substances' excretion was determined using a log value of ml/min/kg, with values ranging from −0.61 to 1.24. Table 6 displays all the previous detailed data. Subsequently, the toxicity prediction findings showed that none of the investigated substances had AMES toxicity.

Table 6. Results of ADMET properties of potent synthetic compounds **5a–d**, **11**, and Cisplatin.

Compounds	Absorption	Distribution	Metabolism	Excretion (Log mL/min/kg)	Toxicity
	Intestinal Absorption (Human) Numeric (% Absorbed)	*VDss (Log L/kg)*			*AMES Toxicity (Categorical) (Yes/No)*
5a	74.39	−0.5	CYP3A4 Substrate and CYP3A4 inhibitor	0.25	No
5b	65.21	−0.33	CYP1A2 inhibitor	0.13	No
5c	81.9	−0.51	CYP3A4 substrate; CYP1A2, CYP2C19, CYP2C9 and CYP3A4 inhibitor	−0.57	No

Table 6. *Cont.*

Compounds	Absorption	Distribution	Metabolism	Excretion (Log mL/min/kg)	Toxicity
	Intestinal Absorption (Human) Numeric (% Absorbed)	*VDss (Log L/kg)*			*AMES Toxicity (Categorical) (Yes/No)*
5d	81.37	−0.48	CYP3A4 substrate; CYP2C19, CYP2C9 and CYP3A4 inhibitor	−0.61	No
11	83.02	−0.04	CYP2D6, CYP3A4 substrate; CYP1A2, CYP2C19, CYP2C9, and CYP3A4 inhibitor	0.76	No
Cisplatin *	92.60	0.30	-	1.24	No

* Reference drug.

3. Experimental

The melting points of the newly synthesized compounds were determined using an electrothermal Gallenkamp apparatus, Cambridge, UK. IR spectra were obtained with a Pye-Unicam SP300 instrument (Pye, Cambridge, UK) using potassium bromide discs. Mass spectra were acquired using GCMS-Q1000-EX Shimadzu (Kyoto, Japan) and GCMS 5988-A HP spectrometers (HP Inc., Palo Alto, CA, USA), operating at an ionizing voltage of 70 eV. The ^{1}H-NMR and ^{13}C-NMR spectra were recorded using a Jeol-500 spectrometer (JEOL, Tokyo, Japan), operating at 500 MHz for ^{1}H-NMR and 125 MHz for ^{13}C-NMR. Elemental analyses were conducted at the Microanalytical Center of Cairo University, Giza, Egypt.

3.1. *Synthesis of 2-cyano-N′-(2-cyanoacetyl)acetohydrazide* (**3**)

A mixture of 3-(3,5-dimethyl-1*H*-pyrazol-1-yl)-3-oxopropanenitrile **1** (1.63 g, 10 mmol) and 2-cyanoacetohydrazide **2** (1.00 g, 10 mmol) was dissolved in 30 mL of dioxane and heated under reflux for 2 h. The resulting solid, which formed after cooling, was collected by filtration and recrystallized from aqueous ethanol, yielding colorless needles with an 89% yield and a melting point of 207–209 °C (lit., 207–208 °C) [61].

3.2. *Reaction of* **3** *with Aromatic (or Heteroarmatic) Carbonyl Compounds* **4a–e**, **6a,b**, **8**, **10**, *and* **12**

3.2.1. Method A

A mixture of 2-cyano-*N*′-(2-cyanoacetyl)acetohydrazide (**3**, 0.166 g, 1 mmol) and the appropriate carbonyl compounds as substituted benzaldehyde derivatives (**4a–e**), 2-(4-formyl-3-methoxyphenoxy)-*N*-arylacetamide derivatives (**6a,b**), 1-(4-bromophenyl)ethan-1-one (**8**), 1-(1*H*-indol-3-yl)ethan-1-one (**10**), or 1-(pyridin-3-yl)ethan-1-one (**12**) (2 mmol for each) was dissolved in 20 mL of aqueous ethanol (1:1) containing L-proline (0.006 g, 5 mol %). The mixture was refluxed for 3–5 h. After confirming the completion of the reaction by TLC (EtOAc: *n*-hexane, 1:1, *v*/*v*), the reaction mixture was cooled. The insoluble catalyst was separated by filtration, washed, dried, and reused up to four times. The solid was washed with cold water and recrystallized from DMF or DMF-EtOH to obtain pure compounds (**5a–e**, **7a,b**, **9**, **11**, and **13**), respectively.

3.2.2. Method B

A mixture of 2-cyano-*N*′-(2-cyanoacetyl)acetohydrazide (**3**, 0.166 g, 1 mmol) and the appropriate carbonyl compounds as substituted benzaldehyde derivatives (**4a–e**), 2-(4-formyl-3-methoxyphenoxy)-*N*-arylacetamide derivatives (**6a,b**), 1-(4-bromophenyl)ethan-1-one (**8**), 1-(1*H*-indol-3-yl)ethan-1-one (**10**), or 1-(pyridin-3-yl)ethan-1-one (**12**) (each at 2 mmol), was ground in a mortar with a pestle at 25 °C in the presence of L-proline (0.006 g, 5 mol%) and one drop of water. The initially syrupy reaction mixture solidified within

20–30 min. The reaction's completion was monitored by TLC (EtOAc: n-hexane, 1:1, *v/v*). The solid was washed with cold water and recrystallized from DMF or DMF-EtOH to obtain pure compounds (**5a–e**, **7a,b**, **9**, **11**, and **13**), respectively. The spectral data (available in the Supplementary Materials) and physical constants for these products are listed below:

2-Cyano-*N'*-(2-cyano-3-(4-methoxyphenyl)acryloyl)-3-(4-methoxyphenyl) acrylohydrazide (5a).

Orange crystal; mp 216–218 °C (DMF-EtOH); IR(KBr) υ (cm^{-1}) 3277 (NH), 2216 (CN), 1661 (C=O); ^{1}H-NMR (DMSO-d_6): δ 3.79 (s, 6H, 2OCH$_3$), 7.35–7.61 (m, 8H, Ar-H), 8.14 (s, 2H, 2CH=), 10.40 (s, 2H, 2NH) ppm; ^{13}C-NMR (DMSO-d_6): δ = 57.7 (OCH$_3$), 115.4, 121.8, 128.0, 129.3, 132.2, 137.7, 149.6, (Ar-C and C=N), 166.9 (C=O) ppm; MS *m/z* (%): 402 (M$^+$, 60), 328 (71), 232 (100), 165 (50), 63 (83). Anal.Calcd for $C_{22}H_{18}N_4O_4$ (402.41): C, 65.66; H, 4.51; N, 13.92. Found C, 65.47; H, 4.38; N, 13.73%.

2-Cyano-*N'*-(2-cyano-3-(4-hydroxyphenyl)acryloyl)-3-(4-hydroxyphenyl) acrylohydrazide (5b).

Orange crystals; mp 186–188 °C (DMF-EtOH); IR(KBr) υ (cm^{-1}) 3408, 3288 (OH and NH), 2215 (CN), 1679 (C=O); ^{1}H-NMR (DMSO-d_6): δ 8.06 (s, 2H, 2CH=), 6.89–7.73 (m, 8H, Ar-H), 9.76 (s, 2H, 2OH), 10.61 (s, 2H, 2NH) ppm; ^{13}C-NMR (DMSO-d_6): δ = 115.5, 124.6, 128.1, 129.3, 131.1, 134.7, 146.7 (Ar-C and C=N), 165.8 (C=O) ppm; MS *m/z* (%): 374 (M$^+$, 46), 227 (100), 104 (72), 61 (92). Anal. Calcd for $C_{20}H_{14}N_4O_4$ (374.36): C, 64.17; H, 3.77; N, 14.97. Found C, 64.05; H, 3.62; N, 14.85%.

3-(4-Chlorophenyl)-*N'*-(3-(4-chlorophenyl)-2-cyanoacryloyl)-2-cyanoacrylohydrazide (5c).

Orange crystals; mp 241–243 °C (DMF); IR(KBr) υ (cm^{-1}) 3279 (NH), 2218 (CN), 1652 (C=O); ^{1}H-NMR (DMSO-d_6): δ 8.24 (s, 2H, 2CH=), 7.29–7.99 (m, 8H, Ar-H), 10.78 (s, 2H, 2NH) ppm; MS *m/z* (%): 411 (M$^+$, 29), 317 (47), 227 (100), 126 (63), 59 (67). Anal. Calcd for $C_{20}H_{12}Cl_2N_4O_2$ (411.24): C, 58.41; H, 2.94; N, 13.62. Found C, 58.30; H, 2.85; N, 13.49%.

3-(4-Bromophenyl)-*N'*-(3-(4-bromophenyl)-2-cyanoacryloyl)-2-cyanoacrylohydrazide (5d).

Orange crystals; mp 230–232 °C (DMF); IR(KBr) υ (cm^{-1}) 3290 (NH), 2217 (CN), 1657 (C=O); ^{1}H-NMR (DMSO-d_6): δ 8.14 (s, 2H, 2CH=), 7.78–7.88 (m, 8H, Ar-H), 10.76 (s, 2H, 2NH) ppm; ^{13}C-NMR (DMSO-d_6): δ = 115.2, 128.4, 128.8, 129.2, 129.4, 132.1, 149.4 (Ar-C and C=N), 165.8 (C=O) ppm; MS *m/z* (%): 500 (M$^+$, 83), 412 (47), 360 (62), 273 (100), 171 (52), 57 (55). Anal. Calcd for $C_{20}H_{12}Br_2N_4O_2$ (500.15): C, 48.03; H, 2.42; N, 11.20. Found C, 48.15; H, 2.35; N, 11.04%.

2-Cyano-*N'*-(2-cyano-3-(4-nitrophenyl)acryloyl)-3-(4-nitrophenyl)acrylohydrazide (5e).

Yellow solid; mp 237–239 °C (DMF); IR(KBr) υ (cm^{-1}) 3271 (NH), 2221 (CN), 1677 (C=O); ^{1}H-NMR (DMSO-d_6): δ 8.13–8.39 (m, 8H, Ar-H), 8.82 (s, 2H, 2CH=), 10.96 (s, 2H, 2NH) ppm; MS *m/z* (%): 432 (M$^+$, 48), 303 (39), 217 (100), 130 (66), 61 (53). Anal. Calcd for $C_{20}H_{12}N_6O_6$ (432.35): C, 55.56; H, 2.80; N, 19.44. Found C, 55.44; H, 2.69; N, 19.28%.

2,2'-((-Hydrazine-1,2-diylbis(2-cyano-3-oxoprop-1-ene-3,1-diyl))bis(3-methoxy-4,1-phenylene))bis(oxy))bis(*N*-(4-chlorophenyl)acetamide) (7a).

Yellow microcrystals; mp 212–214 °C (DMF-EtOH); IR(KBr) υ (cm^{-1}) 3381, 3218 (2NH), 2215 (CN), 1664 (C=O); ^{1}H-NMR (DMSO-d_6): δ 3.73 (s, 6H, 2OCH$_3$), 4.77 (s, 4H, 2CH$_2$), 6.98–7.83 (m, 14H, Ar-H), 8.13 (s, 2H, 2CH=), 10.23 (s, 2H, 2NH), 11.32 (s, 2H, 2NH) ppm; ^{13}C-NMR (DMSO-d_6): δ = 55.4 (OCH$_3$), 65.2 (CH$_2$O), 112.5, 116.8, 118.9, 120.0, 121.8, 122.1, 123.2, 131.1, 132.6, 135.2, 144.3, 153.1, 158.3 (Ar-C and C=N), 165.7, 168.5 (C=O) ppm; MS *m/z* (%): 769 (M$^+$, 45), 612 (59), 417 (60), 363 (44), 232 (100), 141 (38), 61 (54). Anal. Calcd for $C_{38}H_{30}Cl_2N_6O_8$ (769.59): C, 59.31; H, 3.93; N, 10.92. Found C, 59.19; H, 3.82; N, 10.85%.

2,2'-((-Hydrazine-1,2-diylbis(2-cyano-3-oxoprop-1-ene-3,1-diyl))bis(3-methoxy-4,1-phenylene))bis(oxy))bis(*N*-(4-bromophenyl)acetamide) (7b).

Yellow microcrystals; mp 217–219 °C (DMF-EtOH); IR(KBr) υ (cm^{-1}) 3399, 3247 (2NH), 2216 (CN), 1663 (C=O); ^{1}H-NMR (DMSO-d_6): δ 3.79 (s, 6H, 2OCH$_3$), 4.82 (s, 4H, 2CH$_2$), 7.41–7.56 (m, 14H, Ar-H), 8.10 (s, 2H, 2CH=), 10.30 (s, 2H, 2NH), 11.22 (s, 2H, 2NH) ppm; MS *m/z* (%): 858 (M$^+$, 36), 683 (48), 493 (72), 366 (100), 281 (80), 163 (70), 63 (81). Anal. Calcd for $C_{38}H_{30}Br_2N_6O_8$ (858.50): C, 53.16; H, 3.52; N, 9.79. Found C, 53.04; H, 3.47; N, 9.60%.

3-(4-Bromophenyl)-*N*′-(3-(4-bromophenyl)-2-cyanobut-2-enoyl)-2-cyanobut-2-enehydrazide (9).

Yellow crystals; mp 211–213 °C (DMF); IR(KBr) υ (cm^{-1}) 3248 (NH), 2216 (CN), 1660 (C=O); ^{1}H-NMR (DMSO-d_6): δ 2.32 (s, 6H, $2CH_3$), 6.98–7.96 (m, 8H, Ar-H), 11.32 (s, 2H, 2NH) ppm; ^{13}C-NMR (DMSO-d_6): δ = 15.1 (CH_3), 115.4, 120.2, 126.9, 128.9, 130.2, 138.3, 150.8 (Ar-C and C=N), 165.6 (C=O) ppm; MS m/z (%): 528 (M^+, 27), 440 (51), 294 (100), 171 (50), 63 (79). Anal. Calcd for $C_{22}H_{16}Br_2N_4O_2$ (528.20): C, 50.03; H, 3.05; N, 10.61. Found C, 50.14; H, 3.01; N, 10.49%.

2-Cyano-*N*′-(2-cyano-3-(1*H*-indol-3-yl)but-2-enoyl)-3-(1H-indol-3-yl)but-2-enehydrazide (11).

Orange crystals; mp 233–235 °C (DMF); IR(KBr) υ (cm^{-1}) 3403, 3233 (2NH), 2216 (CN), 1670 (C=O); ^{1}H-NMR (DMSO-d_6): δ 2.41 (s, 6H, $2CH_3$), 7.13–8.27 (m, 10H, Ar-H), 10.46 (s, 2H, 2NH), 11.88 (s, 2H, 2NH) ppm; ^{13}C-NMR (DMSO-d_6): δ = 15.5 (CH_3), 115.0, 122.8, 126.8, 127.4, 128.1, 128.8, 129.2, 130.2, 138.3, 147.9, 158.3 (Ar-C and C=N), 165.6 (C=O) ppm; MS m/z (%): 448 (M^+, 73), 310 (62), 218 (100), 162 (38), 61 (52). Anal. Calcd for $C_{26}H_{20}N_6O_2$ (448.49): C, 69.63; H, 4.50; N, 18.74. Found C, 69.53; H, 4.44; N, 18.68%.

2-Cyano-N′-(2-cyano-3-(pyridin-3-yl)but-2-enoyl)-3-(pyridin-3-yl)but-2-enehydrazide (13).

Brown crystals; mp 250–252 °C (DMF); IR(KBr) υ (cm^{-1}) 3271 (NH), 2217 (CN), 1673 (C=O); ^{1}H-NMR (DMSO-d_6): δ 2.36 (s, 6H, $2CH_3$), 8.07–9.31 (m, 8H, Ar-H), 10.57 (s, 2H, 2NH) ppm; ^{13}C-NMR (DMSO-d_6): δ = 15.4 (CH_3), 115.7, 121.7, 127.9, 129.3, 132.2, 137.8, 142.5, 159.6 (Ar-C and C=N), 167.0 (C=O) ppm; MS m/z (%): 372 (M^+, 100), 277 (63), 185 (82), 63 (74). Anal. Calcd for $C_{20}H_{16}N_6O_2$ (372.39): C, 64.51; H, 4.33; N, 22.57. Found C, 64.43; H, 4.26; N, 22.40%.

3.3. *Cytotoxicity Assay*

In the cytotoxicity and antitumor studies, cells were planted in 96-well plates at 5×10^4 cells/well and incubated for 24 h. Subsequently, different concentrations of the compounds were applied in six replicates per concentration. Control wells contained only medium or 0.5% DMSO. After a 24 h treatment period, the MTT assay was employed to evaluate cell viability [17,56].

3.4. *Docking Study*

Ligand preparation: Compounds with the lowest IC_{50} values in the MTT assay were selected for docking with the target protein. Using CheDraw Professional 16.0, the potent molecules were sketched, and molecular modeling was performed with MOE software (http://structure.bioc.cam.ac.uk/pkcsm, accessed 20 May 2024). Minimizations were performed until a root mean square deviation (RMSD) gradient of 0.5 kcal·mol^{-1}Å^{-1} was achieved using MMFF 94× (Merck Molecular Force Field 94×) [63]. The oriented compounds were then saved in MDB format for docking.

Protein preparation: Protein preparation involved downloading the enzyme's X-ray crystal structure (PDB ID: 6MTU, resolution: 2.14 Å) from the Protein Data Bank [64]. The enzyme was prepared for docking as follows [65]: (1) Retaining the SEP co-crystalline ligand and removing crystallographic water molecules. (2) Adding hydrogen atoms with standard geometry, reconnecting broken bonds, and fixing potentials. (3) Using Alpha Site Finder to identify large sites with dummy atoms. (4) Marking the protein's binding pocket. (5) Saving the identified pocket in Moe format for modeling ligand-enzyme interactions. Figure 3 shows the prepared protein. (6) Analyzing interactions between the ligand and active site amino acids. Following the docking process, both 2D and 3D interactions with amino acid residues were visualized. As per established protocols, all docking procedures and scoring were meticulously documented [66].

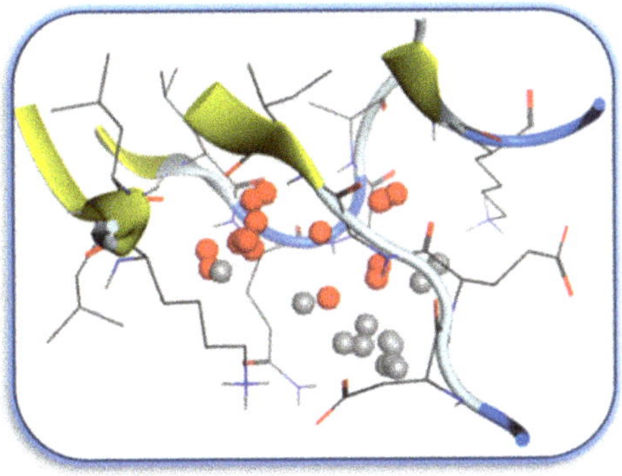

Figure 3. Prepared colon cancer protein (largest pocket).

In silico pharmacokinetic profile (ADMET)

Using pkCSM, the pharmacokinetic profile of Cisplatin and the most significant synthesized compounds were predicted [67]. A web server that is publicly available (http://structure.bioc.cam.ac.uk/pkcsm, accessed 27 July 2024) offers a comprehensive platform for quickly assessing pharmacokinetic [68] and toxicity properties [69]. Predicting the features of new compounds related to ADMET is an effort that requires linking their pharmacokinetic and toxicological characteristics [70].

4. Conclusions

This study demonstrates the effectiveness of green synthesis using L-proline as an organocatalyst, highlighting its reusability, mild conditions, clean reactions, easy workup, high purity, short reaction times, and high yields. The synthesized compounds, including various 2-cyano-*N'*-(2-cyanoacetyl)acetohydrazide derivatives, were structurally characterized and screened for anticancer activities against the HCT-116 colon carcinoma cell line. The reusability of L-proline was confirmed, making the synthesis protocol efficient and sustainable. Molecular docking studies and in silico ADMET investigations revealed that the most potent compounds have favorable binding properties and pharmacokinetic profiles. The docking study showed that compound **11** was the most effective agent against the colon cancer cell line. Additionally, studies using the ADMET tool showed that hydrazide compounds have drug-like properties. These findings suggest that the novel compounds have significant potential as targeted anti-colon cancer agents. The information supports the conclusion that, in $3OCH_3$, the use of advanced technology may result in the development and production of novel, potent colon anticancer medications based on these molecules.

Supplementary Materials: The following supporting information can be downloaded at: https://www.mdpi.com/article/10.3390/catal14080489/s1, File S1: ^{1}H-NMR and ^{13}C-NMR spectra of compounds **3**, **5a–e**, **7a,b**, **9**, **11**, and **13**.

Author Contributions: Conceptualization, S.M.G. and T.Z.A.; methodology, A.H.A., B.F. and W.E.B.; software, B.F.; validation, S.M.G., S.A.A.-H. and M.E.A.Z.; formal analysis, S.M.G., B.F., A.H.A., T.Z.A. and A.M.H.; investigation, S.M.G., B.F. and A.H.A.; resources, A.H.A., S.A.A.-H., T.Z.A. and M.E.A.Z.; data curation, M.E.A.Z. and A.M.H.; writing—original draft preparation, S.M.G. and M.E.A.Z.; writing—review and editing, S.M.G. and W.E.B.; visualization, S.M.G. and A.M.H.; supervision, S.M.G. and T.Z.A. All authors have read and agreed to the published version of the manuscript.

Funding: This research received no external funding.

Data Availability Statement: The data presented in this study are available on request.

Conflicts of Interest: The authors declare no conflicts of interest.

References

1. Anastas, P.; Eghbali, N. Green Chemistry: Principles and Practice. *Chem. Soc. Rev.* **2010**, *39*, 301–312. [CrossRef] [PubMed]
2. Zhou, B.; Liu, Z.; Qu, W.; Yang, R.; Lin, X.; Yan, S.; Lin, J. An environmentally benign, mild, and catalyst-free reaction of quinones with heterocyclic ketene aminals in ethanol: Site-selective synthesis of rarely fused [1,2-a]indolone derivatives via an unexpected anti-Nenitzescu strategy. *Green Chem.* **2014**, *16*, 4359–4370. [CrossRef]
3. Kiyani, H.; Ghorbani, F. Expeditious Green Synthesis of 3,4-disubstituted isoxazole-5(4H)-ones Catalyzed by Nano-MgO. Res. *Chem. Intermed.* **2016**, *42*, 6831–6844. [CrossRef]
4. Kiyani, H.; Darbandi, H.; Mosallanezhad, A.; Ghorbani, F. 2-Hydroxy-5-sulfobenzoic acid: An efficient organocatalyst for the three-component synthesis of 1-amidoalkyl-2-naphthols and 3,4-disubstituted isoxazol-5(4H)-ones. Res. *Chem. Intermed.* **2015**, *41*, 7561–7579. [CrossRef]
5. Brahmachari, G.; Banerjee, B. Facile and One-Pot Access to Diverse and Densely Functionalized 2-Amino-3-cyano-4H-pyrans and Pyran-Annulated Heterocyclic Scaffolds via an Eco-Friendly Multicomponent Reaction at Room Temperature Using Urea as a Novel Organo-Catalyst. ACS Sustain. *Chem. Eng.* **2014**, *2*, 411–422. [CrossRef]
6. Gomha, S.M.; Abdelaziz, M.R.; Abdel-aziz, H.M.; Hassan, S.A. Green Synthesis and Molecular Docking of Thiazolyl-thiazole Derivatives as Potential Cytotoxic Agents. *Mini-Rev. Med. Chem.* **2017**, *17*, 805–815. [CrossRef] [PubMed]
7. Gomha, S.M.; Badrey, M.G.; Arafa, W.A.A. DABCO-catalyzed green synthesis of thiazole and 1,3-thiazine derivatives linked to benzofuran. *Heterocycles* **2016**, *92*, 1450–1461.
8. Khalil, K.D.; Riyadh, S.M.; Gomha, S.M.; Ali, I. Synthesis, characterization and application of copper oxide chitosan nanocomposite for green regioselective synthesis of [1,2,3]triazoles. *Int. J. Biol. Macromol.* **2019**, *130*, 928–937. [CrossRef]
9. Gomha, S.M.; Muhammad, Z.A.; Abdel-aziz, H.M.; Matar, I.K.; El-Sayed, A.A. Green synthesis, molecular docking and anticancer activity of novel 1,4-dihydropyridine-3,5-dicarbohydrazones under grind-stone chemistry. *Green Chem. Lett. Rev.* **2020**, *13*, 6–17. [CrossRef]
10. Gomha, S.M.; Khalil, K.D.; El-Zanate, A.M.; Riyadh, S.M. A facile green synthesis and anti-cancer activity of bis-arylhydrazononitriles, triazolo[5,1-c][1,2,4]triazine, and 1,3,4-thiadiazoline. *Heterocycles* **2013**, *87*, 1109–1120.
11. Kumar, S.; Sharma, M. A grinding-induced catalyst- and solvent-free synthesis of highly functionalized 1,4-dihydropyridines via a domino multicomponent reaction. *Green Chem.* **2011**, *13*, 2017–2020. [CrossRef]
12. Pasha, M.A.; Nizam, A. Amberlite IR-120 catalyzed, microwave-assisted rapid synthesis of 2-aryl-benzimidazoles. *J. Saudi Chem. Soc.* **2012**, *16*, 237–240. [CrossRef]
13. Dhinakaran, V.; Padmini, N.; Bhuvanesh, N.S.P. Chemodivergent, One-Pot, Multi-Component Synthesis of Pyrroles and Tetrahydropyridines under Solvent- and Catalyst-Free Conditions Using the Grinding Method. *ACS Comb. Sci.* **2016**, *18*, 236–242. [CrossRef] [PubMed]
14. Tanaka, F.; Toda, F. Solvent-Free Organic Synthesis. *Chem. Rev.* **2000**, *100*, 1025–1047. [CrossRef] [PubMed]
15. Dolotko, J.W.; Wiench, K.W.; Dennis, K.W.; Pecharsky, V.K.; Balema, V.P. Mechanically induced reactions in organic solids: Liquid eutectics or solid-state processes? *New J. Chem.* **2010**, *34*, 25–28. [CrossRef]
16. Brahmachari, G.; Das, S. L-Proline catalyzed multicomponent one-pot synthesis of gem-diheteroarylmethane derivatives using facile grinding operation under solvent-free conditions at room temperature. *RSC Adv.* **2014**, *4*, 7380. [CrossRef]
17. Edrees, M.M.; Abu-Melha, S.A.; Saad, M.; Kheder, N.A.; Gomha, S.M.; Muhammad, Z.A. Eco-Friendly synthesis, characterization and biological evaluation of some novel pyrazolines containing thiazole moiety as potential anticancer and antimicrobial agents. *Molecules* **2018**, *23*, 2970. [CrossRef] [PubMed]
18. Rashdan, H.R.M.; Gomha, S.M.; El-Gendey, M.S.; El-Hashash, M.A.; Soliman, A.M.M. Eco-friendly one-pot synthesis of some new pyrazolo[1,2-b]phthalazinediones with antiproliferative efficacy on human hepatic cancer cell lines. *Green Chem. Lett. Rev.* **2018**, *11*, 264–274. [CrossRef]
19. Gomha, S.M.; Abdel-aziz, H.M.; Abolibda, T.Z.; Hassan, S.A.; Abdalla, M.M. Green synthesis, molecular docking and pharmacological evaluation of new triazolo-thiadiazepinylcoumarine derivatives as sedative-hypnotic scaffold. *J. Heterocycl. Chem.* **2020**, *57*, 1034–1043. [CrossRef]
20. MacMillan, D.W.C. The advent and development of organocatalysis. *Nature* **2008**, *455*, 304. [CrossRef]
21. Huang, W.; Anwar, S.; Chen, K. Morita–Baylis–Hillman (MBH) reaction derived nitroallylic alcohols, acetates and amines as synthons in organocatalysis and heterocycle synthesis. *Chem. Rec.* **2017**, *17*, 363–381. [CrossRef]
22. Amarante, G.W.; Coelho, F. Quim. p-Sulfonic acid calix[4]arene as a new reusable organocatalyst for the transesterification of vegetable oil. *Nova* **2009**, *32*, 469.
23. Stowe, G.N.; Janda, K.D. Diels-Alder reaction conducted within the parameters of aqueous organocatalysis: Still just smoke and mirrors. *Tetrahedron Lett.* **2011**, *52*, 2085–2087. [CrossRef] [PubMed]
24. Mecadon, H.; Rumum, M.; Kharbangar, I.; Laloo, B.M.; Kharkongor, I.; Rajbangshi, M.; Myrboh, B. L-Proline as an efficicent catalyst for the multi-component synthesis of 6-amino-4-alkyl/aryl-3-methyl-2, 4-dihydropyrano[2,3-c]pyrazole-5-carbonitriles in water. *Tetrahedron Lett.* **2011**, *52*, 3228. [CrossRef]
25. Hernández, J.G.; García-López, V.; Juaristi, E. Solvent-free asymmetric aldol reaction organocatalyzed by (S)-proline-containing thiodipeptides under ball-milling conditions. *Tetrahedron* **2012**, *68*, 92–97. [CrossRef]

26. Yarhosseini, M.; Javanshir, S.; Dekamin, M.G.; Farhadnia, M. Tetraethylammonium 2-(carbamoyl)benzoate as a bifunctional organocatalyst for one-pot synthesis of Hantzsch 1,4-dihydropyridine and polyhydroquinoline derivatives. *Monatsh. Chem.* **2016**, *147*, 1779–1787. [CrossRef]
27. Schettini, R.; De Riccardis, F.; Sala, G.D.; Izzo, I. Enantioselective Alkylation of Amino Acid Derivatives Promoted by Cyclic Peptoids under Phase-Transfer Conditions. *J. Org. Chem.* **2016**, *81*, 2494–2505. [CrossRef] [PubMed]
28. Shaala, L.A.; Youssef, D.T.A.; Badr, J.M.; Harakeh, S.M. Bioactive 2(1H)-Pyrazinones and Diketopiperazine Alkaloids from a Tunicate-Derived Actinomycete Streptomyces sp. *Molecules* **2016**, *21*, 1116. [CrossRef] [PubMed]
29. Schettini, R.; Costabile, C.; Della Sala, G.; Buirey, J.; Tosolini, M.; Tecilla, P.; Vaccaro, M.C.; Bruno, I.; De Riccardis, F.; Izzo, I. Tuning the biomimetic performances of 4-hydroxyproline-containing cyclic peptoids. *Org. Biomol. Chem.* **2018**, *16*, 6708–6717. [CrossRef]
30. Bojarska, J.; Mieczkowski, A.; Ziora, Z.M.; Skwarczynski, M.; Toth, I.; Shalash, A.O.; Parang, K.; El-Mowafi, S.A.; Mohammed, E.H.M.; Elnagdy, S.; et al. Cyclic Dipeptides: The Biological and Structural Landscape with Special Focus on the Anti-Cancer Proline-Based Scaffold. *Biomolecules* **2021**, *11*, 1515. [CrossRef]
31. Arnold, M.; Sierra, M.; Laversanne, M.; Soerjomataram, I.; Jemal, A.; Bray, F. Global patterns and trends in colorectal cancer incidence and mortality. *Gut* **2017**, *66*, 683–691. [CrossRef] [PubMed]
32. Sung, H.; Ferlay, J.; Siegel, R.L.; Laversanne, M.; Soerjomataram, I.; Jemal, A.; Bray, F. *Global Cancer Statistics* 2020: GLOBOCAN Estimates of Incidence and Mortality Worldwide for 36 Cancers in 185 Countries. *CA Cancer J. Clin.* **2021**, *71*, 209–249. [CrossRef] [PubMed]
33. Hull, R.; Francies, F.Z.; Oyomno, M.; Dlamini, Z. Colorectal cancer genetics, incidence and risk factors: In search for targeted therapies. *Cancer Manag. Res.* **2020**, *12*, 9869–9882. [CrossRef]
34. Anand, P.; Kunnumakkara, A.B.; Sundaram, C.; Harikumar, K.B.; Tharakan, S.T.; Lai, O.S.; Sung, B.; Aggarwal, B.B. Cancer is a preventable disease that requires major lifestyle changes. *Pharm. Res.* **2008**, *25*, 2097–2116. [CrossRef] [PubMed]
35. Rejhova, A.; Opattova, A.; Cumova, A.; Slíva, D.; Vodicka, P. Natural compounds and combination therapy in colorectal cancer treatment. *Eur. J. Med. Chem.* **2018**, *144*, 582–594. [CrossRef]
36. Alam, W.; Bouferraa, Y.; Haibe, Y.; Mukherji, D.; Shamseddine, A. Management of colorectal cancer in the era of COVID-19: Challenges and suggestions. *Sci. Prog.* **2021**, *104*, 00368504211010626. [CrossRef]
37. Fang, F.Q.; Guo, H.S.; Zhang, J.; Ban, L.Y.; Liu, J.W.; Yu, P.Y. Anti-cancer effects of 2-oxoquinoline derivatives on the HCT116 and LoVo human colon cancer cell lines. *Mol. Med. Rep.* **2015**, *12*, 8062–8070. [CrossRef] [PubMed]
38. McQuade, R.M.; Bornstein, J.C.; Nurgali, K. Anti-colorectal cancer chemotherapy-induced diarrhoea: Current treatments and side effects. *Int. J. Clin. Med.* **2014**, *5*, 393–406. [CrossRef]
39. Ismail, T.; Donati-Zeppa, S.; Akhtar, S.; Turrini, E.; Layla, A.; Sestili, P.; Fimognari, C. Coffee in cancer chemoprevention: An updated review. *Expert Opin. Drug Metab. Toxicol.* **2020**, *17*, 69–85. [CrossRef]
40. Steward, W.P.; Brown, K. Cancer chemoprevention: A rapidly evolving field. *Br. J. Cancer.* **2013**, *109*, 1–7. [CrossRef]
41. Rollas, S.; Küçükgüzel, Ş.G. Biological activities of hydrazone derivatives. *Molecules* **2007**, *12*, 1910–1939. [CrossRef] [PubMed]
42. Singh, M.; Raghav, N. Biological activities of hydrazones: A review. *Int. J. Pharm. Pharm. Sci.* **2011**, *3*, 26–32.
43. Verma, G.; Marella, A.; Shaquiquzzaman, M.; Akhtar, M.; Rahmat Ali, M.; Mumtaz Alam, M. A review exploring biological activities of hydrazones. *J. Pharm. Bioallied Sci.* **2014**, *6*, 69–80.
44. Corcé, V.; Gouin, S.G.; Renaud, S.; Gaboriau, F.; Deniaud, D. Recent advances in cancer treatment by iron chelators. *Bioorg. Med. Chem. Lett.* **2016**, *26*, 251–256. [CrossRef] [PubMed]
45. Kumar, H.S.; Parumasivam, T.; Jumaat, F.; Ibrahim, P.; Asmawi, M.Z.; Sadikun, A. Synthesis and evaluation of isonicotinoyl hydrazone derivatives as antimycobacterial and anticancer agents. *Med. Chem. Res.* **2014**, *23*, 269–279. [CrossRef]
46. Wirries, A.; Breyer, A.; Quint, K.; Schobert, R.; Ocker, M. Thymoquinone hydrazone derivatives cause cell cycle arrest in p53-competent colorectal cancer cells. *Exp. Ther. Med.* **2010**, *1*, 369–375. [CrossRef] [PubMed]
47. Patil, S.; Kuman, M.M.; Palvai, S.; Sengupta, P.; Basu, S. Impairing powerhouse in colon cancer cells by hydrazide–hydrazone-based small molecule. *ACS Omega* **2018**, *3*, 1470–1481. [CrossRef] [PubMed]
48. Yu, Y.; Gutierrez, E.; Kovacevic, E.; Saletta, F.; Obeidy, P.; Suryo Rahmanto, Y.; Richardson, D.R. Iron chelators for the treatment of cancer. *Curr. Med. Chem.* **2012**, *19*, 2689–2702. [CrossRef] [PubMed]
49. Li, L.-Y.; Peng, J.-D.; Zhou, W.; Qiao, H.; Deng, X.; Li, Z.-H.; Li, J.-D.; Fu, Y.-D.; Li, S.; Sun, K.; et al. Potent hydrazone derivatives targeting esophageal cancer cells. *Eur. J. Med. Chem.* **2018**, *148*, 359–371. [CrossRef]
50. Kitchen, D.B.; Decornez, H.; Furr, J.R.; Bajorath, J. Docking and scoring in virtual screening for drug discovery: Methods and applications. *Nat. Rev. Drug Discov.* **2004**, *3*, 935–949. [CrossRef]
51. Jang, G.R.; Harris, R.Z.; Lau, D.T. Pharmacokinetics and its role in small molecule drug discovery research. *Med. Res. Rev.* **2001**, *21*, 382–396. [CrossRef] [PubMed]
52. Eddershaw, P.J.; Beresford, A.P.; Bayliss, M.K. ADME/PK as part of a rational approach to drug discovery. *Drug Discov. Today* **2000**, *5*, 409–414. [CrossRef] [PubMed]
53. Li, A.P. Screening for human ADME/Tox drug properties in drug discovery. *Drug Discov. Today* **2001**, *6*, 357–366. [CrossRef] [PubMed]
54. Gomha, S.M.; Abdalla, M.A.; Abdelaziz, M.; Serag, N. Eco-friendly one-pot synthesis and antiviral evaluation of pyrazolyl pyrazolines of medicinal interest. *Turk. J. Chem.* **2016**, *40*, 484–498. [CrossRef]

55. Gomha, S.M.; Muhammad, Z.A.; Abdel-aziz, M.R.; Abdel-aziz, H.M.; Gaber, H.M.; Elaasser, M.M. One pot synthesis of new thiadiazolyl-pyridines as anticancer and antioxidant agents. *J. Heterocycl. Chem.* **2018**, *55*, 530–536. [CrossRef]
56. Gomha, S.M.; Riyadh, S.M.; Farag, B.; Al-Hussain, S.A.; Zaki, M.E.A.; Mohamed, M.A. Green Synthesis of Hydrazono-Thiazolones Using Vitamin B1 and Their Antibacterial Implications. *Green Chem. Lett. Rev.* **2024**, *17*, 2380746. [CrossRef]
57. Abdallah, M.A.; Riyadh, S.M.; Abbas, I.M.; Gomha, S.M. Synthesis and biological activities of 7-arylazo-7H-pyrazolo[5,1-c][1,2,4]triazolo-6(5H)-ones and 7-arylhydrazono-7H-[1,2,4]triazolo[3,4-b][1,3,4]thiadiazines. *J. Chin. Chem. Soc.* **2005**, *52*, 987–994. [CrossRef]
58. Gomha, S.M. A facile one-pot synthesis of 6,7,8,9-tetrahydrobenzo[4,5]thieno[2,3-d]-1,2,4-triazolo[4,5-a]pyrimidin-5-ones. *Monatsh. Chem.* **2009**, *140*, 213–220. [CrossRef]
59. Gomha, S.M.; Shawali, A.S.; Abdelhamid, O.A. Convenient method for the synthesis of various fused heterocycles via the utility of 4-acetyl-5-methyl-1-phenyl-pyrazole as a precursor. *Turk. J. Chem.* **2014**, *38*, 865–879. [CrossRef]
60. Gomha, S.M.; El-Sayed, A.A.A.; Alrehaily, A.; Elbadawy, H.M.; Farag, B.; Al-Shahri, A.A.; Alsenani, S.R.; Abdelgawad, F.E.; Zaki, M.E.A. Synthesis, molecular docking, in silico study, and evaluation of bis-thiazole-based curcumin derivatives as potential antimicrobial agents. *Res. Chem.* **2024**, *7*, 101504. [CrossRef]
61. Kurdi, A.A.; Morad, M.; Abumelha, H.M.; Alkhamis, K.; Aljohani, M.M.; Mersal, G.A.M.; El-Metwaly, N. Synthesis and characterization of novel VO(II) complexes from cyanoacetohydrazide-based ligands; Hirshfeld crystals, DFT study and biodiesel catalytic synthesis from waste frying oil. *J. Mol. Struct.* **2021**, *1240*, 130603. [CrossRef]
62. Gomha, S.M.; Abdelaziz, M.R.; Abdel-aziz, H.M.; Hassan, S.A. Green route synthesis and molecular docking of azines using cellulose sulfuric acid under microwave irradiation. *Crystals* **2023**, *13*, 260. [CrossRef]
63. Abolibda, T.Z.; Albalawi, M.; Abdelaziz, M.R.; Abdel-aziz, H.M.; Gomha, S.M. Synthesis and molecular docking of some novel 3-thiazolyl-coumarins as inhibitors of VEGFR-2 kinase. *Molecules* **2023**, *28*, 689. [CrossRef] [PubMed]
64. Caria, S.; Saka, Y.; Gervasio, F.L.; Tanaka, H. Structural analysis of phosphorylation-associated interactions of human MCC with Scribble PDZ domains. *FEBS J.* **2019**, *286*, 4910–4925. [CrossRef] [PubMed]
65. Ibrahim, M.S.; El-Gohary, N.S.; El-Messiry, M.A.; Mohamed, N.A.; Gomha, S.M. Mechanochemical synthesis and molecular docking studies of new azines bearing indole as anticancer agents. *Molecules* **2023**, *28*, 3869. [CrossRef] [PubMed]
66. Nafie, M.S.; Tantawy, M.A.; Elmgeed, G.A. Screening of different drug design tools to predict the mode of action of steroidal derivatives as anti-cancer agents. *Steroids* **2019**, *152*, 108485. [CrossRef] [PubMed]
67. Pires, D.E.; Blundell, T.L.; Ascher, D.B. pkCSM: Predicting small-molecule pharmacokinetic and toxicity properties using graph-based signatures. *J. Med. Chem.* **2015**, *58*, 4066–4072. [CrossRef] [PubMed]
68. Moda, T.L.; Bender, A.; Williams, A.J.; Ekins, S. PK/DB: Database for pharmacokinetic properties and predictive in silico ADME models. *Bioinformatics* **2008**, *24*, 2270–2271. [CrossRef] [PubMed]
69. Cao, D.; Sun, X.; Zhang, L.; Miao, R.; Zhang, H.; Jiang, H.; Wang, J. ADMET evaluation in drug discovery. 11. PharmacoKinetics Knowledge Base (PKKB): A comprehensive database of pharmacokinetic and toxic properties for drugs. *J. Chem. Inf. Model.* **2012**, *52*, 1132–1137. [CrossRef]
70. Cumming, J.G.; Davis, A.M.; Muresan, S.; Haeberlein, M.; Chen, H. Chemical predictive modelling to improve compound quality. *Nat. Rev. Drug Discov.* **2013**, *12*, 948–962. [CrossRef]

catalysts

MDPI

Article

Eco-Friendly Synthesis of Thiazole Derivatives Using Recyclable Cross-Linked Chitosan Hydrogel Biocatalyst Under Ultrasonic Irradiation as Anti-Hepatocarcinogenic Agents

Sobhi M. Gomha [1,*], Nahed A. Abd El-Ghany [2], Manal S. Ebaid [3], Tariq Z. Abolibda [1], Magdi E. A. Zaki [4,*], Mohammad Alhilal [5], Suzan Alhilal [6] and Nadia A. Mohamed [7]

1 Department of Chemistry, Faculty of Science, Islamic University of Madinah, Madinah 42351, Saudi Arabia; t.z.a@iu.edu.sa
2 Department of Chemistry, Faculty of Science, Cairo University, Giza 12613, Egypt; dr_nahed_055@yahoo.com
3 Department of Chemistry, College of Science, Northern Border University, Arar 91431, Saudi Arabia; manalshaabanebied@gmail.com
4 Department of Chemistry, Faculty of Science, Imam Mohammad Ibn Saud Islamic University (IMSIU), Riyadh 11623, Saudi Arabia
5 Department of Nursing, Faculty of Health Sciences, Mardin Artuklu University, Mardin 47200, Turkey; mohammadalhilal@artuklu.edu.tr
6 Department of Medical Services and Techniques, Vocational School of Health Services, Mardin Artuklu University, Mardin 47200, Turkey; suzanalhilal@artuklu.edu.tr
7 Department of Chemistry, College of Science, Qassim University, Buraidah 51452, Saudi Arabia; na.ahmed@qu.edu.sa
* Correspondence: smgomha@iu.edu.sa (S.M.G.); mezaki@imamu.edu.sa (M.E.A.Z.)

Abstract: In the current study, pyromellitimide benzoyl thiourea cross-linked chitosan (PIBTU-CS) hydrogel, was evaluated as a green biocatalyst for the efficient synthesis of novel thiazole derivatives. The PIBTU-CS hydrogel showcased key advantages, such as an expanded surface area and superior thermal stability, establishing it as a potent eco-friendly catalyst. By employing PIBTU-CS alongside ultrasonic irradiation, we successfully synthesized a series of novel thiazoles through the reaction of 2-(4-((2-carbamothioylhydrazineylidene)methyl)phenoxy)-N-(4-chlorophenyl)acetamide with a variety of hydrazonoyl halides (**6a–f**) and α-haloketones (**8a–c** or **10a**,**b**). A comparative analysis with TEA revealed that PIBTU-CS hydrogel consistently delivered significantly higher yields. This synthetic strategy provided several benefits, including mild reaction conditions, reduced reaction times, and consistently high yields. The robustness of PIBTU-CS was further underscored by its ability to be reused multiple times without a substantial reduction in catalytic efficiency. The structures of the synthesized thiazole derivatives were meticulously characterized using a range of analytical techniques, including IR, ^{1}H-NMR, ^{13}C-NMR, and mass spectrometry (MS), confirming their successful formation. These results underscore the potential of PIBTU-CS hydrogel as a sustainable and recyclable catalyst for the synthesis of heterocyclic compounds. Additionally, all synthesized products were tested for their anticancer activity against HepG2-1 cells, with several new compounds exhibiting good anticancer effects.

Keywords: chitosan; heterogeneous catalysis; hydrazonoyl halides; thiazoles; synthesis; anticancer

1. Introduction

Green chemistry has gained significant attention in recent years due to its potential to reduce human risk, minimize environmental pollution, decrease chemical hazards, and reduce waste generation. In organic synthesis, catalysts play a crucial role by reducing reaction times and enhancing product yields [1]. The development of heterogeneous catalytic systems for liquid-phase chemical and biochemical transformations has become a major research focus. These systems offer the advantage of ease of catalyst separation, recovery, and reusability, which are key for cleaner and more efficient processes [2].

Chitosan, a biopolymer derived from chitin, has attracted considerable attention in both scientific and engineering fields [3]. As a renewable and biodegradable material, chitosan-based catalysts are seen as promising candidates for organic catalysis due to their environmental friendliness [4]. Numerous studies have focused on modifying chitosan to enhance its hydrophilicity, chemical stability, solubility, and biological activity, making it suitable for a variety of applications [5].

Hydrophilic gels, known as hydrogels, are cross-linked materials that absorb significant amounts of water without dissolving. Their softness, smartness, and water retention capabilities make hydrogels unique materials. The water absorption capacity arises from hydrophilic functional groups on the polymer backbone, while their resistance to dissolution results from cross-links between polymer chains. In acidic environments, the amino groups of cross-linked chitosan are protonated, forming a cationic hydrogel that swells within the network. Various strategies, including cross-linking, grafting, and blending, have been employed to prepare chitosan hydrogels [6]. Reagents such as dialdehydes [7] and diisothiocyanates [8] are commonly used as chemical cross-linkers for chitosan.

Moreover, chitosan hydrogels have been successfully utilized as cationic catalysts in organic transformations [4,9]. Chitosan hydrogel offers several advantages as a catalyst, including the ability to retain reactant molecules within its cavities, facilitating their interactions and leading to higher product yields. In this work, a chitosan hydrogel was prepared by cross-linking with a functionalized crosslinker containing thiourea and imide moieties. These moieties, when protonated in acidic conditions, enhance the cationic nature of the polymeric matrix.

Ultrasound-assisted reactions offer several benefits, including shorter reaction times, simple experimental setups, high yields, improved selectivity, and cleaner processes [10,11]. Ultrasonic irradiation plays a pivotal role in such reactions, especially in cases where conventional methods require harsh conditions or prolonged reaction times [12,13].

The liver plays a vital role in regulating essential metabolic functions, including bile acid, fatty acid, cholesterol, and glucose metabolism [14–16]. Cancer is a complex and varied group of diseases that can affect almost any organ or tissue in the body, characterized by uncontrolled and accelerated cell growth, resulting in malignant tumors [17]. Globally, cancer remains one of the leading causes of death, posing a significant barrier to improving life expectancy across many countries [18,19]. According to GLOBOCAN 2020, around 19.3 million deaths worldwide were attributed to cancer [20]. Although chemotherapy is one of the primary treatments for cancer, its effectiveness is often compromised by a lack of selectivity and the development of drug resistance [21,22]. As a result, the development of safer and more effective anticancer therapies has become a major focus of global research. While several anticancer drugs, including the antitumor antibiotic doxorubicin, are currently in clinical use, recent research has increasingly concentrated on developing new chemically synthesized compounds to enhance cancer treatment.

The 1,3-thiazole core has been extensively explored as a key structure in the search for new lead compounds in drug discovery. Various thiazole-containing therapies have gained recognition, including tiazofurin, an inhibitor of IMP dehydrogenase [23]; dasatinib, a Bcr-Abl tyrosine kinase inhibitor [24]; dabrafenib, which inhibits the enzyme B-RAF [25]; ixabepilone, known for stabilizing microtubules [26]; and epothilone, which disrupts microtubule function (Figure 1) [27]. These compounds either became top-selling drugs or served as foundational lead structures in drug development. Several research teams have also investigated the thiazole scaffold's potential for designing anticancer drugs [28–31].

Considering the favorable results achieved in our previous studies on green synthesis [32–34], we now introduce an eco-friendly protocol for the synthesis of 1,3-thiazole derivatives using PIBTU-CS hydrogel as an efficient and recyclable base catalyst. This method utilizes ultrasonic irradiation, offering a mild and energy-efficient alternative to conventional heating while maintaining high product yields and reaction efficiency. Building on our earlier work, this study further emphasizes the eco-friendly characteristics and catalytic efficiency of the PIBTU-CS hydrogel. The reusability of the hydrogel, coupled with its minimal environmental impact, highlights its potential as a sustainable catalyst for green chemistry applications.

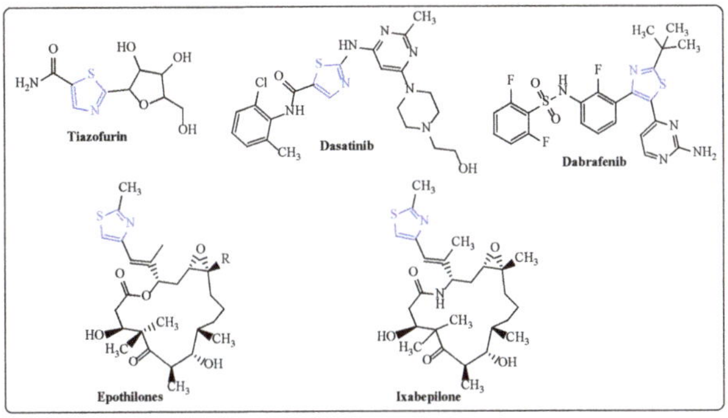

Figure 1. Examples of thiazole-bearing anticancer drugs.

2. Results and Discussion

2.1. Synthesis and Characterization of PIBTU-CS Hydrogel

Using a previously reported three-step method [8], N,N′-bis [4-(chlorocarbonyl)phenyl] pyromellitimide was synthesized and further reacted to form PIBTU-CS hydrogel. Pyromellitic anhydride and p-aminobenzoic acid were dissolved in DMF, followed by sequential reactions with acetic anhydride and thionyl chloride, yielding N,N′-bis[4-(isothiocyanate carbonyl)phenyl]pyromellitimide. This compound was then reacted with chitosan, producing PIBTU-CS hydrogel (Scheme 1; details in SI).

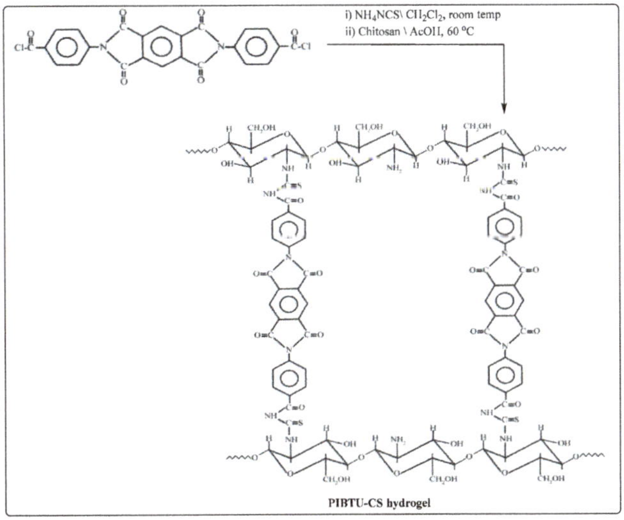

Scheme 1. Synthesis of PIBTU-CS hydrogel.

The hydrogel's structure was confirmed using elemental analysis, FTIR, ^{1}H-NMR, XRD, and SEM, aligning with previous findings [8]. Elemental analysis showed increased nitrogen (8.95%) and sulfur (3.13%), confirming successful preparation. FTIR indicated NH, C=O, and N–C–S groups, while ^{1}H-NMR revealed characteristic signals for glucose and aromatic rings. XRD showed reduced crystallinity due to crosslinking, and SEM displayed a porous and channel-like surface, enhancing water permeation compared to smooth chitosan.

PIBTU-CS-2 was selected for this study based on its superior properties observed in previous research [12]. Among the four hydrogels synthesized with varying degrees of crosslinking, PIBTU-CS-2 demonstrated an optimal balance of porosity, thermal stability, and catalytic efficiency, making it the most suitable candidate for the synthesis of thiazole derivatives. These properties, combined with its reusability, highlight PIBTU-CS-2 as a robust and eco-friendly catalyst for green synthetic applications.

2.2. Thermogravimetric (TG) Analysis

The application of chitosan as an eco-friendly green biocatalyst for organic synthesis needs to reinforce its stabilization against thermal deterioration. There are many strategies suggested to improve the stability of chitosan against elevated temperatures, including the formation of chitosan derivatives, chitosan copolymers, chitosan blends, and chitosan metal nanocomposites. The chemical cross-linking process has already been established as an effective way to boost chitosan for high heat resistance. In the current study, we used chitosan that has previously been crosslinked with N,N′-bis [4-(isothiocyanate carbonyl)phenyl]pyromellitimide to obtain chitosan/pyromellitimide benzoyl thiourea super-porous hydrogel (PIBTU-CS).

The alteration that arose in the thermal stability of chitosan subsequent to crosslinking could be evaluated employing thermogravimetric analysis (TG) mensuration. The degradation demeanor and the heat stability of PIBTU-CS hydrogel were inspected via temperatures with a scope of 25 and 500 °C, in a nitrogen atmosphere at a flow rate of 30 mL min^{-1}, and at a rate of heating of 10 °C min^{-1}.

Table 1 and Figure 2 showed the loss in mass that was observed during the measuring of TG of chitosan and PIBTU-CS hydrogel. Both of them showed a similar degradation pattern, encompassing double-steps event at which remarkable losses in mass were noticed.

In the early stage, at temperatures between 25 and 150 °C, the recorded first losses in mass of PIBTU-CS was 1.00% versus 9.00% for parent chitosan (Table 1). During this stage, the adsorbed water was vaporized from the inspected samples. The losses in mass at temperatures over 100 °C could be ascribed to breaking down the hydrogen bonding that formed between H_2O and the functional groups of the examined samples.

Table 1. TG measurements data of chitosan and PIBTU-CS hydrogel.

Temperature (°C)	Weight Loss (%) [a]	
	Chitosan	PIBTU-CS
25	0.41	0.00
50	1.66	0.00
100	6.04	0.00
150	9.00	1.00
200	10.48	2.50
250	12.53	4.70
260	14.26	7.59
270	17.53	8.51
280	22.20	10.00
290	27.29	11.35
300	31.41	13.63
310	33.80	16.51
320	35.49	19.44
330	37.07	22.00
340	38.30	24.65
350	39.36	27.00
400	42.60	29.00
450	44.77	33.99
500	47.06	37.16

[a] *Weight loss (%) indicates thermal stability under test conditions.*

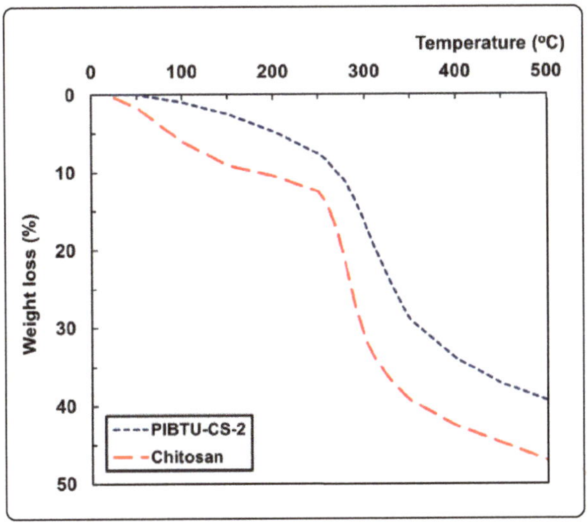

Figure 2. TG analysis of chitosan and PIBTU-CS hydrogel.

In the subsequent stage, a steep thermal deterioration operation was noticed which initiated at around 250 °C. At 500 °C, PIBTU-CS showed a 37.16% loss in mass in comparison to 47.06% of the virgin chitosan (Table 1). The decomposition process is proceeded by multi-steps reaction, comprising dehydrating and indiscriminate splitting of the saccharide nuclei as well as the emancipation and fumigation of the decomposition products. The results confirmed that PIBTU-CS has better thermal stability than that of chitosan. This was clarified not only from its lesser mass losses but also from its greater mass residues at all the inspected temperatures. This is associated with the possession of the cross linkages in the imide moieties, thiourea groups, and aromatic nuclei that are known by their high ability to tolerate the high heat.

2.3. *Synthesis of Thiazole Derivatives Using PIBTU-CS Hydrogel as Basic Heterogeneous Catalyst*

In this study, we present a simple and efficient method for the synthesis of novel thiazoles. The key intermediate, thiosemicarbazone derivative **5**, was initially prepared by condensing *N*-(4-chlorophenyl)-2-(4-formylphenoxy)acetamide **3** [35] with thiosemicarbazide **5** in acidic ethanol solution following the reported method [36] (Scheme 2). The structure of thiosemicarbazone derivative **5** was confirmed through spectroscopic data, including IR, ^{1}H NMR, and MS.

Subsequently, new thiazole derivatives **7a–f** were synthesized by reacting equimolar amounts of thiosemicarbazone derivative **5** with 2-oxo-N-arylpropanehydrazonoyl halides **6a–f** in ethanol, using either triethylamine (TEA) or PIBTU-CS hydrogel as a base catalyst (Scheme 2). The structures of the newly synthesized thiazoles **7a–f** were validated through various spectral techniques (IR, MS, ^{1}H-NMR, ^{13}C-NMR) and elemental analysis. For example, the ^{1}H-NMR spectrum of compound **7a** displayed the expected signals, including five singlets at 2.46 ppm (thiazole-CH_3), 4.79 ppm (CH_2O), 8.10 ppm (CH=N), and two D_2O-exchangeable NH signals at 10.24 and 11.31 ppm, along with thirteen multiplet aromatic proton signals between 6.98 and 7.95 ppm. The infrared spectrum of **7a** also showed characteristic bands at 3413, 3306, and 1681 cm^{-1}, corresponding to the two NH and C=O groups, respectively.

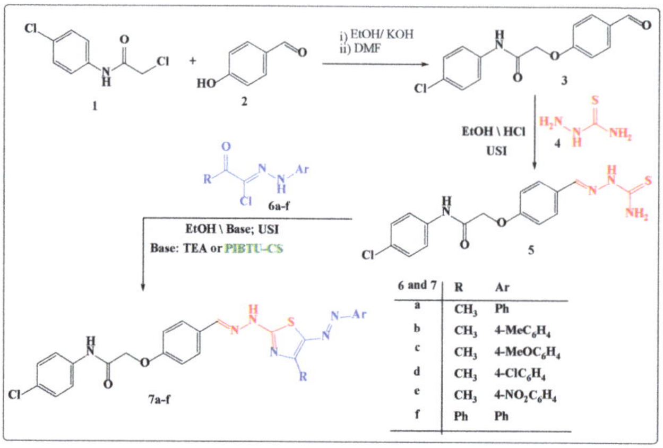

Scheme 2. Synthesis of arylazothiazoles **7a–f**.

The heightened catalytic activity of PIBTU-CS can be attributed to the presence of amine groups (CS–NH and CS–NH_2), which facilitate proton abstraction from the thiol form of the thiosemicarbazone derivative **5**. This process enables the formation of a thiolate intermediate that reacts with hydrazonoyl halide, producing the S-alkylated product. The S-alkylated product subsequently undergoes dehydrative cyclization, yielding the desired thiazole derivatives (**7a–f**) as shown in Scheme 3. This mechanism highlights the critical role of PIBTU-CS's functional groups in promoting efficient and selective catalysis under mild conditions.

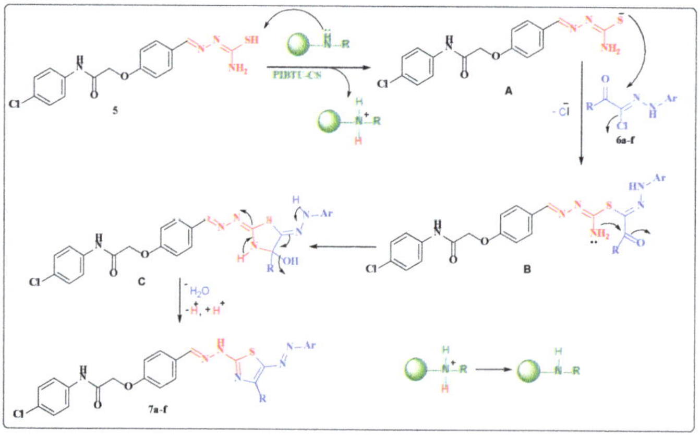

Scheme 3. Proposed mechanism for synthesis of arylazothiazoles **7a–f**.

Similarly, the reaction of compound **5** with 2-bromo-1-arylethan-1-ones **8a–c**, 3-chloro-pentane-2,4-dione **10a**, and ethyl 2-chloro-3-oxobutanoate **10b**, using TEA and PIBTU-CS hydrogel as basic catalysts under the same conditions described in Scheme 4, led to the formation of thiazoles **9a–c** and **11a,b** as the final products. The chemical structures of the target compounds **9a–c** and **11a,b** were verified through comprehensive analysis of their IR, ^{1}H-NMR, ^{13}C-NMR, and MS spectra, as outlined in the Experimental section. The progress of all reactions was monitored using thin-layer chromatography (TLC).

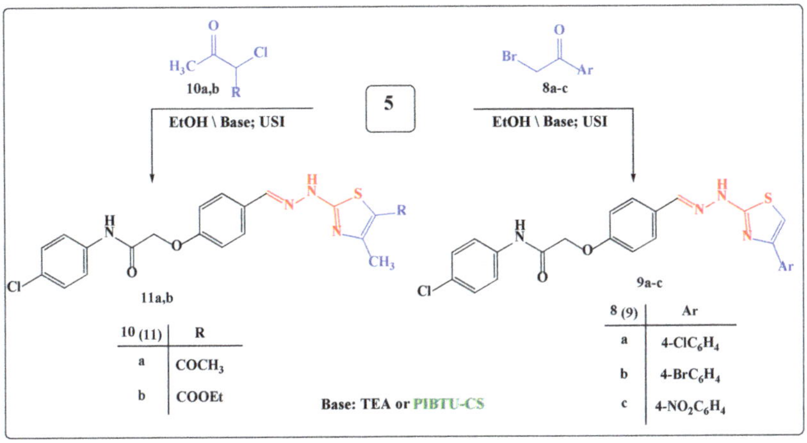

Scheme 4. Synthesis of thiazoles **9a–c** and **11a,b**.

The yields of products **7a–f**, **9a–c**, and **11a,b** were evaluated and are summarized in Table 2.

Table 2. The reaction time and yield percentages of thiazoles **7a–f**, **9a–c**, and **11a,b** were compared using two different basic catalysts under ultrasonic irradiation.

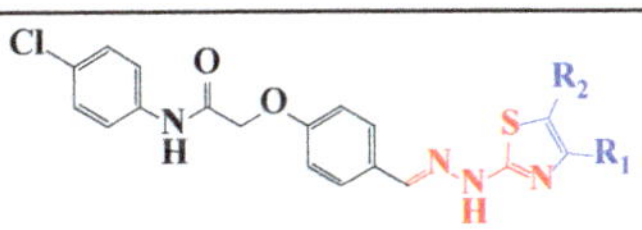

7a-f, 9a-c, and **11a,b**

Compound.	R_1	R_2	TEA		PIBTU-CS Hydrogel	
			Time (min)	(%) Yield [b]	Time (min)	(%) Yield [b]
7a	Me	C_6H_5	48	69	25	87
7b	Me	4-MeC_6H_4	51	71	27	89
7c	Me	4-$MeOC_6H_4$	57	72	35	86
7d	Me	4-ClC_6H_4	47	70	22	90
7e	Me	4-$NO_2C_6H_4$	32	73	20	87
7f	Ph	Ph	49	74	30	90
9a	4-ClC_6H_4	H	53	71	26	88
9b	4-BrC_6H_4	H	39	72	28	87
9c	4-$NO_2C_6H_4$	H	44	68	21	89
11a	Me	$COCH_3$	49	70	29	87
11b	Me	COOEt	60	67	31	85

[b] *Yield (%) represents the isolated product after purification.*

The investigation into the most effective basic catalyst commenced at the start of the study (Table 2).

As demonstrated in Table 2, PIBTU-CS hydrogel outperformed traditional TEA as a basic catalyst under ultrasonic irradiation. Replacing triethylamine with PIBTU-CS hydrogel resulted in higher yields of the desired products (**7a–f**, **9a–c**, and **11a,b**) and shorter reaction times, all under the same reaction conditions.

To identify the optimal conditions and variables for the reaction of **5** with **6a** to synthesize thiazole derivative **7a**, we explored factors such as catalyst loading, temperature, solvent, and reaction time, using PIBTU-CS hydrogel under ultrasonic irradiation (USI).

We first investigated the effect of catalyst concentration on the yield of **7a** (Table 3, entries 1–3). The highest yield (87%) was obtained with a catalyst loading of 15 mol% (Table 3, entry 3), while reducing the catalyst amount resulted in lower yields (Table 3). Next, we examined the performance of various solvents under ultrasonic irradiation (USI)

(Table 3, entries 3, 4, and 5). Ethanol proved to be the most effective, providing the highest yield and the fastest reaction rate (Table 3, entry 3). We then optimized the reaction time (Table 3, entries 3, 6, and 7), identifying 25 min as the most efficient for producing **7a** (Table 3, entry 3). Lastly, we explored the impact of temperature (Table 3, entries 3, 8, 9, and 10). As the temperature increased from 25 °C to 35 °C and 50 °C, the yields improved from 75% to 82% and ultimately 87%. The optimal temperature for the reaction was determined to be 40 °C (Table 3, entry 3).

Table 3. The synthesis of compound **7a** was optimized by adjusting catalyst loading, solvent, reaction time, and temperature.

Entry	Catalyst (Wt%)	Solvent	Time (min)	Temperature (°C)	Yield (%)
1	5	EtOH	25	40	61
2	10	EtOH	25	40	83
3 [a]	15	EtOH	25	40	87
4	15	DMSO	25	40	78
5	15	Dioxan	25	40	80
6	15	EtOH	20	40	81
7	15	EtOH	30	40	87
8	15	EtOH	25	25	75
9	15	EtOH	25	35	82
10	15	EtOH	25	50	87

[a] *The optimal reaction condition for synthesizing compound **7a**.*

The recyclability of PIBTU-CS hydrogel as a basic catalyst was evaluated by assessing its performance over multiple reaction cycles. After each use, the catalyst was cleaned with distilled water and dried at 60 °C for 30 min before being reused. The catalyst maintained high catalytic activity over three consecutive cycles with only a slight decrease in efficiency, demonstrating its potential for reuse under optimal conditions, as detailed in Table 4.

Table 4. Recyclability of PIBTU-CS hydrogel as basic catalyst.

State of Catalyst	Fresh Catalyst	Run 1	Run 2	Run 3	Run 4
Product **7a** (% Yield) [b]	87	85	82	79	55

[b] *Yield (%) represents the isolated product after purification.*

Based on the data in Table 4, the catalyst retained approximately 90.8% of its performance after three runs, with an average reduction of about 2.6% per run. However, in the fourth run, it lost around 30.4% of its performance, dropping to a 55% yield.

The optimal reaction conditions for synthesizing product **7a**, as shown in Table 3, involve the reaction of **5** with **6a** in ethanol under ultrasonic irradiation (USI) with PIBTU-CS hydrogel (15% wt) at 40 °C for 25 min. Under these conditions, the irradiation of **5** with **6b–f** resulted in the successful formation of thiazole derivatives **7b–f** (Scheme 2), along with **9a–c** and **11a,b** (Scheme 3).

2.4. Antitumor Activity

The synthesized compounds (**7a–f**, **9a–c**, and **11a,b**) were evaluated for cytotoxicity against the liver carcinoma cell line (HEPG2-1) using the MTT assay, with doxorubicin serving as the control drug. Data from these tests were used to generate dose–response curves, which helped determine the concentration (μM) of each compound required to achieve a 50% reduction in cell viability (IC_{50}). The cytotoxicity findings were expressed as mean IC_{50} values, derived from three independent experiments (Table 5, Figure 3).

Table 5. Cytotoxic activity of the tested compounds against HEPG2-1.

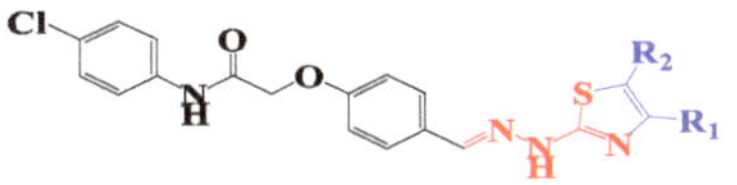

7a-f , 9a-c, and 11a,b

Compound	R_1	R_2	IC_{50} (μM) [a]
7a	Me	$C_6H_5N{=}N$-	2.31 ± 0.57
7b	Me	4-$MeC_6H_4N{=}N$-	1.25 ± 0.67
7c	Me	4-$MeOC_6H_4N{=}N$-	0.73 ± 0.48
7d	Me	4-$ClC_6H_4N{=}N$-	9.04 ± 0.77
7e	Me	4-$NO_2C_6H_4N{=}N$-	14.36 ± 0.53
7f	Ph	Ph	35.27 ± 0.75
9a	4-ClC_6H_4	H	8.08 ± 0.96
9b	4-BrC_6H_4	H	7.06 ± 0.48
9c	4-$NO_2C_6H_4$	H	17.35 ± 0.94
11a	Me	$COCH_3$	1.42 ± 0.71
11b	Me	COOEt	1.03 ± 0.29
Doxorubicin	-	-	0.31 ± 0.48

[a] *IC_{50} (μM) shows the concentration needed for 50% inhibition of cell viability.*

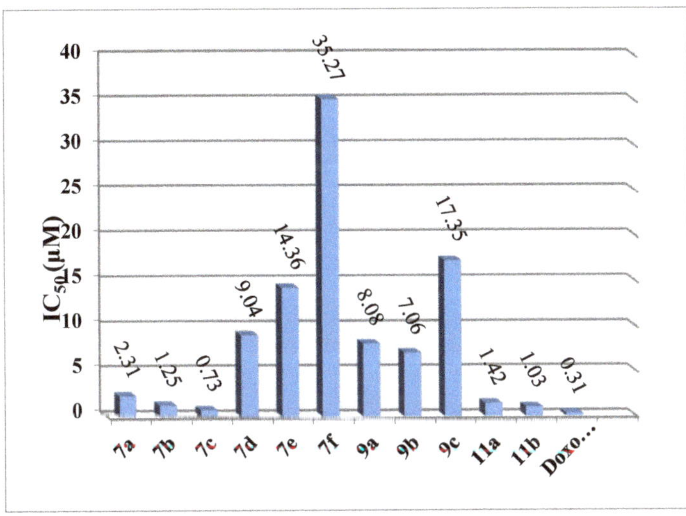

Figure 3. Cytotoxic effects of the tested compounds on HEPG2-1 cells.

The structure-activity relationship (SAR) analysis for the synthesized compounds (**7a–f**, **9a–c**, and **11a**,**b**) against the HEPG2-1 liver carcinoma cell line reveals key trends in cytotoxic activity based on the nature of the substituents at R_1 and R_2.

For compounds **7a–f**, electron-donating groups like methoxy (-OMe) and methyl (-Me) at the para position significantly enhance cytotoxicity. Compound **7c**, with a methoxy group, showed the highest potency (IC_{50} = 0.73 μM), while **7b** with a methyl group followed closely (IC_{50} = 1.25 μM). In contrast, the unsubstituted **7a** (IC_{50} = 2.31 μM) had moderate activity, and electron-withdrawing groups like chloro (-Cl) in **7d** (IC_{50} = 9.04 μM) and nitro (-NO_2) in **7e** (IC_{50} = 14.36 μM) reduced potency. Bulky phenyl groups in **7f** led to the lowest activity (IC_{50} = 35.27 μM).

In compounds **9a–c**, halogen substituents, such as chlorine (-Cl) in **9a** (IC_{50} = 8.08 μM) and bromine (-Br) in **9b** (IC_{50} = 7.06 μM), maintained moderate activity. However, the nitro group in **9c** resulted in lower potency (IC_{50} = 17.35 μM), reinforcing the negative impact of electron-withdrawing groups.

Compounds **11a** and **11b**, with carbonyl substituents, showed strong cytotoxicity. Compound **11b**, containing an ethyl ester (-COOEt), exhibited an IC_{50} of 1.03 μM, while **11a**, with an acetyl group (-$COCH_3$), showed an IC_{50} of 1.42 μM, indicating that carbonyl groups enhance activity, with esters being slightly more favorable.

Compared to doxorubicin (IC_{50} = 0.31 μM), compounds **7c** and **11b** showed good cytotoxicity, though not exceeding the reference drug's potency. The SAR analysis indicates that electron-donating groups (e.g., methoxy, methyl) and carbonyl-containing substituents enhance cytotoxicity, while electron-withdrawing groups and bulky substituents reduce activity.

3. Materials and Methods

3.1. Apparatus and Instrumentations

Details Were Inserted in SI.

3.2. Synthesis and Characterization of (PIBTU-CS) Hydrogel

N,N′-Bis[4-(chlorocarbonyl)phenyl]pyromellitimide was reacted with ammonium thiocyanate in methylene chloride using polyethylene glycol-400 as a catalyst. The resulting N,N′-bis[4-(isothiocyanate carbonyl)phenyl]pyromellitimide was combined with a chitosan solution in acetic acid and stirred at 60 °C to form the PIBTU-CS hydrogel (Scheme 1). The hydrogel was neutralized, dewatered, dried, and labeled as PIBTU-CS, corresponding to PIBTU-CS-2 in previous research [12]. The structure and morphology were confirmed using elemental analysis, FTIR, ^{1}H-NMR, XRD, and SEM.

*3.3. Synthesis of 2-(4-((2-Carbamothioylhydrazineylidene)Methyl)Phenoxy)-N-(4-Chlorophenyl)Acetamide (**5**)*

A mixture of *N*-(4-chlorophenyl)-2-(4-formylphenoxy)acetamide **3** (2.89 g, 10 mmol) and thiosemicarbazide **4** (0.91 g, 10 mmol) was dissolved in 30 mL of EtOH and treated with a catalytic amount of concentrated hydrochloric acid. The reaction mixture was subjected to ultrasonic irradiation in a water bath at 50 °C for 20 min. After cooling, the resulting precipitate was filtered, washed with ethanol, and recrystallized from acetic acid to yield thiosemicarbazones **5** as a yellowish-white solid with a 77% yield: m.p. 258–260 °C (Lit. m.p. 256–260 °C [37]).

*3.4. Synthesis of Thiazole Derivatives **7a–f**, **9a–c**, and **11a,b***

Method A: An equivalent amount of triethylamine (0.101 g, 1 mmol) was added to a solution containing equimolar amounts of thiosemicarbazone derivative **5** (0.362 g, 1 mmol) and the corresponding hydrazonoyl halides **6a–f** or α-haloketones **8a–c** or **10a,b** (1 mmol each) in 20 mL of ethanol. The mixture was irradiated in a water bath at 40 °C for 30–60 min using an ultrasonic generator operating at a frequency of 40 kHz and a power of 250 W. The reaction progress was monitored by thin-layer chromatography (TLC), and irradiation was continued until the starting materials were fully consumed and the desired product formed. Upon completion, the reaction mixture was allowed to cool, and the resulting red precipitate was filtered, washed with ethanol, and dried. The product was then recrystallized from the proper solvent to afford the thiazole derivatives **7a–f**, **9a–c**, and **11a,b**. The physical constants for the synthesized thiazoles **7a–f**, **9a–c**, and **11a,b** are provided below.

Method B: A PIBTU-CS hydrogel (15% wt) was added to a solution of equimolar amounts of thiosemicarbazone derivative **5** (0.362 g, 1 mmol) and the appropriate hydrazonoyl halides **6a–f** or α-haloketones **8a–c** or **10a,b** (1 mmol each) in 20 mL of ethanol. The mixture was subjected to ultrasonic irradiation using a generator operating at 40 kHz and 250 W in a water bath at 40 °C for 20–35 min. After completion, the PIBTU-CS hydrogel was removed by filtration, and the excess solvent was evaporated under reduced pressure. The resulting residue was treated with methanol to remove impurities and facilitate the isolation of the crude products. The solid products were then filtered, recrystallized from the appropriate solvent, and their purity was confirmed by melting point analysis and microanalysis.

***N*-(4-Chlorophenyl)-2-(4-((*E*)-(2-(4-methyl-5-((*E*)-phenyldiazenyl)thiazol-2-yl)hydrazineylidene)methyl)phenoxy)acetamide (7a).**

Red solid, mp. 195–197 °C (EtOH); IR (KBr): *υ* 3413, 3306 (2NH), 3052, 2924 (CH), 1681 (C=O), 1607 (C=N) cm^{-1}; ^{1}H NMR (DMSO-d_6): *δ* 2.46 (s, 3H, CH_3), 4.79 (s, 2H, -CH_2O), 6.98–7.95 (m, 13H, Ar-H), 8.10 (s, 1H, CH=N), 10.24 (br s, 1H, NH), 11.31 (br s, 1H, NH) ppm; MS *m*/*z* (%): 504 (M^+, 38). Anal. Calcd for $C_{25}H_{21}ClN_6O_2S$ (504.99): C, 59.46; H, 4.19; N, 16.64. Found: C, 59.37; H, 4.06; N, 16.55%.

***N*-(4-Chlorophenyl)-2-(4-((*E*)-(2-(4-methyl-5-((*E*)-*p*-tolyldiazenyl)thiazol-2-yl)hydrazineylidene)methyl)phenoxy)acetamide (7b).**

Red solid, mp. 181–183 °C (EtOH); IR (KBr): *υ* 3403, 3326 (2NH), 3044, 2927 (CH), 1679 (C=O), 1609 (C=N) cm^{-1}; ^{1}H NMR (DMSO-d_6): *δ* 2.29 (s, 3H, CH_3), 2.46 (s, 3H, CH_3), 4.73 (s, 2H, -CH_2O), 6.98–7.96 (m, 12H, Ar-H), 8.09 (s, 1H, CH=N), 10.24 (br s, 1H, NH), 11.32 (br s, 1H, NH) ppm; MS *m*/*z* (%): 519 (M^+, 33). Anal. Calcd for $C_{26}H_{23}ClN_6O_2S$ (519.02): C, 60.17; H, 4.47; N, 16.19. Found: C, 60.01; H, 4.33; N, 16.03%.

***N*-(4-Chlorophenyl)-2-(4-((*E*)-(2-(5-((*E*)-(4-methoxyphenyl)diazenyl)-4-methylthiazol-2-yl)hydrazineylidene)methyl)phenoxy)acetamide (7c).**

Red solid, mp. 175–177 °C (EtOH); IR (KBr): *υ* 3417, 3273 (2NH), 3058, 2929 (CH), 1683 (C=O), 1603 (C=N) cm^{-1}; ^{1}H NMR (DMSO-d_6): *δ* 2.46 (s, 3H, CH_3), 3.77 (s, 3H, OCH_3), 4.72 (s, 2H, -CH_2O), 6.98–7.96 (m, 12H, Ar-H), 8.09 (s, 1H, CH=N), 10.23 (br s, 1H, NH), 11.32 (br s, 1H, NH) ppm; MS *m*/*z* (%): 535 (M^+, 47). Anal. Calcd for $C_{26}H_{23}ClN6O_3S$ (535.02): C, 58.37; H, 4.33; N, 15.71. Found: C, 58.21; H, 4.25; N, 15.57%.

***N*-(4-Chlorophenyl)-2-(4-((*E*)-(2-(5-((*E*)-(4-chlorophenyl)diazenyl)-4-methylthiazol-2-yl)hydrazineylidene)methyl)phenoxy)acetamide (7d).**

Red solid, mp. 207–209 °C (Dioxan); IR (KBr): *υ* 3402, 3298 (2NH), 3052, 2927 (CH), 1680 (C=O), 1609 (C=N) cm^{-1}; ^{1}H NMR (DMSO-d_6): *δ* 2.46 (s, 3H, CH_3), 4.73 (s, 2H, -CH_2O), 6.98–7.96 (m, 12H, Ar-H), 8.11 (s, 1H, CH=N), 10.22 (br s, 1H, NH), 11.32 (br s, 1H, NH) ppm; MS *m*/*z* (%): 539 (M^+, 56). Anal. Calcd for $C_{25}H_{20}Cl_2N_6O_2S$ (539.441): C, 55.66; H, 3.74; N, 15.58. Found: C, 55.53; H, 3.71; N, 15.47%.

***N*-(4-Chlorophenyl)-2-(4-((*E*)-(2-(4-methyl-5-((*E*)-(4-nitrophenyl)diazenyl)thiazol-2-yl)hydrazineylidene)methyl)phenoxy)acetamide (7e).**

Brown solid, mp. 193–195 °C (EtOH); IR (KBr): *υ* 3415, 3299 (2NH), 3055, 2931 (CH), 1688 (C=O), 1613 (C=N) cm^{-1}; ^{1}H NMR (DMSO-d_6): *δ* 2.46 (s, 3H, CH_3), 4.79 (s, 2H, -CH_2O), 6.98–7.95 (m, 12H, Ar-H), 8.16 (s, 1H, CH=N), 10.22 (br s, 1H, NH), 11.32 (br s, 1H, NH) ppm; MS *m*/*z* (%): 549 (M^+, 100). Anal. Calcd for $C_{25}H_{20}ClN_7O_4S$ (549.99): C, 54.60; H, 3.67; N, 17.83. Found: C, 54.39; H, 3.44; N, 17.72%.

***N*-(4-Chlorophenyl)-2-(4-((*E*)-(2-(4-phenyl-5-((*E*)-phenyldiazenyl)thiazol-2-yl)hydrazineylidene)methyl)phenoxy)acetamide (7f).**

Orange solid, mp. 219–241 °C (DMF); IR (KBr): *υ* 3409, 3302 (2NH), 3039, 2927 (CH), 1681 (C=O), 1603 (C=N) cm^{-1}; ^{1}H NMR (DMSO-d_6): *δ* 4.73 (s, 2H, -CH_2O), 6.98–7.96 (m, 18H, Ar-H), 8.12 (s, 1H, CH=N), 10.22 (br s, 1H, NH), 11.32 (br s, 1H, NH) ppm; MS *m*/*z* (%): 567 (M^+, 58). Anal. Calcd for $C_{30}H_{23}ClN_6O_2S$ (567.06): C, 63.54; H, 4.09; N, 14.82. Found: C, 63.42; H, 4.01; N, 14.73%.

(*E*)-*N*-(4-Chlorophenyl)-2-(4-((2-(4-(4-chlorophenyl)thiazol-2-yl)hydrazineylidene) methyl)phenoxy)acetamide (9a).

Red solid, mp. 254–256 °C (DMF); IR (KBr): *υ* 3399, 3297 (2NH), 3038, 2979 (CH), 1674 (C=O), 1602 (C=N) cm^{-1}; ^{1}H NMR (DMSO-d_6): *δ* 4.73 (s, 2H, -CH_2O), 6.98–7.96 (m, 13H, 12Ar-H and thiazole-H5), 8.09 (s, 1H, CH=N), 10.23 (br s, 1H, NH), 11.33 (br s, 1H, NH) ppm; MS *m*/*z* (%): 497 (M^+, 32). Anal. Calcd for $C_{24}H_{18}Cl_2N_4O_2S$ (497.39): C, 57.95; H, 3.65; 11.26. Found: C, 57.72; H, 3.49; 11.17%.

(*E*)-2-(4-((2-(4-(4-Bromophenyl)thiazol-2-yl)hydrazineylidene)methyl)phenoxy)-*N*-(4-chlorophenyl)acetamide (9b).

Brown solid, mp. 241–243 °C (DMF); IR (KBr): v 3413, 3305 (2NH), 3051, 2926 (CH), 1679 (C=O), 1601 (C=N) cm^{-1}; ^1H NMR (DMSO-d_6): δ 4.73 (s, 2H, -CH_2O), 7.34–7.96 (m, 13H, 12Ar-H and thiazole-H5), 8.10 (s, 1H, CH=N), 10.23 (br s, 1H, NH), 11.32 (br s, 1H, NH) ppm; MS m/z (%): 541 (M^+, 60). Anal. Calcd for $C_{24}H_{18}BrClN_4O_2S$ (541.85): C, 53.20; H, 3.35; N, 10.34. Found: C, 53.13; H, 3.31; N, 10.25%.

(*E*)-*N*-(4-Chlorophenyl)-2-(4-((2-(4-(4-nitrophenyl)thiazol-2-yl)hydrazineylidene)methyl)phenoxy)acetamide (9c).

Brown solid, mp. 247–249 °C (DMF); IR (KBr): v 3422, 3325 (2NH), 3050, 2932 (CH), 1682 (C=O), 1618 (C=N) cm^{-1}; ^1H NMR (DMSO-d_6): δ 4.74 (s, 2H, -CH_2O), 7.35–8.08 (m, 13H, 12Ar-H and thiazole-H5), 8.23 (s, 1H, CH=N), 10.22 (br s, 1H, NH), 11.33 (br s, 1H, NH) ppm; MS m/z (%): 507 (M^+, 92). Anal. Calcd for $C_{24}H_{18}ClN_5O_4S$ (507.95): C, 56.75; H, 3.57; N, 13.79. Found: C, 56.68; H, 3.52; N, 13.61%.

(*E*)-2-(4-((2-(5-Acetyl-4-methylthiazol-2-yl)hydrazineylidene)methyl)phenoxy)-N-(4-chlorophenyl)acetamide (11a).

Yellow solid, mp. 213–215 °C (DMF); IR (KBr): v 3419, 3283 (2NH), 3037, 2924 (CH), 1707, 1680 (2C=O), 1604 (C=N) cm^{-1}; ^1H NMR (DMSO-d_6): δ 2.37 (s, 3H, CH_3), 2.46 (s, 3H, CH_3), 4.73 (s, 2H, -CH_2O), 7.00–7.96 (m, 8H, Ar-H), 8.09 (s, 1H, CH=N), 10.23 (br s, 1H, NH), 11.32 (br s, 1H, NH) ppm; MS m/z (%): 442 (M^+, 63). Anal. Calcd for $C_{21}H_{19}ClN_4O_3S$ (442.92): C, 56.95; H, 4.32; N, 12.65. Found: C, 56.64; H, 4.27; N, 12.49%.

Ethyl (*E*)-2-(2-(4-(2-((4-chlorophenyl)amino)-2-oxoethoxy)benzylidene)hydrazineyl)-4- methylthiazole-5-carboxylate (11b).

Yellow solid, mp. 191–193 °C (EtOH); IR (KBr): v 3426, 3305 (2NH), 3044, 2927 (CH), 1729, 1680 (2C=O), 1606 (C=N) cm^{-1}; ^1H NMR (DMSO-d_6): δ 1.25 (t, 3H, CH_3CH_2), 2.37 (s, 3H, CH_3), 4.27 (q, 2H, CH_3CH_2), 4.72 (s, 2H, -CH_2O), 6.98–7.96 (m, 8H, Ar-H), 8.09 (s, 1H, CH=N), 10.22 (br s, 1H, NH), 11.32 (br s, 1H, NH) ppm; MS m/z (%): 472 (M^+, 33). Anal. Calcd for $C_{22}H_{21}ClN_4O_4S$ (472.94): C, 55.87; H, 4.48; N, 11.85. Found: C, 55.71; H, 4.39; N, 11.77%.

3.5. Anticancer Activity

The assessment of cytotoxicity for the synthesized compounds was measured following the method reported in reference [33,34,38].

4. Conclusions

In conclusion, the PIBTU-CS hydrogel proved to be an effective and eco-friendly biocatalyst for the synthesis of novel thiazole derivatives, offering numerous advantages such as high yields, mild reaction conditions, and reusability without significant loss in efficiency. The use of ultrasonic irradiation further enhanced the reaction process, making it a sustainable approach. The synthesized thiazole derivatives were successfully characterized and showed good anticancer activity against HepG2 cells, indicating potential for future development as hepatocarcinogenic agents.

Supplementary Materials: Supporting information is available for download at https://www.mdpi.com/article/10.3390/catal14120840/s1. This includes Figure S1, demonstrating the synthesis of N,N′-bis[4-(chlorocarbonyl)phenyl] pyromellitimide; Figure S2, presenting the FTIR spectra of chitosan and PIBTU-CS hydrogel; Figure S3, showcasing the ^{1}H-NMR spectrum of PIBTU-CS hydrogel; Figure S4, illustrating the powder X-ray diffraction of chitosan and PIBTU-CS hydrogel; and Figure S5, featuring SEM images of chitosan and PIBTU-CS hydrogel. Additionally, detailed descriptions of the Chemicals and Materials and Measurements are included.

Author Contributions: Conceptualization: S.M.G. and M.E.A.Z.; Methodology: S.M.G., N.A.A.E.-G. and M.S.E.; Software: M.S.E. and T.Z.A.; Validation: N.A.M., S.M.G. and M.E.A.Z.; Formal analysis: M.A., N.A.M. and T.Z.A.; Investigation: M.S.E., T.Z.A., M.A. and S.A.; Resources: M.A. and M.E.A.Z.; Data curation: M.A. and N.A.M.; Writing—original draft preparation: M.E.A.Z. and S.A.; Writing—review and editing: S.M.G., T.Z.A. and M.E.A.Z.; Visualization: M.S.E. and N.A.M.; Supervision: S.M.G. and N.A.M.; Funding acquisition: M.S.E. All authors have read and agreed to the published version of the manuscript.

Funding: This research was funded by the Deanship of Scientific Research at Northern Border University, Arar, KSA, through project number "NBU-FFR-2024-80-04".

Data Availability Statement: The data presented in this study are available on request from the corresponding author.

Acknowledgments: The authors extend their appreciation to the Deanship of Scientific Research at Northern Border University, Arar, KSA for funding this research through the project number "NBU-FFR-2024-80-04".

Conflicts of Interest: The authors declare no conflict of interest.

References

1. Castiello, C.; Junghanns, P.; Mergel, A.; Jacob, C.; Ducho, C.; Valente, S.; Rotili, D.; Fioravanti, R.; Zwergel, C.; Mai, A. Green med chem: The challenge in the next decade toward eco-friendly compounds and processes in drug design. *Green Chem.* **2023**, *25*, 2109–2169. [CrossRef]
2. Miceli, M.; Frontera, P.; Macario, A.; Malara, A. Recovery/Reuse of Heterogeneous Supported Spent Catalysts. *Catalysts* **2021**, *11*, 591. [CrossRef]
3. Jiménez-Gómez, C.P.; Cecilia, J.A. Chitosan: A natural biopolymer with a wide and varied range of applications. *Molecules* **2020**, *25*, 3981. [CrossRef]
4. Al-Humaidi, J.Y.; Gomha, S.M.; Abd El-Ghany, N.A.; Farag, B.; Zaki, M.E.A.; Abolibda, T.Z.; Mohamed, N.A. Green synthesis and molecular docking study of some new thiazoles using terephthalohydrazide chitosan hydrogel as eco-friendly biopolymeric catalyst. *Catalysts* **2023**, *13*, 1311. [CrossRef]
5. Elmehbad, N.Y.; Mohamed, N.A.; Abd El-Ghany, N.A. Synthesis of novel antimicrobial and food-preserving hydrogel nanocomposite films based on carboxymethylcellulose. *Starch-Stärke* **2024**, *76*, 2300258. [CrossRef]
6. Ekapakul, N.; Lerdwiriyanupap, T.; Siritanon, T.; Choochottiros, C. Double network structure via ionic bond and covalent bond of carboxymethyl chitosan and poly(ethylene glycol): Factors affecting hydrogel formation. *Carbohydr. Polym.* **2023**, *318*, 121130. [CrossRef]
7. Drzymalska, K.W.; Tomaszewska, E.; Sadowska, M.; Opielak, M.; Kucinska-Lipka, J.; Janik, H. Crosslinking of chitosan with dialdehyde chitosan as a new approach for biomedical applications. *Materials* **2020**, *13*, 3413. [CrossRef] [PubMed]
8. Mohamed, N.A.; Abd El-Ghany, N.A.; Fahmy, M.M. Novel antimicrobial superporous cross-linked chitosan/pyromellitimide benzoyl thiourea hydrogels. *Int. J. Biol. Macromol.* **2016**, *82*, 589–598. [CrossRef]
9. Zhong, S. Incorporation of palladium catalyst inside cross-linked chitosan hybrid nanofibers for the Sonogashira reaction. *Kinet. Catal.* **2020**, *61*, 480–485. [CrossRef]
10. Xu, H.; Liao, W.M.; Li, H.F. A mild and efficient ultrasound-assisted synthesis of diaryl ethers without any catalyst. *Ultrason. Sonochem.* **2007**, *14*, 779–782. [CrossRef]
11. Jarag, K.J.; Pinjari, D.V.; Pandit, A.B.; Shankarling, G.S. Synthesis of chalcone (3-(4-fluorophenyl)-1-(4-methoxyphenyl)prop-2-en-1-one): Advantage of sonochemical method over conventional method. *Ultrason. Sonochem.* **2011**, *18*, 617–623. [CrossRef] [PubMed]
12. Cravotto, G.; Fokin, V.V.; Garella, D.; Binello, A.; Boffa, L.; Barge, A. Ultrasound-promoted copper-catalyzed azide-alkyne cycloaddition. *J. Comb. Chem.* **2010**, *12*, 13–15. [CrossRef] [PubMed]
13. Pizzuti, L.; Martins, P.L.G.; Ribeiro, B.A.; Quina, F.H.; Pinto, E.; Flores, A.F.C.; Venzke, D.; Pereira, C.M.P. Efficient sonochemical synthesis of novel 3,5-diaryl-4,5-dihydro-1H-pyrazole-1-carboximidamides. *Ultrason. Sonochem.* **2010**, *17*, 34–37. [CrossRef]
14. Watt, M.J.; Miotto, P.M.; De Nardo, W.; Montgomery, M.K. The liver as an endocrine organ—Linking NAFLD and insulin resistance. *Endocr. Rev.* **2019**, *40*, 1367–1393.
15. Mukherji, A.; Bailey, S.M.; Staels, B.; Baumert, T.F. The circadian clock and liver function in health and disease. *J. Hepatol.* **2019**, *71*, 200–211. [CrossRef]
16. Macpherson, A.J.; Heikenwalder, M.; Ganal-Vonarburg, S.C. The liver at the nexus of host-microbial interactions. *Cell Host Microbe* **2016**, *20*, 561–571. [CrossRef]
17. Wang, X.; Zhang, H.; Chen, X. Drug resistance and combating drug resistance in cancer. *Cancer Drug Resist.* **2019**, *2*, 141. [CrossRef] [PubMed]
18. Bray, F.; Laversanne, M.; Weiderpass, E.; Soerjomataram, I. The ever-increasing importance of cancer as a leading cause of premature death worldwide. *Cancer* **2021**, *127*, 3029–3030. [CrossRef] [PubMed]
19. Siegel, R.; DeSantis, C.; Virgo, K.; Stein, K.; Mariotto, A.; Smith, T.; Cooper, D.; Gansler, T.; Lerro, C.; Fedewa, S.; et al. Cancer treatment and survivorship statistics, 2012. *CA Cancer J. Clin.* **2012**, *62*, 220–241. [CrossRef]
20. Sung, H.; Ferlay, J.; Siegel, R.L.; Laversanne, M.; Soerjomataram, I.; Jemal, A.; Bray, F. Global cancer statistics 2020: GLOBOCAN estimates of incidence and mortality worldwide for 36 cancers in 185 countries. *CA Cancer J. Clin.* **2021**, *71*, 209–249. [CrossRef]

21. Wu, Q.; Yang, Z.; Nie, Y.; Shi, Y.; Fan, D. Multi-drug resistance in cancer chemotherapeutics: Mechanisms and lab approaches. *Cancer Lett.* **2014**, *347*, 159–166. [CrossRef]
22. Varmus, H. The new era in cancer research. *Science* **2006**, *312*, 1162–1165. [CrossRef]
23. Franchetti, P.; Cappellacci, L.; Grifantini, M.; Barzi, A.; Nocentini, G.; Yang, H.; O'Connor, A.; Jayaram, H.N.; Carrell, C.; Goldstein, B.M. Furanfurin and thiophenfurin: Two novel tiazofurin analogues. Synthesis, structure, antitumor activity, and interactions with inosine monophosphate dehydrogenase. *J. Med. Chem.* **1995**, *38*, 3829–3837. [CrossRef]
24. Li, X.; He, Y.; Ruiz, C.H.; Koenig, M.; Cameron, M.D. Characterization of dasatinib and its structural analogs as CYP3A4 mechanism-based inactivators and the proposed bioactivation pathways. *Drug Metab. Dispos.* **2009**, *37*, 1242–1250. [CrossRef] [PubMed]
25. Hu-Lieskovan, S.; Mok, S.; Homet Moreno, B.; Tsoi, J.; Robert, L.; Goedert, L.; Pinheiro, E.M.; Koya, R.C.; Graeber, T.G.; Comin-Anduix, B.; et al. Improved antitumor activity of immunotherapy with B-RAF and MEK inhibitors in BRAF(V600E) melanoma. *Sci. Transl. Med.* **2015**, *7*, 279ra41. [CrossRef] [PubMed]
26. Yao, Y.; Chen, S.; Zhou, X.; Xie, L.; Chen, A. 5-FU and ixabepilone modify the microRNA expression profiles in MDA-MB-453 triple-negative breast cancer cells. *Oncol. Lett.* **2014**, *7*, 541–547. [CrossRef] [PubMed]
27. Altmann, K.H. Epothilone B and its analogs—A new family of anti-cancer agents. *Mini-Rev. Med. Chem.* **2003**, *3*, 149–158. [CrossRef] [PubMed]
28. Aliabadi, A.; Shamsa, F.; Ostad, S.N.; Emami, S.; Shafiee, A.; Davoodi, J.; Foroumadi, A. Synthesis and biological evaluation of 2-phenylthiazole-4-carboxamide derivatives as anti-cancer agents. *Eur. J. Med. Chem.* **2010**, *45*, 5384–5389. [CrossRef] [PubMed]
29. Banimustafa, M.; Kheirollahi, A.; Safavi, M.; Ardestani, S.K.; Aryapour, H.; Foroumadi, A.; Emami, S. Synthesis and biological evaluation of 3-(trimethoxyphenyl)-2(3H)-thiazole thiones as combretastatin analogs. *Eur. J. Med. Chem.* **2013**, *70*, 692–702. [CrossRef]
30. Ayati, A.; Emami, S.; Asadipour, A.; Shafiee, A.; Foroumadi, A. Recent applications of 1,3-thiazole core structure in the identification of new lead compounds and drug discovery. *Eur. J. Med. Chem.* **2015**, *97*, 699–718. [CrossRef]
31. Das, D.; Sikdar, P.; Bairagi, M. Recent developments of 2-aminothiazoles in medicinal chemistry. *Eur. J. Med. Chem.* **2016**, *109*, 89–98. [CrossRef]
32. Gomha, S.M.; Abolibda, T.Z.; Alruwaili, A.H.; Farag, B.; Boraie, W.E.; Al-Hussain, S.A.; Zaki, M.E.A.; Hussein, A.M. Efficient green synthesis of hydrazide derivatives using L-proline: Structural characterization, anticancer activity, and molecular docking studies. *Catalysts* **2024**, *14*, 489. [CrossRef]
33. Rashdan, H.R.M.; Gomha, S.M.; El-Gendey, M.S.; El-Hashash, M.A.; Soliman, A.M.M. Eco-friendly one-pot synthesis of some new pyrazolo[1,2-b]phthalazinediones with antiproliferative efficacy on human hepatic cancer cell lines. *Green Chem. Lett. Rev.* **2018**, *11*, 264–274. [CrossRef]
34. Rashdan, H.R.M.; Abdelmonsef, A.H.; Shehadi, I.A.; Gomha, S.M.; Soliman, A.M.M.; Mahmoud, H.K. Synthesis, molecular docking screening and anti-proliferative potency evaluation of some new imidazo[2,1-b]thiazole linked thiadiazole conjugates. *Molecules* **2020**, *25*, 4997. [CrossRef]
35. Ma, L.; Xie, C.; Ma, Y.; Liu, J.; Xiang, M.; Ye, X.; Zheng, H.; Chen, Z.; Xu, Q.; Chen, T.; et al. Synthesis and biological evaluation of novel 5-benzylidenethiazolidine-2,4-dione derivatives for the treatment of inflammatory diseases. *J. Med. Chem.* **2011**, *54*, 2060–2068. [CrossRef]
36. Rashid, K.O.; Mohamed, K.S.; Salam, M.A.E.; Abdel-Latif, E.; Fadda, A.A.; Elmorsy, M.R. Synthesis of novel phenoxyacetamide derivatives as potential insecticidal agents against the cotton leafworm, Spodoptera littoralis. *Polycycl. Aromat. Compd.* **2023**, *43*, 356–369. [CrossRef]
37. Eddya, M.; Tbib, B.; El-Hami, K. A comparison of chitosan properties after extraction from shrimp shells by diluted and concentrated acids. *Heliyon* **2020**, *6*, e03486. [CrossRef]
38. Mosmann, T. Rapid colorimetric assay for cellular growth and survival: Application to proliferation and cytotoxicity assays. *J. Immunol. Methods* **1983**, *65*, 55–63. [CrossRef]

catalysts

MDPI

Review

Catalyst Development for Biogas Dry Reforming: A Review of Recent Progress

Wei Hu [1,2], Jundao Wu [3], Zeai Huang [1,2,4,*], Hao Tan [2], Yifan Tang [2], Zilong Feng [2], Rui Deng [2], Hongwei Zhang [5], Rustem Zairov [3,6] and Zhicheng Pan [4,7,*]

1 State Key Laboratory of Oil and Gas Reservoir Geology and Exploitation, Southwest Petroleum University, Chengdu 610500, China; huw0712@163.com
2 School of New Energy and Materials, Southwest Petroleum University, Chengdu 610500, China; tdsswup@163.com (H.T.); tyfswpu@163.com (Y.T.); fengzl2001@163.com (Z.F.); drswpu@163.com (R.D.)
3 Institute of Chemistry, Kazan Federal University, Kremlevskaya Str. 18, 420008 Kazan, Russia; rustem02@yandex.ru (R.Z.)
4 National Postdoctoral Research Station, Haitian Water Group, Chengdu 610213, China
5 Chengdu Biogas Institute, Key Laboratory of Development and Application of Rural Renewable Energy, Ministry of Agriculture and Rural Affairs, Chengdu 610041, China; zhw9590@163.com
6 Arbuzov Institute of Organic and Physical Chemistry, FRC Kazan Scientific Center, Russian Academy of Sciences, 8 Arbuzov Str., 420088 Kazan, Russia
7 National Engineering Laboratory of Circular Economy, Sichuan University of Science and Engineering, Zigong 643000, China
* Correspondence: zeai.huang@swpu.edu.cn (Z.H.); panzhicheng107@126.com (Z.P.)

Abstract: Amidst the rapid expansion of the global economy, the demand for energy has escalated. The depletion of traditional energy sources coupled with environmental pollution concerns has catalyzed a shift towards the development and utilization of clean, renewable energy. Biogas, as a renewable energy source, provides diverse applications and holds the potential to alleviate energy shortages. Recently, biogas dry reforming technology has garnered substantial attention as a significant pathway for renewable energy utilization, particularly in the development and optimization of catalysts. Contemporary research predominantly focuses on enhancing the activity and stability of catalysts, with particular emphasis on their resistance to coking and sintering. This review delineates the classification of biogas dry reforming catalysts, their catalytic activity, and issues related to carbon deposition, contrasting biogas dry reforming with traditional dry reforming in catalyst design. It synthesizes numerous studies from recent years aimed at mitigating carbon deposition during the biogas dry reforming process and boosting catalytic activity via active components, carriers, and promoters in both precious and non-precious metal catalysts. Furthermore, it discusses the current challenges of biogas dry reforming technology and outlines prospective future development trends. This discussion provides an in-depth understanding of biogas dry reforming technology and catalyst design, offering insights and recommendations for future research and industrial applications.

Keywords: biogas; renewable energy; dry reforming; catalysts; carbon deposition

1. Introduction

Biogas, a combustible gas, is produced from organic wastes (including crop residues, animal manure, discarded fruits and vegetables, kitchen waste, and agricultural by-products) via microbial fermentation under anaerobic conditions, requiring specific temperatures and pH levels [1–4]. Despite biogas's origin from biodegradable materials classifying it as a renewable energy source, its potential is frequently underestimated [5]. Even with the advancement of the shale gas revolution, traditional shale gas may have cost advantages in certain regions. However, utilizing biogas can create additional economic value for the agriculture and waste management industries, promoting their sustainable development.

For example, in 2019, the biogas industry created approximately 335,000 temporary construction jobs and 23,000 full-time operational positions. In China, the biogas industry employs 209,000 workers [6]. And recent years have seen biogas technology's rising potential in addressing global energy shortages, evidenced by a surge in global production from 9519 MW in 2010 to 20,108 MW in 2020, more than doubling the installed capacity [7]. Therefore, from the perspective of sustainable development and environmental impact, biogas possesses distinct advantages over traditional shale gas for utilization.

Biogas primarily finds application in electricity generation, heating, fueling, transportation, chemical production, and biogas upgrading [8–10]. However, biogas's direct combustion may contribute to environmental pollution by emitting NO_x and particulates [11]. Enhancing biogas's value involves converting its methane (CH_4) content into chemical raw materials (like methanol, ethanol, and formaldehyde) or into synthesis gas (H_2 and CO) via dry reforming [12,13]. This technology not only mitigates greenhouse gas emissions but also yields significant economic benefits, underscoring its substantial application potential [14,15]. Unlike classical dry reforming, which relies heavily on natural gas as a feedstock, biogas dry reforming leverages renewable resources, making it more environmentally friendly. Additionally, the integration of biogas, which is often produced from organic waste, can help mitigate greenhouse gas emissions while providing a consistent and localized energy source. As the shale gas revolution has indeed made classical dry reforming more competitive, biogas dry reforming offers a unique advantage by transforming waste into valuable energy, aligning with global efforts to promote circular economy practices and reduce reliance on fossil fuels.

Current review literature primarily examines the synthesis, characterization, and reactivity of catalysts for biogas dry reforming, notably nickel-based catalysts, while often neglecting a detailed categorization of catalysts and the carbon deposition issues encountered during the reaction [16,17]. This review systematically categorizes catalysts for biogas dry reforming into noble and non-noble metal catalysts, delves into the catalytic activity of different catalyst types, and emphasizes addressing carbon deposition issues. It concludes by outlining the current challenges biogas dry reforming technology faces and discussing potential future development trends.

2. Basis of Biogas Dry Reforming Technology

Biogas consists mainly of CH_4 and CO_2, with minor traces of impurities including H_2S, H_2, N_2, NH_3, H_2O, and O_2 [18]. Dry reforming of biogas primarily targets CH_4 and CO_2, with other impurities effectively removed by current technologies [19]. Extensive research has focused on enhancing the conversion rates of reactants and product selectivity. The high bond dissociation energies of CH_4 and CO_2 (435 and 526 kJ mol^{-1}, respectively) make the dry reforming process energy-intensive [17]. Thermodynamic analyses show that higher reaction temperatures promote biogas dry reforming's forward conversion, activating CO_2 and CH_4 at temperatures above 700 °C (Equation (1)) [20]. Several side reactions accompany this process, including the reverse water–gas shift reaction that converts CO_2 and H_2 into CO and H_2O (Equation (2)) [21]. Furthermore, below 800 °C, CH_4 decomposition (Equation (3)) and CO disproportionation (Equation (4)) can lead to carbon formation and catalyst deactivation [22,23].

$$CH_4 + CO_2 \rightleftarrows 2CO + 2H_2 \rightarrow \Delta H_{298K} = +247 \text{ kJ mol}^{-1} \quad (1)$$

$$CO_2 + H_2 \rightleftarrows CO + H_2O \rightarrow \Delta H_{298K} = +41 \text{ kJ mol}^{-1} \quad (2)$$

$$CH_4 \rightleftarrows C(s) + 2H_2 \rightarrow \Delta H_{298K} = +75 \text{ kJ mol}^{-1} \quad (3)$$

$$2CO \rightleftarrows C(s) + CO_2 \rightarrow \Delta H_{298K} = -171 \text{ kJ mol}^{-1} \quad (4)$$

In dry reforming of biogas, CH_4 and CO_2 initially dissociate into reactive intermediates, CH_4 on the active metal and CO_2 at the metal–support interface [24]. CH_4 activation results in C-H bond breakage, yielding active methyl (•CH_3) and hydrogen (H) atoms; CO_2

activation splits the C=O double bond to generate oxygen (O) and carbon monoxide (CO) [25,26]. These intermediates, arising from carbon and hydrogen atom separation, form on the catalyst surface and are crucial for transforming the reaction into the final products, CO and H_2, through complex steps [27,28].

Microbial anaerobic digestion, driven mainly by methanogens, converts organic matter's carbon and hydrogen into CH_4, yielding a biogas CH_4-to-CO_2 ratio of about 1.5 [29,30]. Excessive CH_4 may worsen side reactions like methane decomposition, causing carbon deposition on the catalyst, obstructing active sites, and altering surface structure and morphology [31,32]. This reduction in specific surface area and reactant adsorption and activation capacity slows the reaction rate, leading to catalyst deactivation [33]. Traditional dry reforming typically employs pure CH_4 and CO_2 as feedstock in a 1:1 ratio, which lessens the impact of carbon deposition [34]. Consequently, this necessitates the use of different catalysts. Biogas dry reforming catalysts emphasize addressing carbon deposition, while traditional dry reforming catalysts, benefiting from feedstock purity and an adjustable CH_4-to-CO_2 ratio, prioritize enhancing catalytic activity and selectivity.

Dry reforming of biogas catalysts comprises active components, supports, and promoters. Active components, anchored on stable supports, enhance surface area and stability [35]. Promoters optimize catalyst performance, enhancing reaction stability and efficiency [35,36]. Catalysts are classified into noble and non-noble metal types based on their active components [37]. Noble metal catalysts, with superior activity and selectivity, face limited application due to high costs [38]. Non-noble metal catalysts show significant activity when well dispersed but suffer from carbon deposition and thermal instability, causing deactivation and sintering [39]. To reduce carbon deposition and enhance activity, specific catalyst supports and promoters are selected. Researchers have explored metal oxides like Al_2O_3, SiO_2, La_2O_3, CeO_2, and ZrO_2 as supports [40,41] and used transition (e.g., Fe, Co, Sn) [37,42], noble (e.g., Ag, Pt, Pd) [43], rare earth (e.g., La, Ce, Pr) [44,45], and alkaline earth metals (e.g., Sr, Ca, Ba) [46] as promoters, creating a carrier-mediated system to enhance the reaction [47,48].

3. Advances in Catalyst Research

3.1. Noble Metal Catalysts

Noble metal catalysts in biogas dry reforming show high-temperature stability and resistance to carbon deposition, extending their lifespan [49]. Researchers improve noble metal utilization and efficiency by designing catalyst structures and compositions, including single-atom and supported variants [50]. Selecting appropriate support materials and optimizing their morphology and surface attributes enhances the catalysts' thermal stability and anti-sintering features [51,52]. The influence of reactant gas ratios on biogas dry reforming efficiency in different catalyst systems has been studied to boost syngas production and quality.

A.L.A. Marinho and his colleagues [53] compared the performance of Pt-encapsulated CeO_2 catalysts, synthesized in a single step via the sol–gel method, against Pt/CeO_2 catalysts prepared by the traditional impregnation method in DRM reactions. Figure 1a,b show the changes in the number of exchanged oxygen atoms during TPOIE and IOIE processes at 400 °C for $Pt@CeO_2$ and $Pt@CeZrO_2$ catalysts. The results showed that embedding Pt nanoparticles in $Pt@CeO_2$ and $Pt@CeZrO_2$ catalysts balanced carbon's formation and gasification rates due to the high active oxygen content at the metal–support interface, enhancing carbon deposition resistance. Compared to supported Pt, embedded Pt nanoparticles interact more strongly with Ce, increasing active lattice oxygen on the catalyst surface and facilitating carbon gasification. Stability tests simulated biogas reactions at 800 °C for 24 h under CH_4/CO_2 molar ratios of 1.0 and 1.5. At a CH_4/CO_2 molar ratio of 1.5, all catalysts showed significant stability. Faris Jasim Al-Doghachi et al. [54] enhanced MgO-based catalysts by doping them with varying proportions of NiO to examine the resultant effects on catalyst performance. As demonstrated in Figure 1c through FT-IR analysis, the altered $Pt/Mg_{1-x}Ni_xO$ catalysts revealed the presence of Pt-O, Ni-O, and MgO bonds.

The Pt-O and Ni-O bonds likely serve as catalytic active sites, essential for the methane dry reforming reaction's CH_4 and CO_2 conversion. These bonds are posited to facilitate critical reaction steps, such as methane dissociation and CO_2 activation, thus accelerating the reaction rate. All catalysts achieved CO_2 and CH_4 conversion rates over 95%, with the following efficiency rankings: Pt/$Mg_{0.85}Ni_{0.15}O$ > Pt/$Mg_{0.93}Ni_{0.07}O$ > Pt/$Mg_{0.97}Ni_{0.03}O$ > Pt/MgO. The pore volume to specific surface area ratio (SBET) aligned with the catalytic activity order, indicating that a higher ratio improves performance. Additionally, at a CH_4/CO_2 molar ratio of 2, the catalysts showed enhanced resistance to deactivation from carbon formation and superior H_2 and CO selectivity.

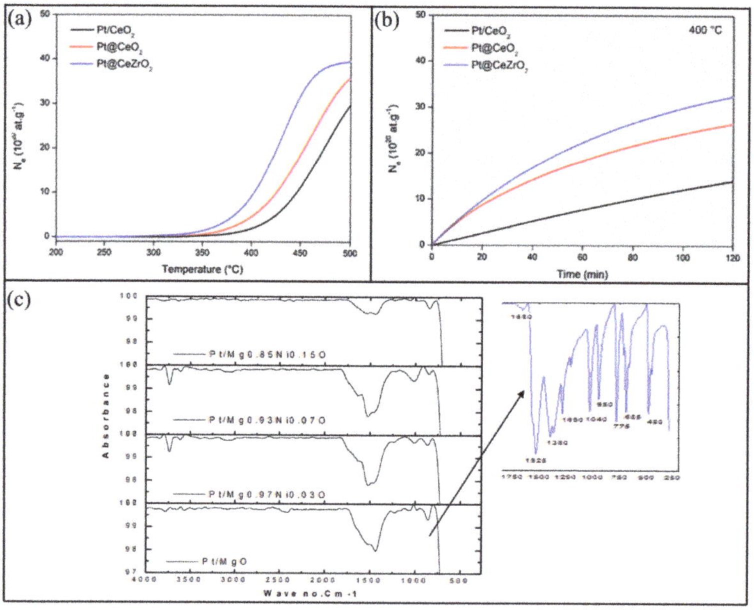

Figure 1. (**a**) The evolution of the number of exchanged oxygen atoms during the TPOIE process on Pt/CeO_2, Pt@CeO_2, and Pt@$CeZrO_2$ and (**b**) the evolution of the number of exchanged oxygen atoms during the IOIE process at 400 °C. Reprinted with permission from Ref. [53]. Copyright 2021, Elsevier. (**c**) FT-IR for different catalysts (Pt/MgO, Pt/$Mg_{0.85}Ni_{0.15}O$, Pt/$Mg_{0.97}Ni_{0.03}O$, and Pt/$Mg_{0.93}Ni_{0.07}O$). Reprinted with permission from Ref. [54]. Copyright 2015, Elsevier.

Ainara Morala et al. [49] investigated Rh-based catalysts utilizing γ-Al_2O_3, SiO_2, and CeO_2 as supports, comparing their activities across various gas hourly space velocities (GHSV) and highlighting CH_4 conversion rates. The findings revealed that both laboratory-prepared and commercial Rh/Al_2O_3 catalysts surpassed others in performance, with a catalytic activity ranking of Rh/Al_2O_3 > Rh/SiO_2 > Rh/CeO_2. Specifically, the Rh/Al_2O_3 catalyst showed the highest metal dispersion (approximately 36%), offering superior sintering resistance.

At a GHSV of 30 N L $CH_4/(g_{cat}\cdot h)$ and 700 °C, dry reforming of biogas with a commercial Rh catalyst reaches equilibrium after 2 h. With higher GHSV values (150 and 300 N L $CH_4/(g_{cat}\cdot h)$), catalytic activity declines over time, particularly at the highest flow rates. Even at elevated flow rates and temperatures, coke production stays minimal, underscoring the substantial benefits of noble metal catalysts like Rh in this process.

Recent research has significantly enhanced the carbon gasification process by increasing the active lattice oxygen on catalyst surfaces, which is crucial for boosting long-term stability and catalytic efficiency. Additionally, the optimization of the catalyst's pore volume and surface area ratio, alongside improvements in metal dispersibility and sintering resistance through the use of diverse supports or element doping, has been achieved. However, the prohibitive cost of precious metals constrains the large-scale industrial deployment of these catalysts. Future studies should focus on ways to sustain catalytic performance while minimizing costs, such as by identifying alternatives to precious metals or refining manufacturing processes. Moreover, the stability and long-term performance of these catalysts in actual applications still require further validation and enhancement. Through these initiatives, we aim to develop commercially viable advanced catalytic technologies that offer practical solutions to global energy challenges.

3.2. Non-Noble Metal Catalysts

Noble metal catalysts are known for their excellent resistance to carbon deposition, activity, and stability, yet their development is constrained by limited resources. Non-noble metal catalysts, like Ni, Co, and Fe, offer economic benefits and are more apt for large-scale industrial use compared to noble metals [45,55]. However, nickel-based catalysts often rapidly deactivate due to carbon buildup and active phase sintering [56]. Cobalt-based catalysts, in comparison to nickel-based ones, exhibit lower CH_4 conversion activity and are costlier, diminishing their economic benefit [57]. Iron-based catalysts are the least expensive, yet they may not match the activity and stability of nickel- or cobalt-based alternatives [38]. Addressing non-noble metal catalysts' issues involves using diverse preparation methods and modification techniques to optimize performance under various conditions or refining the catalyst composition by incorporating promoters or alloying to boost activity and stability [58,59]. Table 1 summarizes some typical non-noble metal catalysts for biogas dry reforming.

Table 1. Reaction conversion rate and product selectivity on non-noble metal catalysts in biogas dry reforming.

Entry	Catalysts	Reaction Conditions	CH_4 Conversion (%)	CO_2 Conversion (%)	Product Selectivity (H_2/CO Ratio)	References
1	Ni/MgO	$CH_4/CO_2/N_2$ = 2:2:1, 750 °C, 200 mg	96.4	91.7	≈1.06	[60]
2	$Co(NO_3)_2 + Al(NO_3)_3$	CH_4/CO_2 = 1:1, 900 °C	100	86.2	≈1.16	[61]
3	NiLaTi-I	CH_4/CO_2 = 1.5:1, 800 °C, 200 mg	79	≈79	>1	[62]
4	$Ni_{0.11}/Ce_{0.20}$ (Al_2O_3-TiO_2)	CH_4/CO_2 = 1.5:1, 850 °C, 150 mg	84.9	95.6	1.06	[55]
5	$LaNiO_3$	CH_4/CO_2 = 1.5:1, 850 °C, 0.5 g	97.2	93.8	1.6	[34]
6	0.2Ca-10Co/Al_2O_3	CH_4/CO_2 = 1:1, 973 K, 100 mg	84	89	0.74	[63]
7	Gd.Ni/$Mg_{1.3}AlO_x$	CH_4/CO_2 = 2:1, 750 °C, 50 mg	49	95	>0.9	[64]
8	$Ni_{0.10}/(Zn_{0.1}\ Ce_{0.9})$	CH_4/CO_2 = 1.5:1, 900 °C, 150 mg	83.1	97	1.04	[65]
9	Ni/Al_2O_3	$CH_4/CO_2/N_2$ = 2:2:1, 750 °C, 200 mg	82.7	/	>0.9	[66]
10	NiO/ZrO_2/MgO	CH_4/CO_2 = 1.5:1, 900 °C, 150 mg	97	100	1.7	[67]
11	Ni-SiO_2@SiO_2	CH_4/CO_2 = 1.5:1, 700 °C, 100 mg	72.7	93	0.95	[68]
12	$Fe_{5\%}Ni_{5\%}Al_2O_3$	CH_4/CO_2 = 1.8:1, 700 °C, 100 mg	22	49	≈1.16	[69]
13	Fe/ZrO_2	CH_4/CO_2 = 1.5:1, 900 °C, 3.5 g	92	89	2.25	[70]
14	NiMo/MgO	CH_4/CO_2 = 1.5:1, 900 °C, 1 g	95	100	3.1	[71]

3.2.1. Nickel-Based Catalysts

Nickel-based catalysts activate CH_4 and CO_2 efficiently at lower temperatures, accelerating product formation [72]. As an active component, Ni lowers the energy needed for C-H bond breaking, thus easing CH_4 decomposition [73]. Its surface sites enhance CO_2 adsorption and activation, simplifying reactions with C and H. However, nickel particle sintering and surface carbon formation during reactions reduce active sites, diminishing catalytic activity. Recently, researchers have improved the catalyst's performance by reducing its size, adjusting its structure, and modifying the carrier [74,75].

Developing nickel-based catalysts with strong metal–support interaction to enhance CO_2 adsorption and conversion is crucial for mitigating metal sintering and minimizing carbon deposition [76]. Pali Rosha et al. [55] synthesized CeO_2-promoted Ni catalysts on TiO_2 and Al_2O_3 supports using the wet impregnation method to investigate biogas dry reforming. Figure 2a shows that H_2-TPR results, after impregnating 11wt% Ni on TiO_2

support, reveal a shift in the NiO reduction peak to a higher temperature, indicating a strong metal–support interaction. Enhanced binding force between the metal and support hinders active metal migration, thereby slowing metal sintering. Additionally, CeO_2's inclusion boosts CH_4 and CO_2 conversion rates, thanks to its high oxygen storage capacity that facilitates timely C and O combination during reactions, thereby somewhat preventing carbon formation. At 850 °C and a CH_4/CO_2 ratio of 1.5, the $Ni_{0.11}/Ce_{0.20}$ (Al_2O_3-TiO_2) catalyst achieved the highest CH_4 and CO_2 conversion rates (84.9% and 95.5%, respectively), operating stably for 7 h with a carbon deposition rate of 8.8 wt%. Danhua Mei et al. [77] employed Dielectric Barrier Discharge (DBD) technology for plasma catalytic dry reforming of biogas using nickel-based bimetallic catalysts. Figure 2b illustrates that, compared to 10Ni, 10Ni3Co has the highest CO_2 desorption peak temperature. Furthermore, the base site quantity ranks as follows: 10Ni3Co > 10Ni3Mn > 10Ni > 10Ni3Cu. Catalysts with higher basicity more effectively adsorb and dissociate CO_2. Absorption at these sites yields more surface oxygen, facilitating carbon gasification and thus increasing resistance to carbon deposition. After a 150-min reaction at 50 W, 10Ni3Co showed minimal carbon deposition (2.9%). At 60 W, it achieved the highest CH_4 (49.1%) and CO_2 (29.8%) conversion rates.

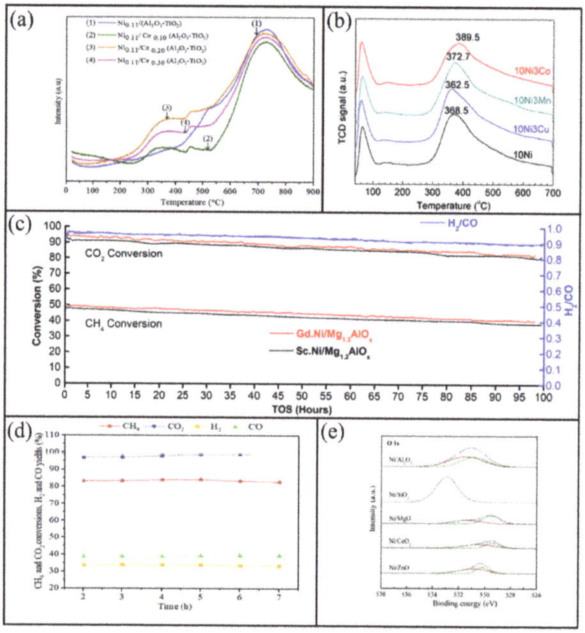

Figure 2. (**a**) H_2-TPR spectra of different catalysts. Reprinted with permission from Ref. [55]. Copyright 2018, Elsevier. (**b**) CO_2-TPD spectra of different catalysts. Reprinted with permission from Ref. [77]. Copyright 2023, Elsevier. (**c**) Biogas dry reforming conversions and H_2/CO ratios of Gd.Ni/$Mg_{1.3}AlO_x$ and Sc.Ni/$Mg_{1.3}AlO_x$ at 100 h. Reprinted with permission from Ref. [64]. Copyright 2022, Elsevier. (**d**) $Ni_{0.10}/(Zn_{0.1}$-$Ce_{0.9})$ catalyst stability test at 900 °C. Reprinted with permission from Ref. [65]. Copyright 2019, Elsevier. (**e**) O 1s XPS spectra of the reduction catalyst. Reprinted with permission from Ref. [66]. Copyright 2020, Elsevier.

Rare-earth-metal-promoted nickel-based catalysts enhance stability by improving CO_2 adsorption and activation capabilities. Quan Luu Manh Ha and colleagues examined low-nickel-load (2.5 wt%) catalysts modified by Gd, Sc, or La, noting that Gd and Sc modifications maintained CH_4 and CO_2 conversion rates during a 100 h stability test, as depicted in Figure 2c. This stability results from the strong metal–support interaction that stabilizes micro-nickel particles and boosts CO_2 adsorption and activation, thus increasing resistance to carbon accumulation [64].

Optimizing support materials enhances metal–support interactions and supplies active oxygen to the catalyst surface. Pali Rosha et al. [65] explored the impact of 10 wt% nickel catalysts on CeO_2 and ZnO supports in methane dry reforming. For the $Ni_{0.10}/CeO_2$ catalyst, adding 10 wt% Zn improved metal–support interaction, Ni particle dispersion at high temperatures, and reduced carbon deposition. CeO_2, offering numerous oxygen vacancies and releasing oxygen, curbs carbon buildup and fortifies metal–support interaction, thereby enhancing the catalyst's activity and selectivity in the reforming reaction. Zinc oxide's high melting point (1975 °C) preserves surface area at elevated temperatures, aids nickel particle dispersion, and boosts active sites, thus improving CH_4 and CO_2 conversion rates. Figure 2d demonstrates that the $Ni_{0.10}/(Zn_{0.1}\text{-}Ce_{0.9})$ catalyst, with lower carbon deposition, attained high CH_4 conversion rates (83.1%) and H_2 yield (33.6%). Similarly, Yuchen Gao and colleagues synthesized nickel catalysts supported by Al_2O_3, SiO_2, MgO, CeO_2, and ZnO. Ni/Al_2O_3, characterized by its smaller particle size and strong metal–support interaction, offers more active sites, thereby enhancing the adsorption and surface reaction rates of CH_4 and CO_2, resulting in the highest CH_4 conversion efficiency of 82.7%. The O 1s XPS spectrum, illustrated in Figure 2e, reveals an O_α/O_β ratio, calculated from peak areas, indicative of oxygen activity. Ni/MgO, with an O_α/O_β ratio of 1.46, suggests a surface rich in active oxygen. This active oxygen interacts with CO_2 to form intermediates such as CO_3^{2-}, which enhances CO_2 adsorption and activation, speeds up the dry reforming reaction, and helps oxidize surface carbon deposits, thereby reducing carbon accumulation and maintaining catalyst activity and stability. A 12 h stability test indicated a slight decrease in the CH_4 conversion rate of Ni/MgO from 68% to 64%; however, unlike the other four catalysts, it did not experience any blockage [66].

Smaller catalyst particle sizes yield a greater number of required active sites. Studies show that incorporating precious metals reduces catalyst particle size [78], boosting the quantity of highly efficient active sites. S. Carrasco-Ruiz et al. developed a stable Ni-Rh bimetallic catalyst on CeO_2-Al_2O_3 mixed oxides. Adding a small amount of Rh decreased the active metal's average particle size, increased active sites, and enhanced reducibility, showcasing superior redox capabilities. This improved CH_4 and CO_2 adsorption and conversion, minimized carbon deposition, and prolonged the catalyst's lifespan, as depicted in Figure 3a [79].

The preparation method significantly affects the particle size of Ni mixed catalysts. Santiago Veiga et al. [62] synthesized Ni-La-Ti catalysts through impregnation, co-precipitation, and an improved Pechini method. Significant differences in Ni particle size were observed among catalysts prepared by different methods, as illustrated in Figure 3b. The improved Pechini method yielded catalysts with smaller active metal particle sizes and stronger metal support interactions, which mitigated metal sintering and increased the number of active sites. Figure 3c shows that the catalyst prepared using the improved Pechini method demonstrated excellent stability and activity during a 10 h stability test at a CH_4/CO_2 ratio of 1.5.

The stability of the methane dry reforming reaction relies heavily on the catalyst's structure. Matinee Chuenjai et al. [67] synthesized $NiO/MgO/ZrO_2$ catalysts using urea carbon surface modification technology. Figure 4a illustrates that adding MgO stabilizes ZrO_2's tetragonal phase and enhances ZrO_2-MgO interaction, which prevents phase transformation, boosts catalyst stability, and increases alkalinity. This process enhances CO_2 adsorption and reduction on the catalyst surface, leading to CO production. Surendar Moogi et al. [34] prepared perovskite catalysts ($LaCoO_3$ and $LaNiO_3$) using both the citrate–gel combustion and co-precipitation methods, observing more severe carbon deposition under CH_4/CO_2 = 1.5 than CH_4/CO_2 = 1.0. Figure 4b shows TEM-EDS images of the reduced catalyst. TEM analysis of perovskite catalysts indicated that synthesis methods influenced particle sizes; notably, the citrate–gel method produced smaller particles than the co-precipitation method, resulting in more active sites. The citrate–gel combustion method enhances internal porosity of the catalysts, facilitating methane diffusion and product desorption at active sites, crucial for minimizing carbon deposition. Consequently, catalysts synthesized via the citrate–gel combustion method showed reduced carbon content.

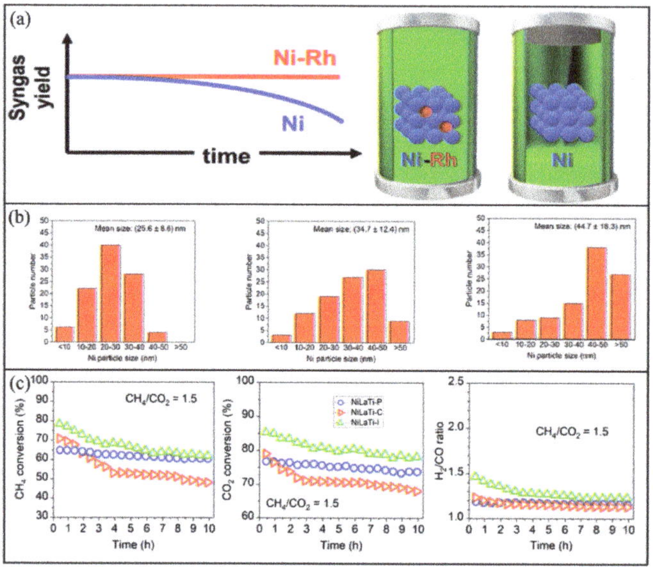

Figure 3. (**a**) Schematic diagram of the structure and properties of bimetallic Ni-Rh catalysts. Reprinted with permission from Ref. [79]. Copyright 2023, Elsevier. (**b**) Histogram of the distribution of nickel particles; (**c**) stability test carried out for 10 h at CH_4/CO_2 = 1.5 feed ratio. Reprinted with permission from Ref. [62]. Copyright 2023, Elsevier.

In previous studies, Maryam Kaviani et al. [68] employed the ammonia evaporation method to synthesize SiO_2-coated Ni-SiO_2@SiO_2 core–shell catalysts, enhancing stability and coking resistance at a CH_4/CO_2 ratio of 1.5, as depicted in Figure 4c. At 700 °C and with 10 wt% Ni loading, Ni-SiO_2@SiO_2 exhibited stable and high CH_4 and CO_2 conversion rates with negligible carbon formation, as Figure 4d shows. The core–shell catalyst's high stability is due to its silica-coated core, which limits nickel particle mobility and imparts anti-migration and anti-sintering properties. The catalyst's unique interlayer structure encapsulates Ni particles between the SiO_2 core and shell. During the reduction process, this leads to a Ni-SiO_2 core forming from layered nickel silicate, enhancing Ni particle dispersion and Ni-SiO_2 interaction, thus increasing stable active sites during the reaction.

Employing diverse synthesis methods and designing catalysts with core–shell configurations can effectively restrict nickel particle migration, conferring anti-migration and anti-sintering properties that enhance catalyst stability. Optimizing metal–support interactions boosts CO_2 adsorption and conversion, increases active sites, and reduces metal sintering and carbon deposition. These innovative strategies are essential for developing efficient and environmentally friendly catalysts, especially in the methane dry reforming process for sustainable energy conversion. Continued research and optimization of these methods could enable future industrial applications of these high-performance catalysts, which would significantly lower energy consumption and carbon emissions, guiding the energy sector towards a more sustainable and greener path.

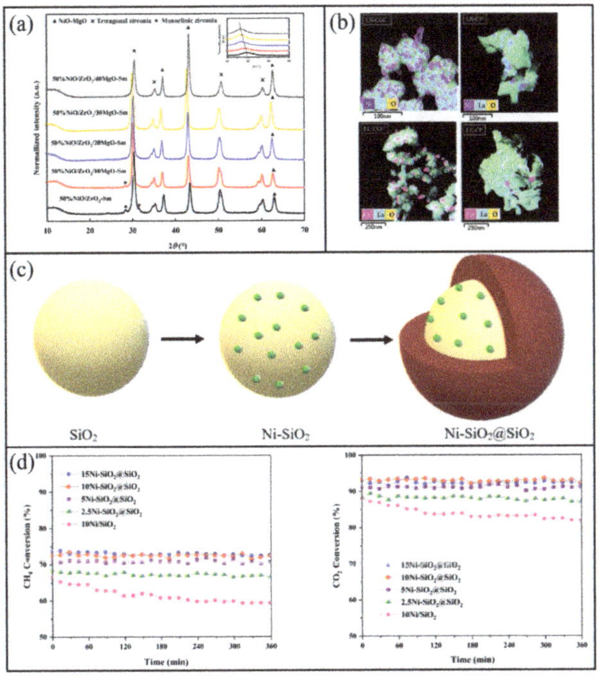

Figure 4. (**a**) XRD spectra of $NiO/ZrO_2/MgO$ catalysts synthesized with different MgO dosages using surface modification technique. Reprinted with permission from Ref. [67]. Copyright 2022, Elsevier. (**b**) TEM-EDS patterns of the reduced catalysts. Reprinted with permission from Ref. [34]. Copyright 2022, Elsevier. (**c**) Schematic structure of Ni-SiO_2@SiO_2 catalysts and (**d**) stability test of Ni-SiO_2@SiO_2. Reprinted with permission from Ref. [68]. Copyright 2022, Elsevier.

3.2.2. Cobalt-Based Catalysts

The key distinction between cobalt-based catalysts in conventional and biogas dry reforming is the minor variations in CH_4/CO_2 feed ratios and the reactants' complexity in biogas dry reforming. Research into the application of cobalt-based catalysts is vital given their significance in these domains, yet the challenges they present are non-trivial [80]. Catalyst deactivation primarily results from Co oxidation or carbon deposition, with smaller Co particles more susceptible to oxidation and larger particles predominantly losing activity due to carbon formation [81]. Cobalt-based catalysts exhibit lower conversion rates compared to precious metals and nickel-based catalysts [82]. Adding promoters is an effective strategy to boost Co-based catalysts' activity and stability [83]. Promoters enhance active metal particle dispersion, electron transfer, carbon gasification, and catalyst reducibility, thereby increasing activity while reducing the reaction's temperature requirement [84,85]. Promoters encompass alkanes, alkaline earth metals, transition metals, and rare earth metals, with alkanes and alkaline earth metals offering notable advantages in availability and cost [86,87].

Zhang et al. [88] synthesized a series of cobalt-based catalysts using γ-Al_2O_3 as the support and rare earth metals (Y, Tb, Gd) as promoters to mitigate carbon deposition. They introduced a carbon elimination mechanism involving rare earth metal oxides, where a bidentate carbonate intermediate is formed by a reaction with CO_2, as depicted in Figure 5a. This intermediate interacts with carbon deposits on the catalyst surface, yielding CO and metal oxides. Employing an oxidation–reduction cycle process significantly improves the catalyst's resistance to carbon accumulation. Additionally, the influence of alkalinity on coke formation has not been extensively investigated. Anh Ngoc T. Cao et al. [63] employed γ-Al_2O_3 as a support with the impregnation method to synthesize cobalt catalysts. The reaction mechanism, depicted in Figure 5b, involves the adsorption and subsequent dissociation of CH_4 into hydrogen and hydrocarbons (CH_x). These CH_x intermediates then react with activated oxygen species (O*) from CO_2 to form CH_xO^* radicals, which further react to produce CO and H_2. Illustrated in Figure 6c, calcium addition notably enhances the catalyst's surface alkalinity, thereby improving its CO_2 adsorption capacity and modifying the reaction pathways to prevent carbon buildup at primary active sites. With low calcium dosages, the average crystallite size of Co_3O_4 is reduced, suggesting an improved dispersion of cobalt. This discovery opens new avenues for using earth-abundant catalysts to enhance the performance of simulated biogas dry reforming. Mabkhoot Alsaiari et al. [82] incorporated cost-effective iron into cobalt-based catalysts, as shown in Figure 5d, significantly affecting the material's alkalinity due to the iron load. The alkalinity peaked at 705.7 μmol CO_2 g^{-1} in the 2Fe-5Co/Al_2O_3 formulation. They established a correlation between the catalytic activity and the physicochemical properties of the materials. The 0.8Fe-5Co/Al_2O_3 sample demonstrated superior performance, improving the dispersion, reducibility, and alkalinity of cobalt on the support. This additive facilitated carbon gasification and effectively minimized coke deposition.

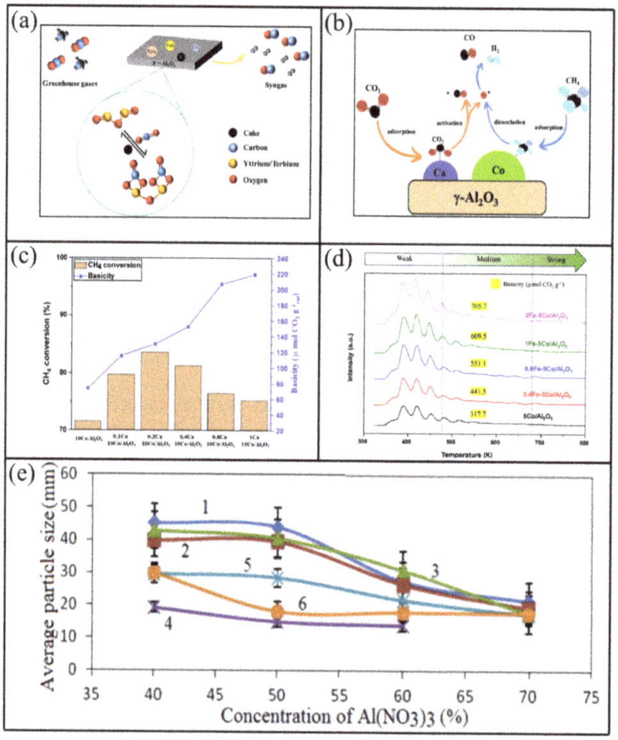

Figure 5. (**a**) The proposed reaction mechanism of carbon elimination. Reprinted with permission from Ref. [88]. Copyright 2023, Elsevier. (**b**) Proposed mechanism for methane dry reforming over Ca-Co/Al_2O_3 catalyst. (**c**) The effects of calcium loading on methane conversion and total basicity of catalyst. Reprinted with permission from Ref. [63]. Copyright 2022, Elsevier. (**d**) The material basicity measurement using the CO_2− TPD method of the pristine and iron-promoted catalyst. Reprinted with permission from Ref. [82]. Copyright 2024, Elsevier. (**e**) Dependence of the crystallite size on the concentration of $Al(NO_3)_3$; 1—$CoAl_2O_4$ (511), 2—Co_2AlO_4 (511), 3—Co_3O_4 (511), 4—CoO (200), 5—AlCo (200), 6—Al (311). Reprinted with permission from Ref. [61]. Copyright 2021, Authors of MDPI.

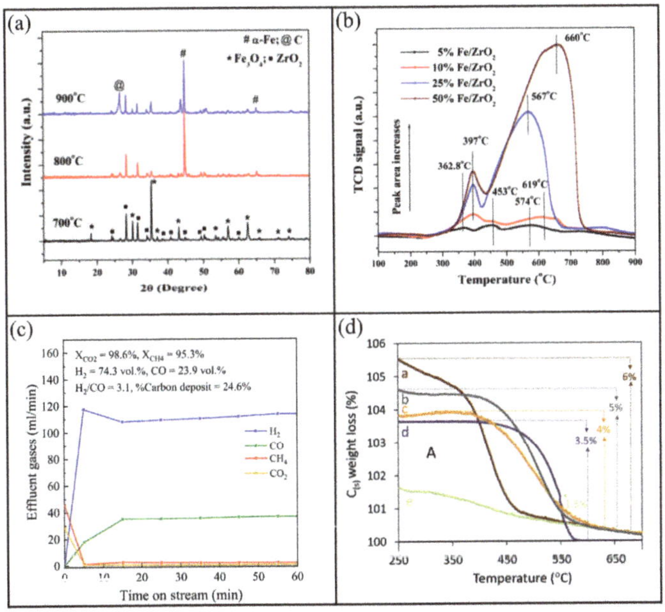

Figure 6. (**a**) XRD patterns of the spent catalysts: effect of reaction temperature; (**b**) TPR profiles of the iron-loaded catalysts. Reprinted with permission from Ref. [70]. Copyright 2020, Elsevier. (**c**) The composition of effluent gases over FeMo/MgO catalysts, GHSV = 4500 mL/(g·h). Reprinted with permission from Ref. [71]. Copyright 2022, Elsevier. (**d**) TGA profiles of spent catalysts: a $Fe_{10\%}Al_2O_3$, b $Fe_{7.5\%}Ni_{2.5\%}Al_2O_3$, c $Fe_{5\%}Ni_{5\%}Al_2O_3$, d $Fe_{2.5\%}Ni_{7.5\%}Al_2O_3$, and e $Ni_{5\%}Al_2O_3$. Reprinted with permission from Ref. [69]. Copyright 2019, Elsevier.

Manapkhan Zhumabek et al. [61] synthesized syngas using catalysts prepared from $Co(NO_3)_2$–$Al(NO_3)_3$–urea mixtures with varied compositions. The crystallite sizes of the catalysts, estimated from the widths of the XRD peaks as shown in Figure 5e, decreased as the $Al(NO_3)_3$ content increased, enhancing the catalysts' thermal stability and resistance to carbon deposition—attributes vital for prolonged catalytic operations. For CO_2 conversion, the Co_xAl_y phase, with a lattice parameter of 1.423 Å, closely matches the C-O bond length of 1.43 Å in CO_2, suggesting that Co_xAl_y could serve as an active site for C-O bond activation. Additionally, incorporating metals such as Zn is seen as an effective strategy to prevent metal oxidation and manage carbon deposition on catalysts. Previous research has indicated that Zn addition prevents the oxidation of Co in the Co/ZrO_2 catalyst during ethanol steam reforming. Jung-Hyun Park et al. [81] examined the effects of Zn addition on the catalytic performance of $CoZn(x)/ZrO_2$ in methane dry reforming. Introducing 2 wt% Zn to Co/ZrO_2 notably improved catalyst stability; however, exceeding this amount reversed the effect, with higher oxidation rates and extents observed in $CoZn(2)/ZrO_2$ compared to Co/ZrO_2. Catalysts enhanced by Zn demonstrated increased stability, as the formation of a Co-Zn alloy impeded Co oxidation.

Adding promoters significantly enhances cobalt-based catalysts' performance in dry reforming reactions by improving their activity, selectivity, stability, and carbon deposition resistance. Mesoporous carriers, like γ-Al_2O_3, have been shown to improve active metal dispersion and thermal stability in traditional dry reforming [89]. However, their application research in methane dry reforming remains limited. Adding promoters not only optimizes the catalyst's physical and chemical properties but also boosts catalytic efficiency, opening new avenues for optimizing methane dry reforming and its industrial applications. Nonetheless, further research is needed to determine the optimal amounts, methods of doping, and the mechanisms by which promoters affect catalysts.

3.2.3. Iron-Based Catalysts

Iron, as the second most abundant metal in the Earth's crust, offers substantial economic benefits, being approximately 250 times cheaper than Ni or Co [69]. With appropriate preparation and use, iron-based catalysts show excellent thermal stability and poison resistance [90]. Despite Fe's economic advantages in methane conversion, related research remains limited. Additionally, under high temperatures and certain conditions, iron-based catalysts can experience structural changes like particle growth or phase transformation. While iron-based catalysts resist poisoning, trace impurities like sulfur and chlorine in methane can lead to their deactivation [29]. Currently, the deactivation mechanisms and reasons for monometallic iron catalysts remain unclear. Studies on bimetallic catalysts highlight iron's potential to enhance the performance of other metals [37]. Research has also concentrated on using secondary metals to boost iron's reactivity, aiming to develop cost-effective, primarily iron-based catalysts for industrial use [91].

Zahira Yaakob et al. [70] synthesized a Fe/ZrO catalyst via a urea-assisted solid-state combustion technique, applied in the thermal catalytic decomposition of undiluted biogas. This approach promotes the uniform dispersion of catalysts. As evidenced in Figure 6a, at elevated reaction temperatures, such as 800 °C and 900 °C, the peak intensity of iron nanoparticles reaches its maximum, suggesting a more thorough reduction of Fe, thereby enhancing catalytic activity. Concurrently, Figure 6b illustrates that the area of the TPR peaks increases substantially with added Fe content, indicating that higher reaction temperatures and greater iron content can improve CH_4 and CO_2 conversion and the generation of hydrogen-rich syngas by augmenting the number of active sites and enhancing metal–carrier interactions. This methodology, in contrast to the absence of a catalyst or the use of alternative catalyst types, allows for more efficient exploitation of biogas as a feedstock, converting it into higher-value products. Thunyathon Kludpantanapan et al. [71] investigated monometallic Fe and bimetallic Fe-based catalysts on MgO supports for synthesizing carbon nanotubes and hydrogen from biogas. The Fe/MgO catalyst showed poor catalytic activity due to the CO_2-induced oxidation of iron, resulting in significant deactivation. Nonetheless, the addition of Mo significantly improved metal dispersion and strengthened metal–support interactions, as evidenced in Figure 6c, leading to substantial increases in conversion rates and syngas production, thereby boosting the yields of H_2 and CNTs. Additionally, Karam Jabbour and colleagues developed ordered mesoporous $Fe_{10\%}$ and $Fe_xNi_{(1-x)}$ catalysts within Al_2O_3 using an evaporation-induced self-assembly direct "one-pot" method to generate syngas [69]. Figure 6d illustrates that carbon deposition is directly proportional to the Fe content and inversely proportional to the Ni content in the catalysts, suggesting that higher Ni levels reduce carbon accumulation. The introduction of Ni facilitates the formation of a Fe-Ni alloy, minimizing the segregation of iron species and the accumulation of iron oxides and carbon. Furthermore, Ni catalyzes the activation of CO_2, where the resulting oxygen atoms oxidize deposited carbon (C(s)) to CO, diminishing carbon buildup. Regarding the resistance to poisoning, particularly against sulfide in real-world applications, incorporating additives enhances the resilience of iron-based catalysts and supports their regeneration and reuse. Research indicates that chromium additives notably improve the low-temperature denitrification activity and sulfur resistance of these catalysts.

Adopting innovative synthesis methods or incorporating secondary metals such as Mo significantly enhances metal dispersion and strengthens metal–support interactions. These advancements lead to marked increases in conversion rates and syngas production, consequently enhancing the yields of hydrogen and carbon nanotubes (CNTs). Moreover, introducing nickel facilitates the formation of a Fe-Ni alloy, which minimizes the segregation of iron species and reduces the accumulation of iron oxides and carbon. This improvement promotes CO_2 activation, potentially enhancing the catalyst's resistance to poisoning and its regenerative capabilities in practical applications. These developments not only underscore the potential of iron-based catalysts in the methane dry reforming process but also create essential technological pathways for developing more efficient and environmentally friendly catalytic systems.

4. Challenges and Future Trends

Biogas dry reforming technology enables the conversion of biogas into syngas, effectively using waste resources and decreasing reliance on traditional fossil fuels, thus promoting sustainable energy use. Furthermore, this technology transforms greenhouse gases into high-value syngas, offering significant environmental and economic benefits, with considerable potential for future expansion. This review classifies noble and non-noble metal catalysts for biogas dry reforming, summarizing recent advances through strategies like smaller particle size design and improved metal–support interactions. The findings suggest that current methods effectively reduce carbon deposition and enhance catalyst activity. Despite laboratory progress, biogas dry reforming technology encounters challenges in industrial production and large-scale applications. Primary challenges encompass the long-term catalyst stability meeting industrial standards and issues in process parameter control, equipment design, and optimization, which are crucial for the technology's commercial application.

Future research will primarily concentrate on enhancing the electronic and structural properties of catalyst surfaces. This will be achieved through strategic modifications including structural alterations, the introduction of additives, and alloying to improve both the resistance to carbon deposition and reaction efficiency. Additionally, efforts will be directed towards the development of innovative catalysts, such as those featuring layered, porous, or nanostructured oxides, sulfides, and nitrides. These structures are expected to offer a greater number of active sites and superior surface activity. Designing catalysts for improved light interaction is a promising research direction. Research demonstrates that illumination not only diminishes the thermal demands of reactions but also stimulates catalyst surface electrons, resulting in the activated adsorption of reactants. Consequently, this enhances the reaction rate and significantly reduces carbon deposition. Investigating the effects of illumination on the electronic properties of catalyst surfaces, and the control of reaction pathways through illumination, can potentially increase both the conversion rate of methane and the selectivity of products. Generative artificial intelligence (AI), including Transformer, Diffusion models, and Generative Adversarial Nets (GANs), is increasingly pivotal in driving catalyst design innovations. By learning from extensive datasets, these models produce catalyst structures with enhanced predictive capabilities, overcoming the constraints of traditional design methods. In developing biogas dry reforming catalysts, these technologies enable rapid screening and optimization, simulating material reactivity and stability to predict performance. This approach not only accelerates the experimental process but also reduces research and development costs. Additionally, generative AI deepens understanding of catalytic mechanisms, shows how various structures and components influence efficiency and selectivity, and offers a scientific basis for designing more efficient and eco-friendly catalysts. The interplay between theoretical simulation and experimental validation remains vital, uncovering the methane dry reforming reaction's mechanisms and key steps. Deep insights into the reaction mechanism will inform catalyst design and optimization, enhancing reaction efficiency and reducing carbon deposition. Such research will offer crucial theoretical and technical support for the industrial use of methane dry reforming and sustainable production.

Author Contributions: Investigation, W.H., J.W., H.T., Y.T., H.Z., Z.F. and R.D.; resources, W.H.; writing—original draft preparation, W.H.; writing—review and editing, Z.H., Z.P., R.Z., J.W., H.T., Y.T. and H.Z.; supervision, Z.H., Z.P. and R.Z. All authors have read and agreed to the published version of the manuscript.

Funding: This work has been financially supported by the National Natural Science Foundation of China (No. 22209136), the Sichuan Provincial Science and Technology Plan "Unveiling and Leading" Project (No. 23JBGOV0008), the Key R&D projects in Sichuan Province (No. 2023YFG0106), the Major

Science and Technology Special Projects in Sichuan Province (No. 23ZDZX0008). And the authors also thank a grant from the Ministry of Science and Higher Education of the Russian Federation for large scientific projects in priority areas of scientific and technological development (075-15-2024-646).

Data Availability Statement: No new data were created or analyzed in this study.

Conflicts of Interest: Authors Zeai Huang and Zhicheng Pan were employed by the company Haitian Water Group. The remaining authors declare that the research was conducted in the absence of any commercial or financial relationships that could be construed as a potential conflict of interest.

References

1. Atelge, M.R.; Krisa, D.; Kumar, G.; Eskicioglu, C.; Nguyen, D.D.; Chang, S.W.; Atabani, A.E.; Al-Muhtaseb, A.H.; Unalan, S. Biogas Production from Organic Waste: Recent Progress and Perspectives. *Waste Biomass Valorization* **2020**, *11*, 1019–1040. [CrossRef]
2. Pavi, S.; Kramer, L.E.; Gomes, L.P.; Schiavo Miranda, L.A. Biogas production from co-digestion of organic fraction of municipal solid waste and fruit and vegetable waste. *Bioresour. Technol.* **2017**, *228*, 362–367. [CrossRef] [PubMed]
3. Wu, J.P.; Jiang, X.; Jin, Z.H.; Yang, S.L.; Zhang, J. The performance and microbial community in a slightly alkaline biotrickling filter for the removal of high concentration H_2S from biogas. *Chemosphere* **2020**, *249*, 126127. [CrossRef] [PubMed]
4. Chen, L.; Yuan, J.; Li, T.; Jiang, X.; Ma, S.; Cen, W.; Jiang, W. A regenerable N-rich hierarchical porous carbon synthesized from waste biomass for H_2S removal at room temperature. *Sci. Total Environ.* **2021**, *768*, 144452. [CrossRef] [PubMed]
5. Garg, A.; Basu, S.; Shetti, N.P.; Bhattu, M.; Alodhayb, A.N.; Pandiaraj, S. Biowaste to bioenergy nexus: Fostering sustainability and circular economy. *Environ. Res.* **2024**, *250*, 118503. [CrossRef] [PubMed]
6. Obaideen, K.; Abdelkareem, M.A.; Wilberforce, T.; Elsaid, K.; Sayed, E.T.; Maghrabie, H.M.; Olabi, A.G. Biogas role in achievement of the sustainable development goals: Evaluation, Challenges, and Guidelines. *J. Taiwan Inst. Chem. Eng.* **2022**, *131*, 104207. [CrossRef]
7. Gupta, P.; Kurien, C.; Mittal, M. Biogas (a promising bioenergy source): A critical review on the potential of biogas as a sustainable energy source for gaseous fuelled spark ignition engines. *Int. J. Hydrogen Energy* **2023**, *48*, 7747–7769. [CrossRef]
8. Di Micco, S.; Romano, F.; Jannelli, E.; Perna, A.; Minutillo, M. Techno-economic analysis of a multi-energy system for the co-production of green hydrogen, renewable electricity and heat. *Int. J. Hydrogen Energy* **2023**, *48*, 31457–31467. [CrossRef]
9. Abanades, S.; Abbaspour, H.; Ahmadi, A.; Das, B.; Ehyaei, M.A.; Esmaeilion, F.; Assad, M.E.; Hajilounezhad, T.; Hmida, A.; Rosen, M.A.; et al. A conceptual review of sustainable electrical power generation from biogas. *Energy Sci. Eng.* **2022**, *10*, 630–655. [CrossRef]
10. He, J.; Jin, Z.; Gan, F.; Xie, L.; Guo, J.; Zhang, S.; Jia, C.Q.; Ma, D.; Dai, Z.; Jiang, X. Liquefiable biomass-derived porous carbons and their applications in CO_2 capture and conversion. *Green Chem.* **2022**, *24*, 3376–3415. [CrossRef]
11. Kim, Y.; Kawahara, N.; Tsuboi, K.; Tomita, E. Combustion characteristics and NO_x emissions of biogas fuels with various CO_2 contents in a micro co-generation spark-ignition engine. *Appl. Energy* **2016**, *182*, 539–547. [CrossRef]
12. Wang, A.; Harrhy, J.H.; Meng, S.; He, P.; Liu, L.; Song, H. Nonthermal plasma-catalytic conversion of biogas to liquid chemicals with low coke formation. *Energy Convers. Manag.* **2019**, *191*, 93–101. [CrossRef]
13. Zhang, Z.G.; Bi, G.C.; Zhang, S.L.; Tan, T.; Xie, J. $CeMgO_x$-promoted Ni/boron nitride ultra-long stability bi-reforming for methanol-friendly syngas ($H_2/CO\approx2$): CO_2/H_2O activation and coke elimination mechanism. *Fuel* **2024**, *364*, 131122. [CrossRef]
14. Bac, S.; Keskin, S.; Avci, A.K. Recent advances in sustainable syngas production by catalytic CO_2 reforming of ethanol and glycerol. *Sustain. Energy Fuels* **2020**, *4*, 1029–1047. [CrossRef]
15. Mondal, K.; Sasmal, S.; Badgandi, S.; Chowdhury, D.R.; Nair, V. Dry reforming of methane to syngas: A potential alternative process for value added chemicals-a techno-economic perspective. *Environ. Sci. Pollut. Res.* **2016**, *23*, 22267–22273. [CrossRef]
16. Manfro, R.L.; Souza, M.M.V.M. Overview of Ni-Based Catalysts for Hydrogen Production from Biogas Reforming. *Catalysts* **2023**, *13*, 1296. [CrossRef]
17. Zhao, X.; Joseph, B.; Kuhn, J.; Ozcan, S. Biogas Reforming to Syngas: A Review. *Iscience* **2020**, *23*, 101082. [CrossRef]
18. Kumar, R.; Kumar, A.; Pal, A. Overview of hydrogen production from biogas reforming: Technological advancement. *Int. J. Hydrogen Energy* **2022**, *47*, 34831–34855. [CrossRef]
19. Bak, C.-u.; Lim, C.-J.; Kim, Y.-D.; Kim, W.-S. Multi-stage adsorptive purification process for improving desulfurization performance of biogas. *Sep. Purif. Technol.* **2019**, *227*, 115702. [CrossRef]
20. Charisiou, N.D.; Siakavelas, G.; Tzounis, L.; Sebastian, V.; Monzon, A.; Baker, M.A.; Hinder, S.J.; Polychronopoulou, K.; Yentekakis, I.V.; Goula, M.A. An in depth investigation of deactivation through carbon formation during the biogas dry reforming reaction for Ni supported on modified with CeO_2 and La_2O_3 zirconia catalysts. *Int. J. Hydrogen Energy* **2018**, *43*, 18955–18976. [CrossRef]
21. Charisiou, N.D.; Douvartzides, S.L.; Siakavelas, G.I.; Tzounis, L.; Sebastian, V.; Stolojan, V.; Hinder, S.J.; Baker, M.A.; Polychronopoulou, K.; Goula, M.A. The Relationship between Reaction Temperature and Carbon Deposition on Nickel Catalysts Based on Al_2O_3, ZrO_2 or SiO_2 Supports during the Biogas Dry Reforming Reaction. *Catalysts* **2019**, *9*, 676. [CrossRef]
22. Calgaro, C.O.; Perez-Lopez, O.W. Biogas dry reforming for hydrogen production over Ni-M-Al catalysts (M = Mg, Li, Ca, La, Cu, Co, Zn). *Int. J. Hydrogen Energy* **2019**, *44*, 17750–17766. [CrossRef]

23. Rosset, M.; Feris, L.A.; Perez-Lopez, O.W. Biogas reforming to syngas over cu-modified Ni-Al-LDH catalysts. *Int. J. Energy Res.* **2022**, *46*, 22664–22678. [CrossRef]
24. Aramouni, N.A.K.; Touma, J.G.; Abu Tarboush, B.; Zeaiter, J.; Ahmad, M.N. Catalyst design for dry reforming of methane: Analysis review. *Renew. Sustain. Energy Rev.* **2018**, *82*, 2570–2585. [CrossRef]
25. Zhang, C.; Li, S.; Wang, T.; Wu, G.; Ma, X.; Gong, J. Pt-based core-shell nanocatalysts with enhanced activity and stability for CO oxidation. *Chem. Commun.* **2013**, *49*, 10647–10649. [CrossRef]
26. Tu, W.; Ghoussoub, M.; Singh, C.V.; Chin, Y.-H.C. Consequences of Surface Oxophilicity of Ni, Ni-Co, and Co Clusters on Methane Activation. *J. Am. Chem. Soc.* **2017**, *139*, 6928–6945. [CrossRef]
27. Liu, Z.; Lustemberg, P.; Gutierrez, R.A.; Carey, J.J.; Palomino, R.M.; Vorokhta, M.; Grinter, D.C.; Ramirez, P.J.; Matolin, V.; Nolan, M.; et al. In Situ Investigation of Methane Dry Reforming on Metal/Ceria(111) Surfaces: Metal-Support Interactions and C-H Bond Activation at Low Temperature. *Angew. Chem.-Int. Ed.* **2017**, *56*, 13041–13046. [CrossRef]
28. Niu, J.; Du, X.; Ran, J.; Wang, R. Dry (CO_2) reforming of methane over Pt catalysts studied by DFT and kinetic modeling. *Appl. Surf. Sci.* **2016**, *376*, 79–90. [CrossRef]
29. Chein, R.; Yang, Z.-W. H_2S effect on dry reforming of biogas for syngas production. *Int. J. Energy Res.* **2019**, *43*, 3330–3345. [CrossRef]
30. Acs, N.; Szuhaj, M.; Wirth, R.; Bagi, Z.; Maroti, G.; Rakhely, G.; Kovacs, K.L. Microbial Community Rearrangements in Power-to-Biomethane Reactors Employing Mesophilic Biogas Digestate. *Front. Energy Res.* **2019**, *7*, 132. [CrossRef]
31. Kameya, Y.; Hanamura, K. Variation in Catalytic Activity of Carbon Black during Methane Decomposition: Active Site Estimations from Surface Structural Characteristics. *Catal. Lett.* **2012**, *142*, 460–463. [CrossRef]
32. Oliveira, L.G.; Machado, B.; de Souza, L.P.; Correa, G.C.G.; Polinarski, M.A.; Cavalcanti Trevisan, S.V.; Borba, C.E.; Brackmann, R.; Alves, H.J. Dry reforming of biogas in a pilot unit: Scale-up of catalyst synthesis and green hydrogen production. *Int. J. Hydrogen Energy* **2022**, *47*, 35608–35625. [CrossRef]
33. Fakeeha, A.H.; Vadodariya, D.M.; Alotibi, M.F.; Abu-Dahrieh, J.K.; Ibrahim, A.A.; Abasaeed, A.E.; Alarifi, N.; Kumar, R.; Al-Fatesh, A.S. Pd+Al_2O_3-Supported Ni-Co Bimetallic Catalyst for H_2 Production through Dry Reforming of Methane: Effect of Carbon Deposition over Active Sites. *Catalysts* **2023**, *13*, 1374. [CrossRef]
34. Moogi, S.; Ko, C.H.; Rhee, G.H.; Jeon, B.-H.; Khan, M.A.; Park, Y.-K. Influence of catalyst synthesis methods on anti-coking strength of perovskites derived catalysts in biogas dry reforming for syngas production. *Chem. Eng. J.* **2022**, *437*, 135348. [CrossRef]
35. Gao, Y.; Jiang, J.; Meng, Y.; Aihemaiti, A.; Ju, T.; Chen, X.; Yan, F. A novel nickel catalyst supported on activated coal fly ash for syngas production via biogas dry reforming. *Renew. Energy* **2020**, *149*, 786–793. [CrossRef]
36. Owgi, A.H.K.; Jalil, A.A.; Hussain, I.; Hassan, N.S.; Hambali, H.U.; Siang, T.J.; Vo, D.V.N. Catalytic systems for enhanced carbon dioxide reforming of methane: A review. *Environ. Chem. Lett.* **2021**, *19*, 2157–2183. [CrossRef]
37. Wang, C.; Zhang, Y.; Wang, Y.; Zhao, Y. Comparative Studies of Non-noble Metal Modified Mesoporous M-Ni-CaO-ZrO_2 (M=Fe, Co, Cu) Catalysts for Simulated Biogas Dry Reforming. *Chin. J. Chem.* **2017**, *35*, 113–120. [CrossRef]
38. Pakhare, D.; Spivey, J. A review of dry (CO_2) reforming of methane over noble metal catalysts. *Chem. Soc. Rev.* **2014**, *43*, 7813–7837. [CrossRef]
39. Akansu, H.; Arbag, H.; Tasdemir, H.M.; Yasyerli, S.; Yasyerli, N.; Dogu, G. Nickel-based alumina supported catalysts for dry reforming of biogas in the absence and the presence of H_2S: Effect of manganese incorporation. *Catal. Today* **2022**, *397*, 37–49. [CrossRef]
40. Goula, M.A.; Charisiou, N.D.; Papageridis, K.N.; Delimitis, A.; Pachatouridou, E.; Iliopoulou, E.F. Nickel on alumina catalysts for the production of hydrogen rich mixtures via the biogas dry reforming reaction: Influence of the synthesis method. *Int. J. Hydrogen Energy* **2015**, *40*, 9183–9200. [CrossRef]
41. Vita, A.; Italiano, C.; Ashraf, M.A.; Pino, L.; Specchia, S. Syngas production by steam and oxy-steam reforming of biogas on monolith-supported CeO_2-based catalysts. *Int. J. Hydrogen Energy* **2018**, *43*, 11731–11744. [CrossRef]
42. Sadykov, V.A.A.; Eremeev, N.F.F.; Sadovskaya, E.; Fedorova, J.E.E.; Arapova, M.V.V.; Bobrova, L.N.N.; Ishchenko, A.V.V.; Krieger, T.A.A.; Melgunov, M.S.S.; Glazneva, T.S.S.; et al. Approaches to the design of efficient and stable catalysts for biofuel reforming into syngas: Doping the mesoporous $MgAl_2O_4$ support with transition metal cations. *Dalton Trans.* **2023**, *52*, 8756–8769. [CrossRef]
43. Theofanidis, S.A.; Galvita, V.V.; Poelman, H.; Marin, G.B. Enhanced Carbon-Resistant Dry Reforming Fe-Ni Catalyst: Role of Fe. *ACS Catal.* **2015**, *5*, 3028–3039. [CrossRef]
44. Oemar, U.; Hidajat, K.; Kawi, S. Pd-Ni catalyst over spherical nanostructured Y_2O_3 support for oxy-CO_2 reforming of methane: Role of surface oxygen mobility. *Int. J. Hydrogen Energy* **2015**, *40*, 12227–12238. [CrossRef]
45. Université, S.; University, A.; Université, S.; Université, S.; Université, S. Nickel Supported Modified Ceria Zirconia Lanthanum/Praseodymium/Yttrium Oxides Catalysts for Syngas Production through Dry Methane Reforming. *Mater. Sci. Forum* **2018**, *941*, 2214–2219.
46. Usman, M.; Daud, W.M.A.W.; Abbas, H.F. Dry reforming of methane: Influence of process parameters—A review. *Renew. Sustain. Energy Rev.* **2015**, *45*, 710–744. [CrossRef]
47. Italiano, C.; Balzarotti, R.; Vita, A.; Latorrata, S.; Fabiano, C.; Pino, L.; Cristiani, C. Preparation of structured catalysts with Ni and Ni Rh/CeO_2 catalytic layers for syngas production by biogas reforming processes. *Catal. Today* **2016**, *273*, 3–11. [CrossRef]

48. Bereketidou, O.A.; Goula, M.A. Biogas reforming for syngas production over nickel supported on ceria-alumina catalysts. *Catal. Today* **2012**, *195*, 93–100. [CrossRef]
49. Moral, A.; Reyero, I.; Alfaro, C.; Bimbela, F.; Gandia, L.M. Syngas production by means of biogas catalytic partial oxidation and dry reforming using Rh-based catalysts. *Catal. Today* **2018**, *299*, 280–288. [CrossRef]
50. Nikolaraki, E.; Goula, G.; Panagiotopoulou, P.; Taylor, M.J.; Kousi, K.; Kyriakou, G.; Kondarides, D.I.; Lambert, R.M.; Yentekakis, I.V. Support Induced Effects on the Ir Nanoparticles Activity, Selectivity and Stability Performance under CO_2 Reforming of Methane. *Nanomaterials* **2021**, *11*, 2880. [CrossRef] [PubMed]
51. Xiao, Y.; Xie, K. Active Exsolved Metal-Oxide Interfaces in Porous Single-Crystalline Ceria Monoliths for Efficient and Durable CH_4/CO_2 Reforming. *Angew. Chem.-Int. Ed.* **2022**, *61*, 13079. [CrossRef] [PubMed]
52. Yentekakis, I.V.; Goula, G.; Panagiotopoulou, P.; Katsoni, A.; Diamadopoulos, E.; Mantzavinos, D.; Delimitis, A. Dry Reforming of Methane: Catalytic Performance and Stability of Ir Catalysts Supported on γ-Al_2O_3, $Zr_{0.92}Y_{0.08}O_{2-\delta}$ (YSZ) or $Ce_{0.9}Gd_{0.1}O_{2-\delta}$ (GDC) Supports. *Top. Catal.* **2015**, *58*, 1228–1241. [CrossRef]
53. Marinho, A.L.A.; Rabelo-Neto, R.C.; Epron, F.; Bion, N.; Noronha, F.B.; Toniolo, F.S. Pt nanoparticles embedded in CeO_2 and $CeZrO_2$ catalysts for biogas upgrading: Investigation on carbon removal mechanism by oxygen isotopic exchange and DRIFTS. *J. CO_2 Util.* **2021**, *49*, 101572. [CrossRef]
54. Al-Doghachi, F.J.; Zainal, Z.; Saiman, M.I.; Embong, Z.; Taufiq-Yap, Y.H. Hydrogen Production from Dry-Reforming of Biogas over $Pt/Mg_{1-x}NixO$ Catalysts. *Energy Procedia* **2015**, *79*, 18–25. [CrossRef]
55. Rosha, P.; Mohapatra, S.K.; Mahla, S.K.; Dhir, A. Biogas reforming for hydrogen enrichment by ceria decorated over nickel catalyst supported on titania and alumina. *Int. J. Hydrogen Energy* **2018**, *43*, 21246–21255. [CrossRef]
56. Navarro-Puyuelo, A.; Reyero, I.; Moral, A.; Bimbela, F.; Banares, M.A.; Gandia, L.M. Effect of oxygen addition, reaction temperature and thermal treatments on syngas production from biogas combined reforming using Rh/alumina catalysts. *J. Ind. Eng. Chem.* **2019**, *80*, 217–226. [CrossRef]
57. Itkulova, S.S.; Zakumbaeva, G.D.; Mukazhanova, A.A.; Nurmakanov, Y.Y. Syngas production by biogas reforming over the Co-based multicomponent catalysts. *Cent. Eur. J. Chem.* **2014**, *12*, 1255–1261. [CrossRef]
58. le Sache, E.; Johnson, S.; Pastor-Perez, L.; Horri, B.A.; Reina, T.R. Biogas Upgrading Via Dry Reforming Over a Ni-Sn/CeO_2-Al_2O_3 Catalyst: Influence of the Biogas Source. *Energies* **2019**, *12*, 1007. [CrossRef]
59. Abdullah, B.; Ghani, N.A.A.; Vo, D.-V.N. Recent advances in dry reforming of methane over Ni-based catalysts. *J. Clean. Prod.* **2017**, *162*, 170–185. [CrossRef]
60. Gao, Y.; Golmakani, A.; Nabavi, S.A.; Jiang, J.; Manovic, V. Simulative optimization of catalyst configuration for biogas dry reforming. *Int. J. Hydrogen Energy* **2021**, *46*, 12835–12845. [CrossRef]
61. Zhumabek, M.; Xanthopoulou, G.; Tungatarova, S.A.; Baizhumanova, T.S.; Vekinis, G.; Murzin, D.Y. Biogas Reforming over Al-Co Catalyst Prepared by Solution Combustion Synthesis Method. *Catalysts* **2021**, *11*, 274. [CrossRef]
62. Veiga, S.; Romero, M.; Faccio, R.; Segobia, D.; Apesteguia, C.; Perez, A.L.; Brondino, C.D.; Bussi, J. Biogas dry reforming over Ni-La-Ti catalysts for synthesis gas production: Effects of preparation method and biogas composition. *Fuel* **2023**, *346*, 128300. [CrossRef]
63. Vo, C.-M.; Cao, A.N.T.; Qazaq, A.S.; Pham, C.Q.; Nguyen, D.L.T.; Alsaiari, M.; Vu, T.V.; Sharma, A.; Phuong, P.T.T.; Van, T.T.; et al. Toward syngas production from simulated biogas dry reforming: Promotional effect of calcium on cobalt-based catalysts performance. *Fuel* **2022**, *326*, 125106. [CrossRef]
64. Ha, Q.L.M.; Atia, H.; Kreyenschulte, C.; Lund, H.; Bartling, S.; Lisak, G.; Wohlrab, S.; Armbruster, U. Effects of modifier (Gd, Sc, La) addition on the stability of low Ni content catalyst for dry reforming of model biogas. *Fuel* **2022**, *312*, 122823. [CrossRef]
65. Rosha, P.; Mohapatra, S.K.; Mahla, S.K.; Dhir, A. Catalytic reforming of synthetic biogas for hydrogen enrichment over Ni supported on ZnO-CeO_2 mixed catalyst. *Biomass Bioenergy* **2019**, *125*, 70–78. [CrossRef]
66. Gao, Y.; Aihemaiti, A.; Jiang, J.; Meng, Y.; Ju, T.; Han, S.; Chen, X.; Liu, J. Inspection over carbon deposition features of various nickel catalysts during simulated biogas dry reforming. *J. Clean. Prod.* **2020**, *260*, 120944. [CrossRef]
67. Chuenjai, M.; Wongsakulphasatch, S.; Yong, N.; Maneeprakorn, W.; Sudsakorn, K.; Tongnan, V.; Hartley, U.W.; Ratchahat, S.; Kiatkittipong, W.; Assabumrungrat, S. Synthesis of NiO/MgO/ZrO_2 catalyst for syngas production from partial oxidation and dry reforming of biogas. *Int. J. Hydrogen Energy* **2022**, *47*, 41386–41396. [CrossRef]
68. Kaviani, M.; Rezaei, M.; Alavi, S.M.; Akbari, E. High coke resistance Ni-SiO_2@SiO_2 core-shell catalyst for biogas dry reforming: Effects of Ni loading and calcination temperature. *Fuel* **2022**, *330*, 125609. [CrossRef]
69. Jabbour, K.; Saad, A.; Inaty, L.; Davidson, A.; Massiani, P.; El Hassan, N. Ordered mesoporous Fe-Al_2O_3 based-catalysts synthesized via a direct "one-pot" method for the dry reforming of a model biogas mixture. *Int. J. Hydrogen Energy* **2019**, *44*, 14889–14907. [CrossRef]
70. Pudukudy, M.; Yaakob, Z.; Syahri, K.M.; Jia, Q.; Shan, S. Production of hydrogen-rich syngas and multiwalled carbon nanotubes by biogas decomposition over zirconia supported iron catalysts. *J. Ind. Eng. Chem.* **2020**, *84*, 150–166. [CrossRef]
71. Rattanaamonkulchai, R.; Kludpantanapan, T.; Srifa, A.; Koo-Amornpattana, W.; Chaiwat, W.; Sakdaronnarong, C.; Charinpanitkul, T.; Assabumrungrat, S.; Wongsakulphasatch, S.; Show, P.L.; et al. Simultaneous production of hydrogen and carbon nanotubes from biogas over mono- and bimetallic catalyst. *J. Environ. Chem. Eng.* **2022**, *10*, 107910. [CrossRef]
72. Wang, Y.; Yao, L.; Wang, S.; Mao, D.; Hu, C. Low-temperature catalytic CO_2 dry reforming of methane on Ni-based catalysts: A review. *Fuel Process. Technol.* **2018**, *169*, 199–206. [CrossRef]

73. Lustemberg, P.G.; Ramirez, P.J.; Liu, Z.; Gutierrez, R.A.; Grinter, D.G.; Carrasco, J.; Senanayake, S.D.; Rodriguez, J.A.; Veronica Ganduglia-Pirovano, M. Room-Temperature Activation of Methane and Dry Reforming with CO_2 on Ni-CeO_2(111) Surfaces: Effect of Ce^{3+} Sites and Metal-Support Interactions on C-H Bond Cleavage. *ACS Catal.* **2016**, *6*, 8184–8191. [CrossRef]
74. Qin, Z.; Chen, J.; Xie, X.; Luo, X.; Su, T.; Ji, H. CO_2 reforming of CH_4 to syngas over nickel-based catalysts. *Environ. Chem. Lett.* **2020**, *18*, 997–1017. [CrossRef]
75. Han, J.W.; Kim, C.; Park, J.S.; Lee, H. Highly Coke-Resistant Ni Nanoparticle Catalysts with Minimal Sintering in Dry Reforming of Methane. *Chemsuschem* **2014**, *7*, 451–456. [CrossRef] [PubMed]
76. Li, M.; van Veen, A.C. Tuning the catalytic performance of Ni-catalysed dry reforming of methane and carbon deposition via Ni-CeO_{2-x} interaction. *Appl. Catal. B-Environ.* **2018**, *237*, 641–648. [CrossRef]
77. Mei, D.H.; Shen, X.Q.; Liu, S.Y.; Zhou, R.S.; Yuan, X.C.; Rao, Z.Q.; Sun, Y.F.; Fang, Z.; Du, X.S.; Zhou, Y.; et al. Plasma-catalytic reforming of biogas into syngas over Ni-based bimetallic catalysts. *Chem. Eng. J.* **2023**, *462*, 142044. [CrossRef]
78. De, S.; Zhang, J.; Luque, R.; Yan, N. Ni-based bimetallic heterogeneous catalysts for energy and environmental applications. *Energy Environ. Sci.* **2016**, *9*, 3314–3347. [CrossRef]
79. Carrasco-Ruiz, S.; Zhang, Q.; Gandara-Loe, J.; Pastor-Perez, L.; Odriozola, J.A.; Reina, T.R.; Bobadilla, L.F. H_2-rich syngas production from biogas reforming: Overcoming coking and sintering using bimetallic Ni-based catalysts. *Int. J. Hydrogen Energy* **2023**, *48*, 27907–27917. [CrossRef]
80. Ayodele, B.V.; Khan, M.R.; Cheng, C.K. Catalytic performance of ceria-supported cobalt catalyst for CO-rich hydrogen production from dry reforming of methane. *Int. J. Hydrogen Energy* **2016**, *41*, 198–207. [CrossRef]
81. Park, J.-H.; Yeo, S.; Kang, T.-J.; Shin, H.-R.; Heo, I.; Chang, T.-S. Effect of Zn promoter on catalytic activity and stability of Co/ZrO_2 catalyst for dry reforming of CH_4. *J. CO_2 Util.* **2018**, *23*, 10–19. [CrossRef]
82. Phuong, D.H.L.; Alsaiari, M.; Pham, C.Q.; Hieu, N.H.; Pham, T.-P.T.; Rajamohan, N.; Pham, D.D.; Vo, D.-V.N.; Trinh, T.H.; Setiabudi, H.D.; et al. Carbon dioxide reforming of methane over modified iron-cobalt alumina catalyst: Role of promoter. *J. Taiwan Inst. Chem. Eng.* **2024**, *155*, 105253. [CrossRef]
83. Budiman, A.W.; Song, S.-H.; Chang, T.-S.; Shin, C.-H.; Choi, M.-J. Dry Reforming of Methane Over Cobalt Catalysts: A Literature Review of Catalyst Development. *Catal. Surv. Asia* **2012**, *16*, 183–197. [CrossRef]
84. Fakeeha, A.H.; Al Fatesh, A.S.; Ibrahim, A.A.; Kurdi, A.N.; Abasaeed, A.E. Yttria Modified ZrO_2 Supported Ni Catalysts for CO_2 Reforming of Methane: The Role of Ce Promoter. *ACS Omega* **2021**, *6*, 1280–1288. [CrossRef] [PubMed]
85. Zhang, F.; Liu, Z.; Chen, X.; Rui, N.; Betancourt, L.E.; Lin, L.; Xu, W.; Sun, C.-J.; Abeykoon, A.M.M.; Rodriguez, J.A.; et al. Effects of Zr Doping into Ceria for the Dry Reforming of Methane over Ni/$CeZrO_2$ Catalysts: In Situ Studies with XRD, XAFS, and AP-XPS. *ACS Catal.* **2020**, *10*, 3274–3284. [CrossRef]
86. Wu, P.; Tao, Y.; Ling, H.; Chen, Z.; Ding, J.; Zeng, X.; Liao, X.; Stampfl, C.; Huang, J. Cooperation of Ni and CaO at Interface for CO_2 Reforming of CH_4: A Combined Theoretical and Experimental Study. *ACS Catal.* **2019**, *9*, 10060–10069. [CrossRef]
87. Ngoc Thang, T.; Quyet Van, L.; Nguyen Van, C.; Trinh Duy, N.; Nguyen Huu Huy, P.; Phuong, P.T.T.; Monir, M.U.; Abd Aziz, A.; Quang Duc, T.; Abidin, S.Z.; et al. La-doped cobalt supported on mesoporous alumina catalysts for improved methane dry reforming and coke mitigation. *J. Energy Inst.* **2020**, *93*, 1571–1580. [CrossRef]
88. Cai, Y.; Zhang, G.; Liu, J.; Zhang, Y.; Li, T.; Zhang, X.; Wang, Y.; Zhao, Y.; Li, G. Exploring the role of promoters (Y, Tb and Gd) in the performance of Co-based catalyst for dry reforming of methane. *Int. J. Hydrogen Energy* **2023**, *48*, 38251–38265. [CrossRef]
89. Bahari, M.B.; Setiabudi, H.D.; Trinh Duy, N.; Phuong, P.T.T.; Quang Duc, T.; Jalil, A.A.; Ainirazali, N.; Vo, D.-V.N. Insight into the influence of rare-earth promoter (CeO_2, La2O_3, Y_2O_3, and Sm_2O_3) addition toward methane dry reforming over Co/mesoporous alumina catalysts. *Chem. Eng. Sci.* **2020**, *228*, 115967. [CrossRef]
90. Wang, S.; Chen, X.; Zhang, Y.; Zhang, Y.; Zheng, S. Lithium adsorption from brine by iron-doped titanium lithium ion sieves. *Particuology* **2018**, *41*, 40–47. [CrossRef]
91. Theofanidis, S.A.; Galvita, V.V.; Sabbe, M.; Poelman, H.; Detavernier, C.; Marin, G.B. Controlling the stability of a Fe-Ni reforming catalyst: Structural organization of the active components. *Appl. Catal. B-Environ.* **2017**, *209*, 405–416. [CrossRef]

catalysts

MDPI

Review

Catalytic Applications of Natural Iron Oxides and Hydroxides: A Review

Adriana Jiménez-Vázquez [1,*], Raciel Jaimes-López [2], Carlos Mario Morales-Bautista [1], Samuel Pérez-Rodríguez [2], Yadira Gochi-Ponce [3] and Luis Alberto Estudillo-Wong [2,*]

[1] División Académica de Ciencias Básicas, Universidad Juárez Autónoma de Tabasco, Km. 1 Carretera Cunduacán-Jalpa de Méndez A.P. 24, Cunduacán C.P. 86690, Mexico; carlos.morales@ujat.mx

[2] Laboratorio de Electroquímica Ambiental y de Materiales (LEAyM), Departamento de Biociencias e Ingeniería, CIIEMAD-IPN, Instituto Politécnico Nacional, Calle 30 de Junio de 1520 s/n Barrio la Laguna Ticomán, Alcaldía GAM, Ciudad de México C.P. 07340, Mexico; rajaimesl@ipn.mx (R.J.-L.); sperezro@ipn.mx (S.P.-R.)

[3] Tecnológico Nacional de México, Instituto Tecnológico de Tijuana, Posgrado en Ciencias de la Ingeniería, Blvd. Alberto Limón Padilla s/n, Mesa de Otay, Tijuana C.P. 22500, Mexico; yadira.gochi@tectijuana.edu.mx

* Correspondence: ajv05573@docente.ujat.mx (A.J.-V.); laestudillo-wong@ipn.mx (L.A.E.-W.)

Abstract: Iron oxides and hydroxides (Fe-OH) extracted from natural sources have garnered significant attention for their diverse catalytic applications. This article provides a comprehensive overview of the catalytic potential of naturally occurring Fe-OH, focusing on the influence of natural sources and preparation methods on their morphological characteristics and application in heterogeneous catalysis. The unique physicochemical properties of these catalysts, including their high surface area, redox activity, and tunable surface chemistry, make them promising candidates for various catalytic processes. The review discusses key catalytic reactions facilitated by natural Fe-OH, such as advanced oxidation processes (AOPs), electrochemical applications, catalytic cracking, and biodiesel production. Furthermore, it highlights recent advancements and challenges in utilizing these materials as heterogeneous catalysts. By presenting an analysis of the catalytic potential of natural iron oxides, this review aims to stimulate further research about the use of these materials, which are widely distributed in the Earth's crust.

Keywords: minerals; natural sources; hematite; maghemite; magnetite

1. Introduction

Iron oxides and hydroxides, both referred to here as Fe-OH, are ubiquitous earth materials with high adsorption capacities for toxic elements and degradation ability towards organic contaminants. By mass, iron is the fourth most abundant element in the Earth's crust, being found in 16 Fe-OH [1]. Iron oxides are widely distributed in nature and can be found across various components of the global system, including the lithosphere, hydrosphere, biosphere, pedosphere, and atmosphere. They participate in the complex interactions between these environmental compartments [2]. Iron's different oxidation states (ferric (Fe^{3+}) and ferrous (Fe^{2+})) further influence its interactions within these systems. For instance, iron availability in the oceans affects the growth of phytoplankton, impacting the entire marine food chain.

Geological Abundance of Iron Oxide and Hydroxide Minerals

Iron constitutes a significant portion of minerals on our planet, ranging from 28 to 35% of Earth's total mass. The Earth's inner core primarily comprises solid iron (Fe) and nickel (Ni), while the outer core consists of iron alloyed with 10–20% of other elements, such as oxygen (O), sulfur (S), and potassium (K). In the Earth's structure, the primitive mantle contains roughly 12% iron, whereas the crust holds 4.32% iron, ranking just below oxygen (47.2%), silicon (Si, 28.8%), and aluminum (Al, 7.96%) in abundance. Consequently, iron silicates and iron oxides emerge as the most prevalent minerals in the Earth's crust [3]. Table 1 enumerates the naturally occurring iron oxide and hydroxide minerals. Goethite, hematite, and magnetite are the most prevalent, often occurring as standalone rock formations in nature. Lepidocrocite, ferrihydrite, and maghemite are found in significantly smaller quantities but are still distributed across many locations. On the other hand, wüstite, akaganéite, feroxyhyte, and bernalite are comparatively rare, both in terms of their distribution and abundance [4].

Table 1. Naturally occurring iron oxide minerals. Reproduced with permission from Reference [3]. Copyright© 2016, John Wiley and Sons.

Name	Formula	Crystal Symmetry
Magnetite (Mt)	Fe_3O_4	Cubic spinel
Hematite (Ht)	α-Fe_2O_3	Rhombohedral
Goethite (Gt)	α-FeOOH	Orthorhombic
Lepidocrocite	γ-FeOOH	Orthorhombic
Ferrihydrite (Fh)	$5Fe_2O_3 \cdot 9H_2O$	Hexagonal
Maghemite (Mh)	γ-Fe_2O_3	Cubic spinel
Wüstite	FeO	Isometric-hexoctahedral
Akaganéite	β-FeOOH	Monoclinic
Feroxyhyte	δ'-FeOOH	Hexagonal
Bernalite	$Fe(OH)_3$	Orthorhombic

The primary source of iron is the rocks known as banded iron formations (BIFs). These formations are named for their characteristic composition, featuring alternating layers of iron (hematite or magnetite) and silica. Iron can also be found in geological formations like greenschist, gabbro, basalt, amphibolite, and others [5].

Schwertmann and Fitzpatrick (1993) [6] describe various notable features of iron oxides found on Earth's surface. These minerals are physicochemical environmental indicators with unique mineral characteristics and strong pigmenting abilities, even at low concentrations. They tend to have fine grain sizes (5–100 nm) and large surface areas, allowing functionalized surfaces to absorb a range of molecules. Iron oxides are also highly insoluble and may include other cations like aluminum in their crystal structure. Due to their limited crystallinity, advanced techniques, such as Mössbauer spectroscopy, X-ray absorption near-edge structure (XANES), thermogravimetric analysis (TGA), X-ray diffraction (XRD), infrared spectroscopy (IR), differential thermal analysis (DTA), transmission electron microscopy (TEM), and selective chemical dissolution, are essential for identification [6].

Over the past decade, numerous researchers have investigated the utilization of naturally derived iron oxides as heterogeneous catalysts, driven by their widespread viability, readily accessible sources, low production costs, appealing physicochemical properties, and inherent biocompatibility [7–9]. The success achieved in numerous applications has led to reviews of articles on the subject. However, to the best of our knowledge, these works have focused on specific processes or the use of specific minerals. Therefore, there is a need for a comprehensive and integrative view of the topic. Fenton's reaction, a well-established catalytic process, utilizes ferrous and/or ferric ions to decompose hydrogen peroxide, generating potent oxidizing agents capable of degrading a wide range of organic and inorganic contaminants. This reaction exemplifies a catalytic application of iron oxides. Thus, it is only tangentially addressed, referring readers to the reviews already published. A search on Scopus with the search terms TITLE (* "iron" AND "Catalyst" AND "review") yields 35 results, of which only 14 are about natural minerals of iron oxyhydroxides. Three of these address their use in Fenton, photo-Fenton, and electro-Fenton processes [10–12], with four on water–gas shift, syngas and Fischer–Tropsch synthesis [13–16], three on persulfate activation [17–19], one on peroxides [20], one on nanoparticles applied in different catalysts [21], and two on advanced oxidation [22,23]. Additionally, a review of the performance of magnetite [24] in Fenton-like processes is identified. In just the last two years, by using the search terms 'iron', 'oxide', 'catalysts', and 'review' in article titles, the SciFinder® search platform has yielded additional examples focused on Fischer–Tropsch synthesis [25,26], persulfate activation [27], NOx reduction [28,29], oxygen reduction reaction [30], and selective oxidation of hydrogen sulfide [31]. However, none of these address the use of natural materials. However, other catalytic processes involving iron oxides have also been tested. Among these applications are hydrocracking of coal tar, biodiesel production, hydrogenation, ammonia synthesis, reduction of nitrates, heterogeneous catalytic ozonation, oxidation and reduction of inorganic ions, hydrogen production, and water dissociation, which have not been sufficiently disseminated and analyzed in the literature. The main catalytic application of Fe-OH and the iron species used in them are depicted in Figure 1.

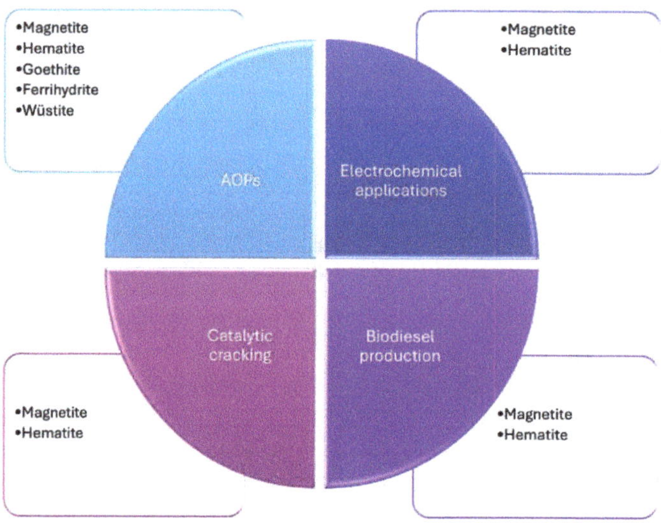

Figure 1. Scheme of main catalytic applications of Fe-OH from natural sources and natural iron species.

To the authors' knowledge, there has been no previous attempt to present, in a single document, most of the applications proposed in the literature for naturally derived Fe-OH. In this review, chemical industry research professionals will find a comprehensive compilation of the identified applications of Fe-OH, especially in the last two decades. The aim is to provide an integrative approach, emphasizing the abundance and versatility of natural iron catalysts, which makes them economically attractive and environmentally benign. This document presents basic information on the relative abundance of Fe-OH in the Earth's crust, their occurrence in nature, pretreatments of the catalysts, and both common and uncommon applications. It is divided into three sections: the first one for the most abundant, the second one for those less abundant but distributed globally, and finally, the third one for those considered difficult to obtain.

2. Most Prevalent Iron Oxides and Hydroxides

Magnetite (Fe_3O_4) and hematite (α-Fe_2O_3) are among the most prevalent iron oxides, while goethite (α-FeOOH) is the most abundant iron hydroxide. In this section, we will explore principal characteristics and both the primary and less common catalytic applications of these iron oxides and hydroxides.

2.1. Magnetite (Fe_3O_4)

Magnetite is a widely distributed accessory mineral in numerous ore deposits and ranks among the most abundant oxide minerals in the continental crust [32]. It occurs globally and can form under diverse temperature and pressure conditions in igneous, metamorphic, and sedimentary settings [33].

At ambient conditions, they crystallize in the inverse spinel cubic structure, which belongs to the space group $Fd\bar{3}m$. Its distinctive structure is characterized by a combination of diverse anions and cations. In a conventional spinel arrangement, divalent cations exclusively occupy tetrahedral positions, while trivalent cations are confined to octahedral coordination. However, within an inverse spinel configuration, the octahedral sites accommodate both divalent and trivalent cations simultaneously [34]. Fe^{3+} ions are positioned in tetrahedral (A) sites coordinated by oxygen, while octahedral (B) sites contain an equal mixture of Fe^{3+} and Fe^{2+} ions. Below the magnetic transition temperature of 850 K, the spins of iron atoms in the A and B sublattices align in a ferrimagnetic configuration [35].

Fe_3O_4 is the most utilized among various iron oxides, including FeO, γ-Fe_2O_3, and $FeTiO_3$, because it exhibits a notable degree of spin polarization at the Fermi level, a high Curie temperature, and electrocatalytic effect, stability, and ease of synthesis. Furthermore, nano-sized Fe_3O_4 shows additional advantageous features beyond those inherited from its bulk form (see Table 2), such as remarkable size and shape controllability, a large specific surface area, and excellent magnetothermal conversion capability. Magnetite nanoparticles have been utilized in a wide range of catalytic applications, such as drug delivery, cancer treatment, organic synthesis, environmental remediation, synthesis of biodiesel, and electrocatalysis [36]. Magnetite is widely used in advanced oxidation processes (AOPs) [24]; however, its catalytic applications include organic synthesis, electrocatalysis, and environmental remediation, among others [36].

Table 2. Characteristics of bulk Fe_3O_4 and Fe_3O_4 NPs. Reproduced with permission from Reference [36].

Characteristics	Bulk Fe_3O_4	Fe_3O_4 NPs
Magnetism	Ferromagnetism	Superparamagnetism or ferrimagnetism
Saturation magnetization (M_s, 300 K, emu/g)	~(84–100)	Depend on size, shape, and coating: ~(0.5–92)
Size controllability	Unachievable	Precisely controllable: ~(2–100) nm
Shape controllability (Spherical, cubic, rod, hollow, and 2D nanoplate)	Unachievable	Precisely controllable
Specific surface area (m^2/g)	0.34 *	Depend on size, shape, and coating: ~(20–300)
Magnetothermal conversion (W/g)	Absent	~(100–2500)
Electrocatalytic activity	Achievable	Achievable

* For an average crystallite size between 38 and 62 μm.

The catalytic performance of magnetite nanoparticles is strongly influenced by their size. In general, smaller nanoparticles exhibit enhanced catalytic efficiency due to their higher surface area and greater number of active sites compared to larger particles. Magnetite-based nanocatalysts are gaining prominence as adaptable tools in heterogeneous catalysis, contributing to the advancement of sustainable reaction methodologies [37].

Numerous methods exist for synthesizing bulk and magnetite nanoparticles [38]. However, the extraction of magnetite nanoparticles from natural sources has received limited attention. While some studies have explored this area [39–43], the primary objective of these syntheses typically differs from catalytic applications. Furthermore, it has been reported that different natural sources of magnetite and preparation methods significantly influence the morphology and size of the particles.

Magnetite prepared from iron sand via coprecipitation shows a more uniform morphology and particle size distribution. In contrast, the ball milling method allows control of particle size based on milling time, with longer times resulting in smaller particles but more irregular shapes (Figure 2A). On the other hand, when the natural source is mineral magnetite, the particles have more regular shapes. Nevertheless, the particle size distribution is broader (Figure 2B,C).

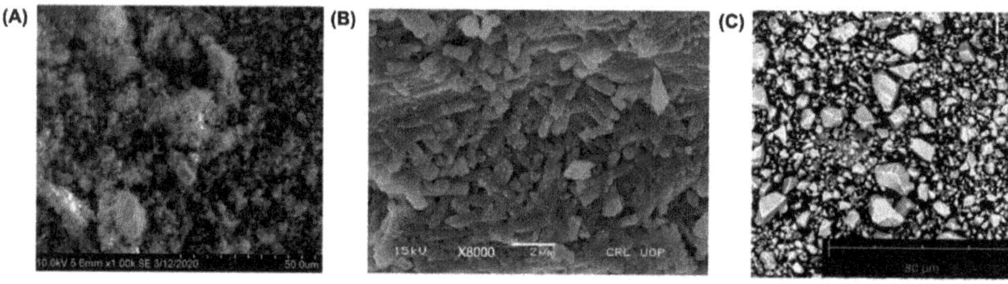

Figure 2. SEM micrographs of magnetite particles prepared from natural sources: (**A**) iron sand via ball milling, (**B**) raw ore after washing and heat treatment, and (**C**) mineral magnetite after magnetic separation and 20 min of ball milling. Reproduced with permission from References (**A**) [44], (**B**) [45], and (**C**) [46]. Copyright© 2020 IOPScience/CC BY 3.0, 2016 and 2021 Elsevier.

Main catalytic applications. Magnetite's capacity to generate hydroxyl radicals enables its utilization in various advanced oxidation processes (AOPs), including Fenton, electro-Fenton, photo-Fenton, and Fenton-like reactions. These processes share the common principle of involving the interaction between iron-based catalysts and H_2O_2 to generate highly reactive hydroxyl radicals for degrading organic pollutants in water. Variations arise in activation methods: in electro-Fenton, H_2O_2 is generated by O_2 reduction at the cathode, photo-Fenton utilizes light to generate $^{\bullet}OH$ and Fe^{2+} from water and Fe^{3+}, and Fenton-like reactions use alternative iron species or transition metals [47]. According to catalytic evaluations and findings performed on various iron oxides in Fenton reactions, magnetite is highly active for both H_2O_2 decomposition and O_2 production at neutral and basic pH [48]. However, due to the magnetic nature of magnetite, the reactor design must consider conditions to prevent particle agglomeration, which could inhibit its catalytic activity by limiting proper dispersion in the solution. To address this issue, magnetic stirring is omitted, using centrifugal [49] or peristaltic pumps [50] for the recirculation of contaminants.

Mineral magnetite has been used to obtain catalysts for the degradation of organic pollutants such as p-nitrophenol [1], textile dye [44,49,51], methylparaben [52], and cefotaxime [46]. The natural sources used for the preparation of these catalysts were mineral magnetite and ferruginous sands. Additionally, the preparation of the catalysts commonly involves magnetic and gravitational separation as well as sieving and ball milling. In some cases, chemical treatment with acid, mainly HCl, is preferred to remove impurities present in the natural magnetite source.

Typical ranges of dosage for the Fenton and Fenton-like reactions are between 0.1 and 0.5 $g \cdot L^{-1}$, and optimal degradation occurs in acidic or neutral conditions, with pH values of 3 to 7. Although H_2O_2 is the primary oxidizing agent used, other species, such as $S_2O_8{}^{2-}$ [49,52], can also be employed. Magnetite obtained from natural sources commonly contains other minerals, including Cr, Mn, Ni, Co, and Ti, among others. However, these impurities can enhance catalytic activity. For example, the comparison between synthetic magnetite and mineral magnetite [1] for p-nitrophenol (p-NP) degradation indicates that the combination of nano-sized magnetite and H_2O_2 does not exhibit significant degradation throughout the entire process, underscoring the superior catalytic efficacy of natural magnetite compared to its synthetic counterpart. This fact was attributed to the presence of exsolved spinel due to the incorporation of transition metals with variable valence.

The preparation method of catalysts from natural sources significantly influences control over particle size, shape, and purity. Mufti et al. [51] and Amiruddin et al. [44] prepared magnetite nanoparticles using different approaches (coprecipitation and ball milling) and tested them in photo-Fenton degradation and Fenton degradation of methylene blue (MB), respectively. The coprecipitation method resulted in smaller particle sizes compared to the ball milling method (11 nm vs. 24–49 nm), with more uniform particle size distribution and a high-purity crystal phase; also, S_{BET} was higher compared to the usual values reported for Fe_3O_4. However, it involved the use of high molar concentrations of HCl and NH_4OH, as well as thermal treatment, which are avoided in the milling method. Although the catalytic performance was evaluated under different conditions (photo-Fenton and Fenton), the results for MB degradation were similar (80% vs. 86.89%). However, the reaction times required to achieve these results differed significantly (30 min vs. 120 min). Similarly, in the Fenton-like degradation of acid orange 7 (AO7) using natural magnetite and persulfate ($S_2O_8{}^{2-}$), high degradation efficiency (90%) was achieved within 120 min under optimized conditions, including 0.5 gL^{-1} of catalyst and 0.2 mM of $S_2O_8{}^{2-}$ under acidic pH [49]. The catalyst exhibited low iron leaching and the potential to be scalable, while the reaction time was longer compared to systems leveraging photo-Fenton activation.

The catalytic degradation of methylparaben (MeP) using a Fenton-like process demonstrated the effectiveness of persulfate (SPS) activation by natural magnetite as a heterogeneous catalyst. Peroxymonosulfate and peroxydisulfate are common ions in aqueous effluents, whose presence can be leveraged to generate strong oxidants ($SO_4{}^{\bullet}$, $^{\bullet}OH$, and $^1O_2$) through appropriate catalysts [53]. The study achieved a high degradation efficiency of 99.5% and mineralization rates of up to 37% within 60 min under optimized conditions, including a catalyst loading of 0.1–0.3 $g \cdot L^{-1}$ and SPS concentrations of 1–5 $mmol \cdot L^{-1}$ under acidic pH. The catalyst exhibited excellent reusability, maintaining over 90% degradation efficiency after five cycles, with minimal decline in performance. This highlights its potential as a cost-effective and robust option for pollutant removal [52].

The preparation of composites of TiO_2/magnetite nanoparticles using natural magnetite and P25 as raw materials resulted in particle sizes of about 20–80 nm [46]. No significant increase in S_{BET} was observed. These particles were tested on Fenton, photo-Fenton, and photocatalysis processes. In the photo-Fenton under UV light and photocatalysis processes, the composites achieved complete degradation of cefotaxime, whereas in the Fenton reaction, the degradation was negligible. The higher degradation rate reached by mediated light degradation was attributed to the fact that Fe^{3+} in the magnetic nanoparticle could capture photo-generated electrons from TiO_2, which accelerated Fe^{2+}/Fe^{3+} cycles in the Fenton reaction, the separation of electron–hole pairs in photocatalysis, the decomposition of H_2O_2, and the production of OH radical, resulting in the effective removal of cefotaxime.

Regarding its application in an electro-Fenton process, it was able to reach a degradation efficiency of about 40% on the degradation of gemcitabine at a pH of 3, which is related to ferrous ions in its structure [54].

In an initial approach to synthesizing nanoparticles from mineral magnetite, Munoz et al. [55] evaluated their catalytic activity in the catalytic wet peroxide oxidation (CWPO) of phenol. In this mode of the Fenton process (heterogeneous), the catalyst used to decompose H_2O_2 into $^{\bullet}OH$ is an insoluble solid [56]. The preparation involved only sieving the raw material, which was obtained from Marphil, Madrid, Spain. This resulted in a Fe_3O_4 catalyst with an S_{BET} of 8 m^2g^{-1} and a saturation magnetization (Ms) of 77.7 emu g^{-1}. Under optimal reaction conditions (H_2O_2 = 500 mg·L^{-1}, Fe_3O_4 = 2 g·L^{-1}, T = 75 °C, pH = 3, and t = 4 h), the enhanced performance of the minerals was confirmed in the CWPO of phenol (100 mg·L^{-1}), achieving complete conversion of the target pollutant with a high degree of mineralization (70–80%). However, even though the process was effective, relatively low oxidation rates were observed, requiring up to 3.5 h of reaction time to reach the complete removal of the pollutant under ambient temperature and circumneutral pH. Therefore, mineral magnetite was modified through thermal treatments of controlled oxidation and reduction to obtain core–shell materials [57]. A slight growth in the magnetization of the material was observed with an increase in the reduction temperature. At a pH of 5, ambient temperature (25 °C), the stoichiometric amount of H_2O_2 for the complete mineralization of sulfamethoxazole (SMX), and a catalyst load of 1 g·L^{-1} were selected. The initial concentration of SMX was fixed at 5 mg·L^{-1}. Reduced iron species exhibited greater activity in the CWPO of SMX, leading to complete removal within 1.5 h. The optimal catalyst was reduced at 400 °C. This material remained active even after five reaction cycles.

Recently, the CWPO of azole pesticides, using natural magnetite as a catalyst, has been investigated [50]. The study utilized a fixed-bed reactor packed with natural magnetite powder to eliminate a representative blend of azole pesticides. System performance was assessed by examining the effects of inlet flow rate (0.25–1 mL·min^{-1}), magnetite loading (2–8 g), H_2O_2 dosage (3.6–13.4 mg·L^{-1}), and initial pollutant concentration (100–1000 μg·L^{-1}) over 300 h of continuous operation. Under optimized conditions (T = 25 °C, pH = 5, flow rate = 0.5 mL·min^{-1}, [H_2O_2] = 6.7 mg·L^{-1}, and [Fe_3O_4] = 8 g), azole pesticide removal exceeded 80%. The catalytic system exhibited high stability over 500 h of operation, with limited iron leaching. Also, the efficiency of the catalyst was proven in the treatment of a real wastewater treatment plant (WWTP) effluent sample, spiked with a mixture of azole pesticides at 500 μg·L^{-1}. The process was less efficient in the WWTP effluent; this was attributed to the presence of organic and inorganic compounds in the treated water since they exhibit OH scavenging properties. On CWPO, H_2O_2 dosage and catalyst loading are the most significant variables that influence catalytic activity; even with low S_{BET} and high particle size, effective degradation can be reached. The reactions are described well by pseudo-first-order rate constants.

Non-common catalytic applications. The use of catalysts based on magnetite obtained from natural sources is not limited to environmental remediation applications. It has also been tested in the conversion of spent engine oil into reusable diesel-like fuel [45]. This process, known as catalytic pyrolysis, involves the use of solid catalysts to break down long organic matter chains into lighter hydrocarbons, promoting specific transformations such as dehydration, dehydrogenation, deoxygenation, and decarboxylation [58]. Natural magnetite was extracted from freshly mined raw ore through magnetic separation. Pyrolysis reactions were conducted at 500 °C with a residence time of 90 min. The overall yield was close to 100% in a thermal non-catalyzed run and decreased linearly with magnetite dosage. However, the results indicated that magnetite was selective to liquid fraction and could promote different product compositions due to cracking and oxidative dehydrogenation of the paraffinic and alkyl aromatics. The resultant catalytically derived liquid product has fuel properties comparable to commercial diesel. Recently, $CaO/MgO/Fe_3O_4$ prepared from iron sand and dolomite was tested as a catalyst for biodiesel production from waste cooking oil [57]. The catalyst calcinated at 1000 °C reached 95.64% yield. The produced biodiesel fulfilled the quality standards for density and viscosity.

Pyrolysis is a promising option in regions with limited petroleum and gas resources because it offers a solution by producing char, tar, and gases simultaneously. Char can be used for combustion or gasification, while tar can be processed into liquid fuels and chemicals, improving economic feasibility. However, coal-derived tar often contains heavy compounds that complicate processing and cause operational challenges. To address this, He et al. [59] proposed a coal pyrolysis method combined with iron ore reduction. Iron facilitates the breaking of C-C and C-O bonds. The study used magnetite with a chemical composition primarily consisting of Fe (62.65%) and minor components like SiO_2 and MgO. Pyrolysis experiments were conducted using a Pt-filament pyrolyzer, with volatiles adsorbed and later desorbed at 300 °C for GC/MS analysis. Coal and catalyst samples were heated to 700 °C under argon. Magnetite showed a low surface area (3.32 $m^2 \cdot g^{-1}$) and primarily produced phenols and CO_2 but performed poorly compared to other iron ores like ilmenite and siderite. Also, the nanohybrid catalyst Fe@C, using natural magnetite as an iron source, was evaluated on Fischer–Tropsch synthesis, a collection of reactions for converting syngas (hydrogen and carbon monoxide) into liquid hydrocarbons [60]. The core–shell structure was confirmed by TEM analysis. The results indicated that CO conversion exceeded 98%, with C_5 hydrocarbon selectivity of 51.7% and stability over 180 h of steam [61].

Table 3 summarizes the catalytic applications of hematite prepared from natural sources. Although its main applications are Fenton-type processes, its use for fuel production is also notable. Furthermore, the analysis of the elimination based on the main parameters in AOPs (Figure 3) shows that the specific surface is not a determining factor in the catalytic activity of magnetite; likewise, acidic or neutral conditions allow better performance. In addition, specific surface area is not a determining factor in the catalytic activity of magnetite, while acidic or neutral pH conditions enable better performance. Even with small specific surface areas and low catalyst concentrations, magnetite enables removal rates close to 100% in advanced oxidation processes.

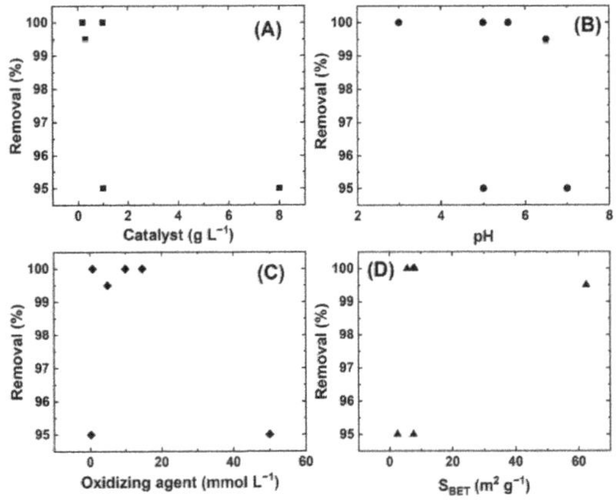

Figure 3. Removal as a function of magnetite catalyst concentration (**A**), pH (**B**), oxidant concentration (**C**), and specific surface area (S_{BET}, (**D**)). Elaborated with data from References [1,46,50,52,55,62].

Table 3. Summary of catalytic applications of natural magnetite (Fe_3O_4).

Catalytic Reaction	Natural Source	Pretreatment Method	S_{BET} ($m^2 \cdot g^{-1}$)	Reaction Conditions	Results of Catalytic Evaluation	Rate Constant	Ref.
Fenton p-NP degradation	Ores	Crushing, grinding, and magnetic and gravity separation	0.54–2.46	Cat = 1.0 $g \cdot L^{-1}$; p-nitrophenol = 10 $mg \cdot L^{-1}$; pH = 7; H_2O_2 = 0.05 $mol \cdot L^{-1}$; NaOH = 0.1 $mol \cdot L^{-1}$	17.8–95.0% degradation	0.24, 0.19, and 0.12 $\mu g \cdot L^{-1} \cdot min^{-1}$	[1]
Photo-Fenton degradation of MB	Natural iron sands	Magnetic separation and coprecipitation	147.12	Cat = 2 $g \cdot L^{-1}$; MB = 10 mL of 20 ppm; UV irradiation	68.52–76.32% degradation	0.000194, 0.0002[illegible]2, and 0.000336 $g \cdot mg^{-1} \cdot min^{-1}$	[51]
Fenton degradation of MB	Beach sand	Drying, magnetic separation, and ball milling	ND	Cat = 1 $g \cdot L^{-1}$; MB = 15 $mg \cdot L^{-1}$; pH = 4.8; H_2O_2 = 150 mL L^{-1}	86.79% degradation	ND	[44]
Photo-Fenton degradation of cefotaxime *	Natural magnetite	Ball milling method	2.92–5.64	Cat = 0.2 $g \cdot L^{-1}$; cefotaxime = 0.4 $mmol \cdot L^{-1}$; pH = 5.6; H_2O_2 = 10 $mmol \cdot L^{-1}$	100% degradation after 60 min	0.09 min^{-1}	[46]
Fenton-like degradation of AO7	Natural magnetite	ND	ND	Cat = 0.5 $mg \cdot L^{-1}$; AO7 = 15 $mg \cdot L^{-1}$; pH = 5; $S_2O_8^{-2}$ = 0.2 mM	75% degradation after 120 min	0.0019 min^{-1}	[49]
Fenton-like degradation of MeP	Natural magnetite	As received	62.5	Cat = 0.3 $g \cdot L^{-1}$; MeP = 10 $\mu mol \cdot L^{-1}$; pH = 6.5; SPS = 5 $mmol \cdot L^{-1}$	90.2–99.5% degradation	0.0066 min^{-1}	[52]
Electro-Fenton removal of Gemcitabine	Natural mineral	Crushing, sieving, and milling	ND	Pt anode; 140 $mL \cdot min^{-1}$ of air; 1 cm separation between electrodes; 0.05 $mol \cdot L^{-1}$ of Na_2SO_4	25–35% degradation	63 $M^{-1} s^{-1}$ min^{-1}	[54]
CWPO of phenol	Natural iron minerals	Sieving	8.0	Cat = 1 $g \cdot L^{-1}$; phenol = 100 $mg \cdot L^{-1}$; pH = 3; H_2O_2 = 500 $mg \cdot L^{-1}$; T = 75 °C; t = 4 h	70–80% mineralization and 100% of conversion	11 L^2 mg^{-1} g_{cat}^{-1} min^{-1}	[62]
CWPO of azole pesticide	Pristine magnetite	As received	7.5	Cat = 8 g; H_2O_2 = 6.7 $mg \cdot L^{-1}$; T = 25 °C; pH = 5; flow rate = 0.5 $mL \cdot min^{-1}$	Conversion of the pollutants > 95%	0.22 $mL \cdot g_{cat}^{-1} \cdot min^{-1}$	[50]
CWPO of SMX	Pristine magnetite	As received	7.5	Cat = 1 $g \cdot L^{-1}$; H_2O_2 = 25 $mg \cdot L^{-1}$; T = 25 °C; pH = 5	50% mineralization and 100% removal	0.0263 $mg \cdot L^{-1} \cdot min^{-1}$	[55]
Fischer–Tropsch synthesis **	Natural magnetite	Sol–gel and high-temperature pyrolysis	73.4–157.7	Cat = 1 g; 2 g of quartz sand; T = 300 °C; P = 2 MPa; GHSV of 3000 $mL \cdot g^{-1} \cdot h^{-1}$; t = 180 h	93–98.5% conversion of CO	ND	[61]
Production of diesel by catalytic pyrolysis	Magnetite ore	Magnetic separation, crushing, grinding, and ball milling	ND	Cat = 3 wt%; T = 500 °C; t = 90 min	Selectivity towards the formation of hydrocarbons having fuel value in conformity with diesel fuel	ND	[45]
Cracking of coal	Mineral magnetite	As received	3.32	Cat = 1 mg; coal; 1 mg; T = 700 °C; Ar atmosphere	Selectivity to aliphatics	ND	[59]
Biodiesel production from waste cooking oil ***	Iron sand	Coprecipitation and wet impregnation	ND	Ratio of methanol to oil = 9:1; Cat = 1 wt%; T = 60 °C; t = 6 h; 1100 rpm	95.64% yield	ND	[63]

* TiO_2/Fe_3O_4, ** Fe@C, and *** $CaO/MgO/Fe_3O_4$; ND = No data; Cat = Catalyst.

2.2. Hematite (α-Fe_2O_3)

Hematite (α-Fe_2O_3) is the oldest recognized form of iron oxide, widely found in soils and rocks. It is more commonly found in sedimentary rocks, either as detrital particles or as precipitates. Hematite is abundant in environments with ample oxygen as the ferric oxyhydroxides initially formed age and transform into hematite. It typically coexists with other iron oxides like goethite. Hematite formation is favored by low water content and elevated temperatures [64]. When finely powdered, hematite exhibits a reddish hue, while in bulk form, it appears black or gray. Known for its exceptional stability, hematite often represents the final stage in the transformation of other iron oxides [2]. Its crystalline structure resembles corundum, with lattice parameters a = 5.0340 Å and c = 13.750 Å.

This mineral displays two magnetic transitions, specifically the Morin transition (T_M) around T = 260 K and the Curie transition at approximately 950 K. The manifestation of its weak ferromagnetic conduct stems from a slight misalignment in spin antiparallelism. Notably, a correlation is established between the T_M and the hematite particle size, with smaller particle sizes correlating to reduced T_M values. However, this relationship is heavily contingent upon preparation methodologies, crystalline lattice defects, and the incorporation of OH groups into the hematite structure [65].

Magnetic properties are intimately tied to nanoparticle morphology and size. For example, the magnetization of saturation decreases and coercivity increases with Mn doping, which are attributed to the presence of more manganese atoms at the grain boundaries [66]. The coercivity interval (H_c) spans from 31 to 530 Oe, while the remanent magnetization (M_r) ranges between 0.6 and 16 $Am^2\ kg^{-1}$ [67–69]. Higher H_c and Mr values are closely linked to larger particle sizes. Conversely, smaller hematite particles of varied morphologies exhibit superparamagnetic behavior beyond the blocking temperature [70]. It is deemed a n-type semiconductor, characterized by a bandgap of 2.2 eV. The valence band comprises occupied 3d orbitals of Fe^{3+} alongside some non-bonding 2p orbitals of oxygen. Meanwhile, the conduction band is composed of vacant 3d orbitals of Fe^{3+} [71]. Regarding its specific surface area, it is recognized that this parameter is influenced by the synthesis method and calcination temperature. Hematite subjected to thermal treatment between 1173 and 1273 K yields a specific surface area below 5 $m^2 \cdot g^{-1}$ due to particle sintering. Broadly, hematite produced at low calcination temperatures (<373 K) possesses specific surface areas ranging from 10 to 90 $m^2 \cdot g^{-1}$, while those generated through Fe^{3+} hydrolysis are dependent on particle size and shape, encompassing values between 2 and 30 $m^2 \cdot g^{-1}$ [2].

Main catalytic applications. While numerous publications document the use of hematite for dye removal, the majority of these studies focus on laboratory-synthesized hematite. However, dye removal using hematite-based catalysts from natural sources has been tested by some research groups. Photo-Fenton [72], Fenton [73,74], Fenton-like [75], and photocatalysis reactions [66,76–78] are the main applications where hematite has been tested. In photocatalysis, photon absorption by a material excites electrons from the valence band to the conduction band. The resulting electrons and holes react with O_2 and H_2O, generating $O_2^{\bullet -}$ and $^{\bullet}OH$ radicals [76]. Although the preparation of hematite catalysts can result in various morphologies, temperature is the key parameter influencing this characteristic. For instance, catalysts prepared from natural hematite (Figure 4A) and from iron ore (Figure 4B) underwent a calcination process at 700 °C, resulting in similar spherical morphologies.

Phenol is naturally found in oilfield-produced waters, and it is commonly used in sectors such as pesticides and pharmaceuticals. Due to its persistence in the environment, much research focuses on its degradation by different catalytic processes. In an interesting approach, a catalyst using natural clay with cassava starch as a porogenic agent and raw material was tested on phenol photodegradation [78]. The catalyst presented around 40%

of α-Fe_2O_3. The maximum phenol photodegradation achieved was 59.15% at a cassava starch concentration of 7%. Even after eight recycling cycles, the degradation efficiency remained above 90%. However, the synthesis method is more complex compared to other preparation methods [74] used to obtain catalysts for similar applications. Iron-bearing mining rejected from the amethyst geode deposit was processed by grinding, ball milling, and sieving to produce a catalyst for the photo-Fenton degradation of phenol. After this processing, the final percentage of Fe_2O_3 was close to 25%. Although the material had low porosity, it demonstrated significant catalytic efficiency in phenol degradation, achieving mineralization and removal efficiencies of 78% and 96.5%, respectively. This performance is largely attributed to its iron content, which effectively facilitates OH radical generation. Both studies demonstrated innovative approaches for preparing Fe_2O_3-based catalysts using low-cost raw materials. Despite differences in synthesis methods and reaction conditions, both highlight sustainable strategies for wastewater treatment.

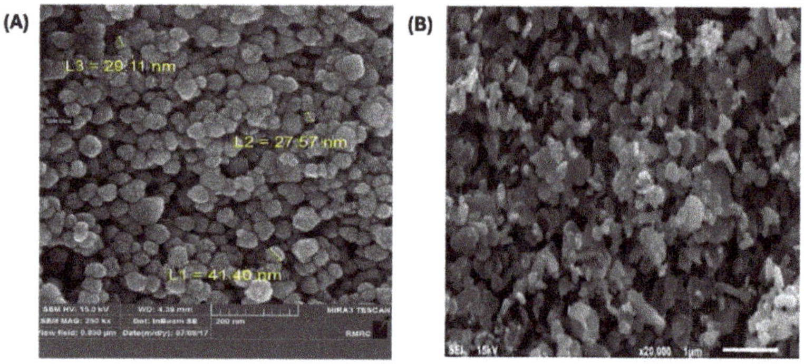

Figure 4. SEM micrographs of α-Fe_2O_3 nanoparticles prepared from natural sources: (**A**) natural hematite via purification and calcination and (**B**) iron ore via magnetic separation, ball milling, and calcination. Reproduced with permission from References (**A**) [73] and (**B**) [76]. Copyright© 2018 Elsevier and 2018 IOPScience/CC BY 3.0, respectively.

A similar organic compound, namely, the low-molecular-mass 4-nitrophenol (4-NP), which is involved in many chemical processes and commonly present in soils and in surface and ground waters, causes severe environmental impacts and health risks. Fe_2O_3 [79] and Cu/Fe_2O_3 [77] catalysts were evaluated on the photocatalytic reduction of 4-NP to 4-aminophenol (4-AP), a less harmful compound. The raw material (iron ore) was crushed, sieved, and thermally treated to prepare the catalysts. Even if goethite was the main phase of iron oxide in the raw material, after heat treatment at 450 °C, only peaks related to hematite were found in an XRD powder pattern. The morphology of the catalyst prepared from natural sources and from iron nitrate is spherical in both cases. TEM analysis revealed a significant difference in particle size, with values of 50 nm (from natural sources) and 20 nm (synthesized). Impregnation with Cu (5 and 20 wt%) and calcination at higher temperatures resulted in a smaller particle size (50 nm) and reduction in S_{BET} (38 $m^2 \cdot g^{-1}$), compared to hematite calcinated at 700 °C (34 nm and 10 $m^2 \cdot g^{-1}$, respectively). The same reaction conditions were applied to 4-NP reduction, and conversion reached >99% in both cases. The rate constant was higher using Cu/Fe_2O_3. This was attributed to the presence of the copper ferrite $CuFe_2O_4$.

Hematite catalysts from natural sources have also been successfully tested in dye degradation. For the oxidation and mineralization of methyl orange (MO), catalysts were prepared from Fe-sand [72]. The sand underwent a series of purification steps to enhance its catalytic properties. Degradation experiments were conducted in a cylindrical batch photoreactor under UV irradiation. A higher rate constant of 0.048 min^{-1} was achieved for a dye concentration of 100 $mg{\cdot}L^{-1}$. One of the major current challenges is the scaling up of catalysts to enable their use in treatment plants [80]. A first step towards this goal is to test them in the laboratory using wastewater samples from real effluents. Under optimized conditions, including 200 mL of textile effluent, 0.3 g of catalyst, pH of 2.5, and an initial H_2O_2 concentration of 200 $mg{\cdot}L^{-1}$, the system achieved an 88% reduction in optical density and a 79% decrease in chemical oxygen demand (COD) after 60 min. One interesting property of using iron oxide catalysts is their magnetic response, which allows for easy recovery and subsequent reuse. In this case, the catalyst was able to reach a degradation of 89% after four cycles, with no obvious leaching. Also, hematite can be used as a catalyst in combination with clay for the photocatalytic degradation of dye using raw iron ore [76]. α-Fe_2O_3/bentonite was further evaluated through photocatalytic degradation on indigo carmine (IC) of a composite prepared by mechanical milling, followed by magnetic separation and purification with HCl. Photocatalytic evaluation was performed under UV and solar light. The α-Fe_2O_3/bentonite composite has high photocatalytic activity on the removal of IC dye compared to extracted α-Fe_2O_3. The maximum removal efficiency (almost 100%) of IC was achieved with the following optimal conditions: pH = 1, 250 mg of catalyst, and 5 $mg{\cdot}L^{-1}$ of IC. The photocatalytic activity under solar light was higher than that of UV irradiation. The removal of 88.88 % of MB was achieved by Mg/Fe_2O_3 prepared from iron sand in a photo-Fenton reaction [66].

The growing interest in the degradation of emergent pollutants results in the evaluation of novel or well-known catalysts in processes related to them. Among these emerging contaminants are herbicides. The simultaneous catalytic degradation of 2,4-dichlorophenoxyacetic acid (2,4-D) and 2-methyl-4-chlorophenoxyacetic acid (MCPA) herbicides over persulfate (PS) activated by mineral hematite via Fenton-like processes resulted in 36% of mineralization under optimum conditions (0.5 $g{\cdot}L^{-1}$ of catalyst; 200 $mg{\cdot}L^{-1}$ of 2,4-D and MCPA; 0.025 M of PS; pH = 3; T = 50 °C; and t = 120 min) [73]. The FESEM analysis indicated that the catalyst was almost porous, providing active sites for the adsorption of both pollutants and oxidant molecules. The synergistic effect between PS and catalysts showed better performance in acidic environments and higher temperatures in the degradation of herbicides.

Due to its abundance on Earth's surface, different sources can be used to obtain hematite. In this line of knowledge, the same research group evaluated the use of Hormuz Red Soil (HRS) as a catalyst for the degradation of diclofenac (DCF) via PS activation [75]. The catalyst beads were prepared by ionic gelation encapsulation using alginate beads. For comparison, HRS in powder form was evaluated too. The catalytic experiments were conducted in two modes: batch experiments for the activation of peroxymonosulfate (PMS) by HRS and supported HRS by alginate beads in flow-through reactors. X-ray fluorescence (XRF) indicated that hematite was the main oxide on HRS (59.30%), with SiO_2 as the second most abundant compound (16.72%). Fresh catalyst particles were spherical, while HRS beads displayed a rougher irregular surface due to HRS crosslinking with alginate, enhancing the surface area and active sites for improved catalytic performance. Additionally, 98.2% of removal (HRS = 5 $mg{\cdot}L^{-1}$; 75 $mg{\cdot}L^{-1}$ of PMS; 50 $mg{\cdot}L^{-1}$ of DCF; and t = 10 min) was achieved with HRS in suspension with PMS. Complete degradation was achieved in HRS supported on alginate beads in batch-mode processes (five granules of HRS beads; 75 $mg{\cdot}L^{-1}$ of PMS; 50 $mg{\cdot}L^{-1}$ of DCF; t = 2 min; and neutral pH). The alginate-supported HRS demonstrated superior performance, achieving complete degradation in significantly shorter reaction times under neutral pH conditions. However, the batch experiment was also efficient, with nearly 100% removal. Also, the catalytic ozonation and peroxone-mediated removal of acetaminophen (ACT) by HRS resulted in the complete degradation of 50 mg·L^{-1} of ACT at 10 min. The calcinated catalyst exhibited high stability and reusability in consecutive catalytic cycles [81].

At the degradation of gemcitabine via the electro-Fenton process, hematite had poor degradation efficiency (20%) [54]. The analysis of parameters in Fenton and Fenton-like reactions catalyzed by hematite indicates that a narrow interval of pH values is available (Figure 5). This trend indicates that higher values of all these parameters cause a decrease in the removal efficiency of different contaminants in reactions catalyzed by hematite.

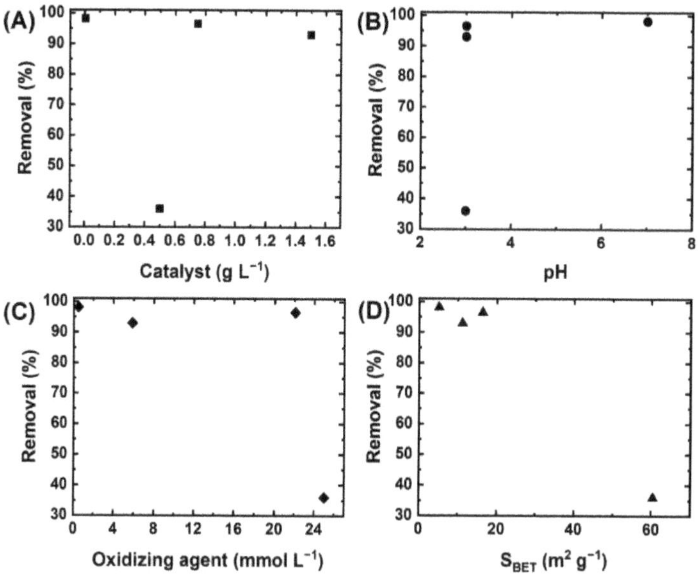

Figure 5. Removal as a function of hematite catalyst concentration (**A**), pH (**B**), oxidant concentration (**C**), and specific surface area (S_{BET}, (**D**)). Elaborated with data from References [72–75].

Non-common catalytic applications. While Fenton and Fenton-like reactions constitute the most prevalent applications of hematite, its utility has also been explored in hydrogen reactions. However, the literature reveals inconsistent results across different research groups in this area. The preparation of catalysts $Mo/Al_2O_3/Fe_2O_3$ using bauxite was evaluated through the hydrocracking of high-temperature coal tar, which was used as the raw material [82]. Hydrocracking involves breaking long C-C chains using a catalyst, hydrogen, and high pressures or temperatures [83]. The higher percentage of hematite was 16.85 wt%, and its catalytic evaluation results demonstrated varying conversion ratios and product yields among different bauxite-derived catalysts. The superior catalytic activity of the best catalyst was attributed to the leaching of Fe_2O_3 and the enrichment of Al_2O_3 in the bauxite modified with HCl, leading to increased acidity and specific surface area. This optimal combination enhanced mass transfer and acidity, resulting in compromised performance in terms of high-temperature catalytic treatment (HTCT) conversion and liquid yield. This result indicates that the presence of Fe_2O_3 is not favorable to the hydrocracking of high-temperature coal tar. In contrast, Kairbekov and coworkers [84] investigated the use of modified iron-containing catalysts and the preliminary ozonization of coal for its hydrogenation. The study focused on catalysts derived from natural bauxite samples with varying Fe_2O_3 contents. Among the tested materials, bauxite samples exhibited superior performance with 23.7% Fe_2O_3, achieving a higher yield of liquid products compared to other bauxite samples and catalysts such as red mud and pyrite concentrate. This result highlights the importance of Fe_2O_3 content in the catalytic hydrogenation process. In catalytic hydrogenation, hydrogen gas is used to reduce unsaturated organic compounds, such as aromatic rings, alkenes, or carbonyl groups, in the presence of a catalyst [85]. Although the specific iron oxide phase was not identified, it is likely that the Fe_2O_3 was in the hematite phase as bauxite typically contains hematite, goethite, and ilmenite. Furthermore, the study noted that the conversion of iron oxides into finely dispersed pyrrhotine during the hydrogenation process, particularly after modification with elemental sulfur, enhanced the catalytic activity of the iron-containing materials.

According to Widayat et al. [86], Indonesia's substantial demand for catalysts is primarily fulfilled through imports, underscoring the critical need to identify and utilize domestic natural resources for catalyst production. Additionally, biofuel, particularly biodiesel, is one of the most widely used energy sources in the country, offering advantages such as being renewable and biodegradable, having lower emission levels, and promoting complete combustion in engines. The preparation of hematite catalysts from sand is of interest to address these issues.

In 2019, Widayat and coworkers prepared magnetic hematite nanoparticles from iron sand and tested them as catalysts for biodiesel production from waste cooking oil. The preparation of catalysts included acid leaching with HCl, chemical coprecipitation, and calcination at varied temperatures (650–800 °C). The optimal catalyst was calcinated at 700 °C. The catalysts were able to support both esterification and transesterification; biodiesel yield reached 86.78%, with 87.88% of fatty acid methyl ester; yield optimization was found to depend on the calcination temperature. Furthermore, Widayat's group explored the potential of bifunctional α-Fe_2O_3/CaO_2 catalysts derived from iron sand for biodiesel production using waste cooking oil [87]. In the study, the catalysts were synthesized by combining hematite obtained through chemical coprecipitation with CaO_2 prepared from sources such as $Ca(OH)_2$, $CaCO_3$, and limestone. The catalytic performance varied depending on the CaO_2 source, with the limestone-derived catalyst yielding the highest biodiesel output at 77.17%. However, the $CaCO_3$-derived catalyst met both Indonesian and European biodiesel standards. It was also discovered that the α-Fe_2O_3 component favored the esterification reaction and formation of diglycerides, while CaO_2 favored the transesterification reaction and formation of methyl esters.

Catalytic cracking is a widely employed chemical process that breaks down complex hydrocarbons into simpler molecules using a catalyst. This process facilities the cleavage of carbon–carbon bonds in large hydrocarbons at lower temperatures and pressures compared to non-catalytic methods, resulting in enhanced energy efficiency and cost-effectiveness. In heavy oil, the increase in lighter hydrocarbons and reduced resin–asphaltene components was achieved by natural hematite containing 41% Fe, 58% Si, and impurities of Al, K, S, Ca, and Cu using a hydrothermal catalytic system. The density of the oil decreased as resin–asphaltene content was reduced [88], and the results demonstrated the possibility of increasing the number of lighter hydrocarbons and reducing their density by a regular decrease in the amount of resin–asphaltene components.

Hematite can be obtained via the heat treatment of other iron oxides. The catalytic cracking performance of hematite derived from thermally treated natural limonite was evaluated using toluene as a model compound for biomass tar [89]. The limonite was crushed, sieved, and calcinated at 700 °C for 2 h under an air atmosphere. The as-prepared hematite achieved 95% toluene conversion at 700 °C. However, at temperatures exceeding 700 °C, the excessive consumption of lattice oxygen led to carbon deposit accumulation, negatively impacting the conversion rate. The evaluation of the stability of the catalysts indicated that toluene conversion was maintained above 80% at 600 min. CO_2 and CO production confirmed the role of hematite as an oxygen carrier. XRD analysis showed the transformation of hematite into magnetite, which occurred due to the consumption of active lattice oxygen and the deposition of carbon; because of this, the spent catalyst was regenerated by heat treatment in an air atmosphere, with the regenerated catalyst maintaining a toluene conversion rate above 85% for the first three cycles.

Natural hematite has also proven to be effective in the catalytic cracking of coal tar [90]. Fe/Al and Fe/Ni composite oxygen carriers (OCs) were prepared through mechanical mixing. The hematite block was first crushed into a fine powder using a jaw crusher and then mixed with γ-Al_2O_3 or NiO powder. After extrusion molding and granulation, the mixture was dried in an oven at 100 °C for 12 h. Compared to natural hematite, the addition of γ-Al_2O_3 helps reduce the carbon deposition, while the inclusion of NiO promotes it. The carbon deposits on the reduced OCs mainly consist of hard coke formed through the graphitization of carbon black at high temperatures. In addition, γ-Al_2O_3 helps improve and preserve the porosity of the OC particles, significantly reducing sintering.

In recent years, catalysts from natural hematite have been tested on catalytic systems, which had not been evaluated before. Iron ore (72% hematite, 22% magnetite, and 6% goethite) was ball-milled as a precursor for the synthesis of a Ni-Fe catalyst for oxygen evolution reaction (OER) [91]. The OER involves the anodic electrooxidation of water in acidic media or OH^- in alkaline media to generate O_2. This reaction is ubiquitous in aqueous environments and typically requires significant overpotentials to proceed, which has driven extensive research into electrocatalysts [92]. Even at a low overpotential (280 mV), this electrocatalyst reaches a current density of 10 mA cm^{-2} and retains its catalytic performance over prolonged electrolysis under alkaline pH. Synthetic hematite has been applied in various structures for the oxygen evolution reaction (OER), including metal–organic frameworks, carbon-supported materials, heterojunctions, and metal–nonmetal/anion-doped Fe_2O_3, primarily in alkaline media (KOH). Catalyst stability varies significantly, with no clear correlation to synthesis methods or composition. One of the cases reporting the lowest stability is Fe_2O_3/g-C_3N_4, which lasts only 10 min and is synthesized by a thermal method [93]. Stability durations of 5–12 h are observed for C-doped $CoFe_2O_4/Fe_2O_3$ [94], Co_3O_4/Fe_3O_4 [95], Fe/Fe_2O_3-Fe-N-doped C [96], FeS/Fe_2O_3 [97], Fe_2O_3/FeS [98], Fe_2O_3/CNT [99], and Fe_2O_3/FeP [100], synthesized by calcination, coprecipitation, pyrolysis, chemical etching/solvothermal, hydrothermal, coprecipitation, and hydrothermal methods, respectively. On the other hand, cases with acceptable operational durations (about 100 h) are reported for RuNi-Fe_2O_3/IF [101], WO_3/Fe_2O_3-NiO [102], and CoMo/Fe_2O_3 [103], synthesized via hydrothermal, chemical etching/decomposition, and hydrothermal/electrodeposition techniques, respectively. Finally, examples with extended durability include IrO_2–Fe_2O_3, which maintained its performance for 600 voltammetry cycles via thermal decomposition [104], and Fe_2O_3–MnO, synthesized via a sol–gel process with 1000 cycles [105].

Twelve minerals, which were washed, dried, milled, and sieved, were evaluated using the catalytic hydrolysis of microcystin-LR (MC-LR). Limonite displayed the best catalytic activity, while hematite resulted in poor yields [106]. Also, Fischer–Tropsch synthesis could be catalyzed by iron oxides. Hematite derived from iron ores allowed CO conversion of up to 80% when H_2 was the reducing gas. This conversion is higher than those achieved by commercial catalysts. However, it was achieved using relatively higher temperature and pressure (270 °C vs. 250 °C and 20.9 vs. 1.85 bar, respectively) [107]. Table 4 provides a summary of the catalytic applications of hematite, which can be derived from various sources. It is the most extensively utilized natural iron oxide in heterogeneous catalysis.

Table 4. Summary of catalytic applications of natural hematite (α-Fe_2O_3).

Catalytic Reaction	Catalyst	Natural Source	Pretreatment Method	S_{BET} ($m^2 \cdot g^{-1}$)	Reaction Conditions	Results of Catalytic Evaluation	Rate Constant	Ref.
Photocatalytic degradation of MB	Mg/α-Fe_2O_3	Natural sand	Magnetic separation, drying, and ball milling	ND	UV irradiation; t = 300 min	Removal of 88.8% of MB	ND	[66]
Photocatalytic degradation of phenol	α-Fe_2O_3/gelcasting porous ceramic	Natural clay	Grinding, sieving, and extraction by 3 M of H_2SO_4 at 30 °C	31.92	UV irradiation; T = 23 °C; pH = 8; t = 3 h; phenol = 10 $mg \cdot L^{-1}$	Removal of 57% of phenol; catalytic activity remains without changes after 8 cycles	ND	[78]
Photo-Fenton degradation of MO	α-Fe_2O_3/SiO_2	Iron sand	Sieving, immersion in HCl solution, washing, and drying	11.16	Cat = 1.5 $g \cdot L^{-1}$; MO = 100 $mg \cdot L^{-1}$; H_2O_2 = 200 $mg \cdot L^{-1}$; pH = 3; UV irradiation	93% of degradation of MO. 89% after 4 cycles of reaction	0.048 min^{-1}	[72]
Fenton degradation of (2,4-D) and MCPA	α-Fe_2O_3	Natural mineral	Crushing, sieving, washing, sonication, and calcination	60.4	Cat = 0.5 $g \cdot L^{-1}$; 2,4-D and MCPA = 200 $mg \cdot L^{-1}$; PS = 0.025 M; pH = 3; T = 50 °C; t = 120 min	36% of mineralization	MCPA = 0.0064; 2,4-D = 0.0059 min^{-1}	[73]
Fenton and photo-Fenton oxidation of phenol in water	Mining reject	Mining reject	Ball milling and sieving	16.35	Cat = 0.75 $g \cdot L^{-1}$; phenol = 50 $mg \cdot L^{-1}$; H_2O_2 = 0.75 $g \cdot L^{-1}$; pH = 3	96.5% degradation at 180 min	0.0411 min^{-1}	[74]
Degradation of DCF via PMS activation	α-Fe_2O_3	HRS	As received	5.17	Cat = 5 $mg \cdot L^{-1}$; DCF = 50 $mg \cdot L^{-1}$; PMS = 75 $mg \cdot L^{-1}$; t = 10 min; neutral pH	98.2% of degradation in 10 min	0.334 min^{-1}	[75]
Photocatalytic degradation of IC	α-Fe_2O_3/bentonite	Iron ore	Magnetic separation, grinding, ball milling, and coprecipitation	ND	Cat = 250 mg; IC = 5 $mg \cdot L^{-1}$; pH = 1; solar light	100% of degradation after 2 h	ND	[76]
Photocatalytic reduction of 4 NP to 4 AP	Cu/Fe_2O_3	Natural iron ore rock	Crushing, sieving, calcination, and impregnation	10–42	Cat = 33.3 $mg \cdot L^{-1}$; 4 NP = 5 × 10^{-5} $mol \cdot L^{-1}$; $NaBH_4$ = 12.5 mL (0.5 M); pH = 11.5; λ = 200–500 nm	>99% conversion in less than 1 min	2.34, 3.36, and 5.4 min^{-1}	[77]
Photocatalytic reduction of 4 NP to 4 AP	α-Fe_2O_3	Natural iron ore rock	Crushing, sieving, heat treatment, and calcination	18–85	Cat = 33.3 $mg \cdot L^{-1}$; 4 NP = 100 mL (0.1 mM); λ = 220–550 nm; $NaBH_4$ = 12.5 mL (0.5 M)	>99% conversion in less than 3 min	1.38 min^{-1}	[79]
Catalytic ozonation and peroxone-mediated removal of ACT	α-Fe_2O_3	HRS	Calcination in air atmosphere	3.63	Cat = 1 $g \cdot L^{-1}$; ACT = 50 $mg \cdot L^{-1}$; O_3 = 1.2 mg/min; t= 10 min; pH = 7	100% degradation	0.40 min^{-1}	[81]
Hydrocracking of high-temperature coal tar	Mo/Al_2O_3-Fe_2O_3	Natural bauxite	Crushing, calcination, purification, washing, and wet impregnation	126.9–237.9	Cat = 1.3 g; sulfur powder = 0.4 g; HTCT = 42 g; T = 430 °C; P = 12.5 MPa; t = 90 min	The presence of Fe_2O_3 is not favorable to the hydrocracking of high-temperature coal tar	ND	[82]
Hydrogenation of coal	Fe_2O_3–SiO_2–Al_2O_3–TiO_2–MnO_2	Natural bauxite	ND	ND	T = 380–440 °C; P = 3–5 MPa; sulfur additive = 0–2%; t = 90 min	Selectivity to liquid products	ND	[84]
Biodiesel production from waste cooking oil	α-Fe_2O_3	Iron sand	Magnetic separation, coprecipitation, and calcination	10.5–22.9	Esterification: methanol: waste cooking oil molar ratio of 5:1; H_2SO_4 = 1 wt%; T = 70 °C; t = 300 min. Transesterification: waste cooking oil: methanol molar ratio of 15:1; Cat = 1 wt%; T = 65 °C; t = 3 h	87.88% biodiesel yield	ND	[86]
Biodiesel production from waste cooking oil	Fe_2O_3/CaO_2	Iron sand	Magnetic separation, coprecipitation, and calcination	ND	Waste cooking oil: methanol molar ratio of 1:15; Cat = 1 wt%; T = 65 °C; t = 3 h	97.04% biodiesel yield with a Fe/Ca ratio of 1:4	ND	[87]
Hydrothermal catalytic conversion of extra-heavy Ashal'chinskoe oil	α-Fe_2O_3	Natural hematite	As received	ND	T = 210, 230, and 300 °C; P = 2–18 MPa; t = 2 h	The possibility of increasing the number of lighter hydrocarbons in heavy oil and reducing its density by a regular decrease in the amount of resin–asphaltene components	ND	[88]

Table 4. *Cont.*

Catalytic Reaction	Catalyst	Natural Source	Pretreatment Method	S_{BET} ($m^2 \cdot g^{-1}$)	Reaction Conditions	Results of Catalytic Evaluation	Rate Constant	Ref.
Catalytic cracking of toluene	α-Fe_2O_3	Natural limonite	Crushing, sieving, and calcination	12.4	Cat = 0.5–1.5 g; toluene = 1000 ppm; T = 500–800 °C; t = 60 min	95% of toluene conversion	ND	[89]
Catalytic cracking of coal tar	α-Fe_2O_3/γ-Al_2O_3 α-Fe_2O_3/NiO	Mineral hematite	Crushing, drying, and mechanical mixing	ND	T = 700–900 °C; consumption of coal tar = 0.3 $kg \cdot h^{-1}$; mass flow rate of coal tar = 5 $g \cdot min^{-1}$	Poor performance of hematite/NiO; the addition of γ-Al_2O_3 could effectively inhibit carbon deposition	ND	[90]
OER	Ni/iron ore	Iron ore	Ball milling and solution-assisted electrode preparation	ND	Cat = 150 mg; NaOH = 1 M	Achieves a current density of 10 mA cm^{-2} at a low overpotential of 280 mV; potentially scalable to industrial applications	ND	[91]
Catalytic hydrolysis of microcystin-LR peptides	α-Fe_2O_3	Mineral hematite	Washing, drying, milling, and sieving	ND	Cat = 20 mg; MC-LR = 1 mL (10 ppm); T = 60 °C	20.7% hydrolysis yield	ND	[106]
Fischer–Tropsch synthesis	α-Fe_2O_3	Raw iron ore	Grinding and sieving	59.0	Syngas at space velocity of 60 $Nml \cdot g_{cat}^{-1}$; T = 270 °C; P = 20 bar	The CO-reduced catalyst exhibited the highest CO conversion of 94.1%, followed by the H_2-reduced catalyst with a CO conversion of 80.1%, while the syngas-reduced catalyst showed the least CO conversion of 54.1%	ND	[107]

2.3. Goethite (α-FeOOH)

Goethite (α-FeOOH) is one of the most thermodynamically stable and abundant crystal structures of iron in minerals, sediments, and soils. It results from various processes (hydroxylation, oxidation, hydration, or decomposition) of iron silicates, magnetite, siderite, and pyrite [108]. Goethite adopts a compact hexagonal arrangement of OH^- and O_2^- anions, with Fe^{3+} at the center of the octahedra aligned parallel to the [001] direction [109]. Goethite is antiferromagnetic, with two Neel temperatures of 63 and 97 °C. Above this, it becomes paramagnetic, transitioning to hematite at temperatures above 400 °C [110]. It often exhibits a weak ferromagnetic component caused by a mismatch in the antiferromagnetic grid, associated with a higher number of defects such as hollows, boundaries, and grain rotations [111].

Main catalytic applications. Goethite has proven to be effective as a catalyst for removing organic contaminants from aqueous effluents, making it a promising natural candidate for water treatment applications due to its widespread abundance in the environment. Liu et al. [109] and Ruan et al. [22] conducted reviews of Fenton-like processes based on the literature on this topic. The direct formation of $^1O_2$ and $^{\bullet}OH$ has been identified, either independently or in the presence of H_2O_2 or UV [112], under slightly acidic pH conditions. Examples such as hydroquinone, 2-chlorophenol, oxalate, succinate, bisphenol A, and benzoate have been treated in this manner [113–116].

As a Fenton reagent, goethite demonstrates remarkable performance. De la Plata and colleagues [117] conducted a comparison between goethite and TiO_2 in heterogeneous photo-Fenton catalysis. The authors employed a spectrophotometric system specifically designed for highly precise measurement of the optical properties of catalysts, demonstrating that goethite has a significantly larger optical path length and the ability to effectively degrade 2-chlorophenol. In a more comprehensive study [118], it was demonstrated that the decomposition of various organic contaminants follows a trend: salicylic acid ≈ m-hydroxybenzoic acid > p-hydroxybenzoic acid ≈ benzoic acid > p-bipthalic acid > phenol > benzenesulfonic acid. This trend is analogous to the sorption pattern of these species on goethite, which, in turn, depends on the structure (substituent groups and number of rings).

Yeh et al. [119] conducted a comparison of the degradation of five chlorinated ethylenes and three aromatic hydrocarbons through Fenton-like processes. They observed that aromatics were more readily degradable than chlorinated compounds. However, removal was achieved within a few minutes in all cases. The type of electrolyte significantly influences $^{\bullet}OH$ generation by goethite–H_2O_2. Lin et al. [120], demonstrated that the maximum generation occurs at pH 3, remaining substantial at pH 6–7. The most suitable electrolyte would be ClO_4^- as it does not impact radical formation, while other tested inorganic anions showed interference tendencies: $H_2PO_4 > SO_4^{2-} > Cl^- > NO_3^-$. Cl^- and SO_4^{2-} more severely suppress $^{\bullet}OH$ generation with an increasing concentration. The reason for this appears to be related to the adsorption and blocking of active sites by these anions. In general, based on the results reported in the literature [22], it can be concluded that for goethite, the photo-Fenton process yields better results than simple Fenton processes, and electro-Fenton is more effective than photo-Fenton. During toluene oxidation, Mn/goethite nanoparticles (Figure 6A) selectively produced CO_2 and water because no significant formation of an incomplete combustion product (CO) was detected [121]. In order to avoid changes in the catalyst structure, the catalytic performance was evaluated at a maximum temperature of 450 °C.

Natural goethite (Figure 6B) has been successfully applied to produce nanofibers for water disinfection under visible light, demonstrating remarkable efficiency in both dye degradation and bacterial inactivation. Specifically, it achieved 90% degradation of MB within 5 h. Its recyclability was confirmed by retaining a 65% degradation rate after five reaction cycles. The goethite used was sourced from an abandoned mine in the Taouz region of Errachidia Province, Morocco. Nanoparticles were synthesized by high-energy ball milling of goethite rock and wet milling of metallic iron rocks in a planetary ball mill. The process utilized stainless steel vials and iron balls, with a ball-to-powder mass ratio of 40:1, operating at a rotational speed of 450 rpm for 5 h [122].

More recently, highly dispersed natural goethite/Fe_3O_4 composites (Figure 6C) were synthesized via a one-step hydrothermal method, demonstrating a potential as catalysts for tetracycline removal. Under optimal conditions (pH of 3.0, tetracycline concentration of 100 $mg \cdot L^{-1}$, catalyst dosage of 0.3 $g \cdot L^{-1}$, and H_2O_2 concentration of 6.0 mM), a removal efficiency of 90.1% was achieved within 30 min. The system also exhibited strong durability, retaining a tetracycline elimination rate of 73.3% after five treatment cycles [123].

Another application of goethite is its role as an efficient dissociator of ozone molecules, making it valuable in heterogeneous catalytic ozonation [124], as demonstrated in the case of nitrobenzene by Zhang, Ma and coworkers [125]. In heterogeneous catalytic ozonation, O_3 is added to a contaminated effluent, along with a solid catalyst that decomposes the ozone molecule into reactive oxygen species, which, in turn, degrade the pollutant [126]. Toxic inorganic agents such as AsO_3^{3-} can also be oxidized (to AsO_4^{3-}) this way, being more effective in the presence of H_2O_2 and UV. Zhang et al. [127] utilized goethite to adsorb tartaric acid and reduce dissolved Cr^{6+} in water through this process. They demonstrated that the presence of goethite is crucial; in its absence, tartaric acid alone promotes only a 12% reduction in Cr^{6+} in 72 h. However, the reduction is 100% within 24 h and can occur in just 4 h when reducing the acid concentration, allowing for the co-adsorption of Cr^{6+}. The authors suggest that goethite's role is to provide efficient channels for electron transfer between the reactants. Similar results were obtained for the reduction and stabilization of Cr^{6+} in soil by using calcium polysulfide catalyzed by goethite and hematite, where the presence of these oxides enhanced the performance. The removal of Cr^{6+} improved with increasing iron oxide content, from 0 to 9 g kg^{-1}, but then declined as the concentration rose from 9 to 12 g kg^{-1} [128]. Pelalak and colleagues synthesized plasma-treated goethite nanoparticles, which exhibited enhanced surface area and a higher density of surface hydroxyl groups. These nanoparticles were derived from natural goethite [129]. They enhanced the heterogeneous catalytic ozonation of the sulfasalazine (SSZ) antibiotic.

Non-common catalytic applications. In a study on chemical synthesis, Fernández-Marchante [130] investigated the catalytic activity of goethite in the Westinghouse process for hydrogen production (HP), specifically focusing on the transformation of H_2SO_4 into SO_2. The study compared the performance of goethite to that of other established catalysts. According to their results, goethite demonstrated superior behavior compared to SiC but was lower than that of Fe_2O_3 and similar to that of CuO. The advantage lies in its resistance to sintering, making catalyst deactivation less likely over the long term. Shehzad et al. [131] used goethite as a catalyst for water dissociation in catalytic bipolar membranes (CBMs), reducing the activation energy (calculated by DFT) of the process by 80% compared to state-of-the-art membranes. This, combined with the natural availability of goethite, significantly enhances the process both economically and energetically. They achieved excellent catalyst stability through a protection strategy involving the in situ growth of conductive polyaniline (PANI). Hu et al. [132] employed goethite nanoflowers as support for Rh nanoparticles in the reduction and hydrogenation of nitrostyrene to vinylaniline. A key advantage over alternative methodologies is the ability to achieve 100% conversion without compromising the integrity of the vinyl group. This is a significant improvement as the co-reduction of other unsaturated groups is a common challenge in heterogeneous catalysts, often necessitating lower conversions to prevent unwanted side reactions. As a hydrogen donor, the use of hydrazine hydrate avoids the challenges of handling H_2 directly, generating only water and N_2 as byproducts in this process. Regarding stability, the authors conducted six cycles of catalyst recovery and reuse, detecting no loss of activity. Composites of Mn, Co, Ni, Cu, and Zn oxides were employed by Kuncser et al. [133] in the oxidation of cyclooctene. Catalytic oxidation occurs in the presence of O_2, using isobutyraldehyde as a reducing agent and CH_3CN as a solvent. In all instances, selectivity towards the formation of the corresponding epoxide exceeded 99% in the composites. However, the conversion varied in the order of Mn > Fe > Co > Ni > Zn > Cu, ranging from 55% to 4%. The concentration of basic sites in the composites was determined by measuring the UV–Vis signal of acrylic acid, irreversibly adsorbed by the materials. The results show a complete correlation between conversion and the ratio of the basic sites to the acidic sites of the composites. In microbial fuel cells (MFCs), goethite has demonstrated its ability to enhance catalysis [134]; the mineral was extracted from mining mud, and its performance was compared to stainless steel alone. Raw goethite tripled the electrical power obtained with stainless steel (11 W m^{-3} compared to 3.5 W m^{-3}). Furthermore, when goethite was treated at 550 °C, the power increased to 17.1 W m^{-3}. Additionally, it substantially improved the current and current efficiency obtained while reducing COD. Goethite exhibited superior performance compared to specularite and hematite as a catalytic bed for the catalytic reforming of lignite-blended co-pyrolysis with corn straw [135]. The process significantly enhanced the formation of light aromatic hydrocarbons. Verma et al. [136] employed goethite for the synthesis of α-aminonitriles from N,N-dimethylaniline using atmospheric oxygen. The authors experimented with various iron salts and minerals, including $Fe(OAc)_2$, $FeCl_2$, $FeSO_4$, $FeCl_3$, $Fe(NO_3)_3$, $Fe_2(SO_4)_3$, $Fe(acac)_2$, $Fe(acac)_3$, FeO@GO, Fe_3O_4, FeO, Fe@g-C_3N_4, and goethite. However, none yielded a performance greater than 17%. Remarkably, when goethite assumed a lychee-like morphology, the yield increased to 96%, indicating a high dependence on maintaining this specific morphology.

Figure 6. SEM micrographs of α-FeOOH particles prepared from natural sources: (**A**) acid mine drainage; (**B**) goethite rocks; (**C**) mineral goethite/Fe_3O_4; and (**D**) mineral goethite. Reproduced with permission from References (**A**) [121], (**B**) [122], (**C**) [123], and (**D**) [137]. Copyright © 2012, 2023 American Chemical Society; 2020 MDPI/CC BY 4.0; and 2023 Elsevier, respectively.

A recent study [137] examined the sunlight-driven activity and wavelength dependency of goethite obtained from Bingzhou, Hunan Province (Figure 6D). Before conducting photochemical experiments, the mineral was crushed, milled, and sieved to attain a particle size of less than 48 μm (<300 mesh). Goethite displayed strong light absorption, extending to approximately 570 nm in the visible spectrum, suggesting a narrow band gap favorable for photocarrier generation. Regarding hydroxyl radical (•OH) production, goethite achieved a rate of 0.648 nM min^{-1}, surpassing both hematite and magnetite. Its performance in H_2O_2 production was still lower than that of hematite.

A comparative study on •OH photochemical production between natural and synthetic iron minerals revealed much higher rates in natural minerals. Specifically, natural goethite, hematite, and magnetite showed 1.7-, 4.1-, and 4.9-fold increases in •OH production, respectively, compared to their synthetic counterparts. Interestingly, the relatively smaller improvement in natural goethite's performance corresponds to its lower impurity levels compared to the other minerals.

Goethite exhibits optimal catalytic performance under slightly acidic to neutral pH conditions, where the generation of hydroxyl radicals (•OH) and reactive oxygen species is enhanced.

Table 5 presents the catalytic applications of goethite prepared from natural sources, with advanced oxidation processes being the most prominent. However, as previously discussed, synthetic goethite has been successfully used in other types of reactions, such as reduction and selective hydrogenation. Therefore, it would be worthwhile to test the efficiency of natural goethite in the same reactions. Nonetheless, its transformation into hematite above 400 °C should be considered.

Table 5. Summary of catalytic applications of natural goethite (α-FeOOH). Unless otherwise stated, the catalyst is goethite.

Catalytic Reaction	Natural Source	Pretreatment Method	S_{BET} ($m^2 \cdot g^{-1}$)	Reaction Conditions	Results of Catalytic Evaluation	Rate Constant	Ref.
Oxidation of toluene *	Acid mine drainage	Sequential precipitation method and wetness impregnation	115.70	Cat = 1 g; T = 250–450 °C; toluene = 0.9 vol%; flow rate = 500 $L \cdot min^{-1}$; GHSV = 18,000 h^{-1}	30% of toluene conversion; selectivity to CO_2	ND	[121]
Photocatalytic degradation of MB **	Goethite rocks	Ball milling, stirring, sonication, and electro-spinning process	ND	Cat = 0.5 g; MB = 100 mL (10 ppm); t = 5 h; visible light and UV	90% of bleaching after 5 h of illumination	0.0141 min^{-1}; similar in UV and visible light	[122]
Fenton-like TC degradation ***	Mineral goethite	Hydrothermal method	20.4	Cat = 0.3 $g \cdot L^{-1}$; pH = 3; TC = 100 $mg \cdot L^{-1}$; H_2O_2 = 6.0 mM	90.1% removal after 30 min of treatment	ND	[123]
Reduction and stabilization of Cr(VI) in soil	Natural goethite	As received	ND	Cat = 3, 6, 9, and 12 g; pH = 8.6 to 9	Goethite increased the apparent rate constant	3.33×10^{-6} $kg \cdot mg^{-1} \cdot min^{-1}$ to 8.33×10^{-6} $kg \cdot mg^{-1} \cdot min^{-1}$	[128]
Ozonation of SSZ ****	Natural goethite	Crushing, milling, washing, and plasma process	29.65, 73.24 and 77.31, respectively	Cat = 1–9 $mg \cdot L^{-1}$; T = 25 °C; P = 1 atm; inlet flow = 1 $L \cdot h^{-1}$; Na_2SO_3 = 1 mL (0.01 M)	96.5% degradation efficiency reached by N_2-goethite after 40 min of reaction	0.076 min^{-1}	[129]
Catalytic reforming of volatiles from co-pyrolysis	Mineral goethite	Dehydration	18.05–140.28	15–20 mg of a mixture of raw lignite and corn straw; T = room temperature to 800 °C; Ar flow = 100 $mL \cdot min^{-1}$	Goethite significantly promotes the production of light aromatic hydrocarbons	ND	[135]
Production of ROS	Natural goethite	Mechanical crushing, milling, and sieving	ND	Cat = 0.1 $g \cdot L^{-1}$; LI = 10 mW cm^{-2}; pH = 7; T = 25 °C	Natural goethite ROS production was 4.9-fold higher than the standard	0.648 nM min^{-1}	[137]

* Mn/goethite; ** PAN/goethite nanofibers; *** Goethite/Fe_3O_4; **** Natural N_2-treated goethite; ND = No data; LI = Light intensity.

Although the most abundant oxides are widely used in catalytic reactions, their preparation from natural sources is less common. In applications where high phase purity is required, their use can be detrimental to the reaction due to impurities commonly found in

the sources from which they are obtained. However, these impurities can enhance catalytic activity in other processes. Additionally, in highly reducing or oxidizing environments, a phase change may occur during the process, potentially leading to its deactivation. While its magnetic properties are widely known and catalysts are magnetically characterized in some research reports, such characterization is carried out with a focus on the recovery and reuse of the catalysts. In spite of this, the effect of magnetic properties on catalytic activity has rarely been studied.

3. Less Abundant but Widely Distributed Fe-OH

While less abundant than magnetite and hematite, iron oxide such as maghemite (γ-Fe_2O_3), lepidocrocite (γ-FeOOH), and ferrihydrite ($Fe_{8.2}O_{8.5}(OH)_{7.4}$ + $3H_2O$) also exhibit notable catalytic properties. This section will explore some of their applications in catalysis.

3.1. Maghemite (γ-Fe_2O_3)

Maghemite is widely recognized as a common component of various soils, particularly in highly weathered toxic soils found in tropical and subtropical regions. Recent environmental magnetic studies have shown that maghemite is distributed globally, having been detected in tropical, subtropical, arid, and tundra regions. The formation of ultrafine maghemite during pedogenesis is considered the primary factor behind magnetic enhancement in paleosols. Studies have detailed the pedogenic processes responsible for maghemite formation, revealing that pedogenic maghemite, often associated with hematite, acts as an intermediate phase in the transformation of ferrihydrite into hematite. Experimental findings demonstrate that when phosphated ferrihydrite is subjected to a temperature of 150 °C for 120 days, it forms hydromaghemite (with grain sizes between 10 and 30 nm). Over time, the grain size of hydromaghemite grows, transitioning from the superparamagnetic to the single-domain range. After four months, hydromaghemite further transforms into the more thermodynamically stable hematite [138]; at nanosizes (36–40 nm), it transforms into hematite at 576 °C [139].

Maghemite (γ-Fe_2O_3) possesses an inverse spinel cubic structure with a lattice parameter a = 8.3500 Å. Its nomenclature derives from the amalgamation of magnetite (Fe_3O_4) and hematite (α-Fe_2O_3), owing to substantial congruence in both structural attributes and chemical composition. Operating as a ferrimagnetic substance at ambient temperature, it exhibits saturation magnetization values as impressive as 90 emu g^{-1}. However, its stability is inferior to that of hematite. Characterized as an n-type semiconductor, it displays a band gap of 2.0 eV. Similarities with magnetite (Fe_3O_4) are clearly observable. The distinctive demarcation lies in the exclusive presence of the Fe^{3+} oxidation state within maghemite, while magnetite encompasses both Fe^{2+} and Fe^{3+}. The distinction between maghemite and magnetite is often complex when using XRD and X-ray photoelectron spectroscopy (XPS). However, some methods have been proposed to differentiate them from one another [140,141]. Determining the Curie temperature of maghemite has been proven to be challenging due to its tendency to transform into hematite, even at temperatures such as 413 K. This transformational behavior is contingent upon precursor selection, synthesis methodology, and external pressure exerted [142]. In addition, its Curie temperature is estimated to be within the range of 820 to 986 K [143,144]; nanoparticles present a considerably lower value of about 545 K [145]. Notwithstanding, since this analysis was conducted via the thermogravimetric analysis (TGA) technique and the observed transition temperature closely corresponds to the maghemite-to-hematite phase transition, assigning it as the Curie temperature remains a significant challenge.

Concerning its band structure, ab initio calculations elucidate that the upper segment of the valence band is dominated by oxygen 2p orbitals, with occupied 3d levels of Fe

situated 6 to 7 eV below the Fermi level. The lower tier of the conduction band is populated by unoccupied 3d levels of Fe (octahedrally coordinated). This classification designates maghemite as a charge transfer semiconductor, with the foremost excitation potentially corresponding to electron transfer from O^{2-} anions to octahedral Fe^{3+} cations [146].

Synthetic maghemite has been successfully tested in hydrogenation reactions, with remarkable yields, especially when used in conjunction with other oxides [85,147,148]. However, to the best of our knowledge, maghemite prepared from natural sources has been poorly explored as a catalyst. Vasanthakumar and Karvembu [149] reported the catalytic evaluation of maghemite, obtained from river sand (Figure 7A), based on the base-free transfer hydrogenation of furfural, levulinic acid, and o-vanillin. River sand was collected from the Cauvery River in Kallanai, Tamil Nadu, India, and γ-Fe_2O_3 was separated using simple magnetic separation with a bar magnet. The γ-Fe_2O_3 powder was then washed with acetone, ethanol, and water to remove impurities and was dried at 80 °C for 3 h. To compare the catalytic activity of natural γ-Fe_2O_3, synthetic γ-Fe_2O_3 was also prepared following existing literature protocols. Catalytic tests were performed under a vacuum at 80 °C for 3 h. The activated catalyst was added to a sealed tube containing 1 mmol of substrate in 2-propanol and placed in a preheated oil bath at 150 °C for a specific duration. After 30–90 min, depending on the substrate, the catalyst was separated with an Nd magnet, and the solution was centrifuged. Magnetic characterization showed that the γ-Fe_2O_3 from river sand had higher saturation and remanent magnetization than synthetic γ-Fe_2O_3, exhibiting stronger magnetic behavior. Catalytic tests revealed that river sand γ-Fe_2O_3 displayed superior catalytic activity, which remained effective even after 10 cycles. The base-free catalytic transformation of levulinic acid to γ-valerolactone, via the 4-hydroxypentanoic acid intermediate, was attributed to the presence of silica in the γ-Fe_2O_3 from river sand.

Figure 7. Micrographs of catalysts from natural sources of less abundant and rare oxides: (**A**) γ-Fe_2O_3 from river sand, (**B**) TEM image of ferrihydrite from AMD, and (**C**) wüstite from Probus reagents. Reproduced with permission from References (**A**) [149], (**B**) [150], and (**C**) [151]. Copyright© 2020, 2007 American Chemical Society and 2022 Elsevier, respectively.

Iron ore raw materials obtained through the catalytic reduction of NO with NH_3 resulted in excellent selective catalytic reduction (SCR) activity (above 80% at 170–350 °C)

and N_2 selectivity (above 90% up to 250 °C) at low temperatures. However, the addition of H_2O and SO_2 in the feed gas showed some adverse effects on SCR activity.

For the preparation of catalysts, the iron ore raw materials were dried at 105 °C for 4 h and ground. Then, these ground samples were dried again at 105 °C for 5 h. Finally, they were calcinated at 250, 350, and 450 °C, respectively, for 6 h [152]. The characterization results indicated that the best catalytic performance was achieved by a catalyst comprised mainly of α-Fe_2O_3 and γ-Fe_2O_3. The utilization of natural maghemite as a catalyst presents challenges due to its inherent impurities. Natural maghemite typically contains substitutional metal cations such as Ti^{4+}, Mg^{2+}, or Al^{+3}, among others, which can significantly influence its catalytic properties. Since γ-Fe_2O_3 is a metastable phase, the methods used to prepare catalysts based on this compound must be adequate to avoid the formation of the most thermodynamically stable hematite (α-Fe_2O_3) [153]. The summary results are presented in Table 6.

Table 6. Summary of catalytic applications of less abundant and rare Fe-OH prepared from natural sources.

Catalytic Reaction	Catalyst	Mineral Source	Pretreatment Method	S_{BET} ($m^2 \cdot g^{-1}$)	Reaction Conditions	Results of Catalytic Evaluation	Ref.
Transfer hydrogenation of furfural, levulinic acid, and o-vanillin	γ-Fe_2O_3	River sand	Magnetic separation, grinding, washing, and drying at 80 °C	22.3	Cat = 20 mg (7 mol %); T = 150 °C; 1 mmol of substrate in 2-propanol (3 mL); t = 30–90 min	γ-Fe_2O_3 recovered from the natural source exhibited superior activity compared to the synthetic counterpart under base-free conditions	[149]
Catalytic reduction of NO with NH_3	α-Fe_2O_3/ γ-Fe_2O_3	Iron ores	Drying, grinding, and calcination	22.8–42.5	NO = 500 ppm; NH_3 = 500 ppm; O_2 = 3 vol %; SO_2 (when used) = 150 ppm; H_2O (when used) = 5 vol%; N_2 = 145 mL·min^{-1}; GHSV = 20,000 h^{-1}	Selectivity catalytic reduction activity above 80% at 170–350 °C and N_2 selectivity (above 90% up to 250 °C) at low temperatures	[152]
Decomposition of hydrogen peroxide	Ferrihydrite	Acid mine drainage	Vacuum drying	270	Cat = 1 g·L^{-1}; H_2O_2 = 0.02 M	Waste ferrihydrite shows the same catalytic activity for H_2O_2 decomposition as the commercial goethite-based catalyst	[150]
Photo-Fenton degradation of ENR and LEV	Wüstite	ND	ND	108.7	Cat = 10 mg·L^{-1}; ENR = LEV = 0.05 mmol·L^{-1}; H_2O_2 = 1.0 mmol·L^{-1}; pH = 6.5; T = 35 °C; t = 180 min	100% and 55 % of antibiotic activity elimination in 180 min for ENR and LEV, respectively; complete antibiotic activity elimination for ENR in the next four recycling cycles	[151]

ND = No data.

3.2. *Lepidocrocite (γ-FeOOH)*

Lepidocrocite is named after its platy crystal shape and orange coloration [2]. It is commonly found in hydromorphic soils with alternating reducing and oxidizing conditions and can biogenically be obtained [154]. It is metastable, serving as a precursor to more magnetic phases like maghemite and magnetite [155], and plays a key role in the geochemical behavior of nutrients, heavy metals, and organic pollutants. It presents an orthorhombic crystal structure with lattice parameters a = 3.072(2), b = 12.516(3), c = 3.873(2) Å, and space group *Cmcm* [156]. In the lepidocrocite structure, ferric iron is coordinated in edge-shared octahedra, which form layers within [001] linked by hydrogen bridges. In contrast to other oxyhydroxides, lepidocrocite exhibits a low Neel temperature, which ranges between 50 and 70 K [157].

Up to the time of writing this article and to the best of the authors' knowledge, lepidocrocite has only been used in its synthetic form in Fenton [158] and Fenton-like processes [159–161], photodegradation [162,163], water splitting [164], CO oxidation [154], arsenic adsorption [165], and the combination of H_2O_2 and PMS to produce ROS and remove chloramphenicol (CAP) [166]. However, no reports were found in the literature regarding its use in mineral form, likely because it is rare to find it in ore deposits [167]. The unique

way to form lepidocrocite is the oxidation of $Fe(OH)_2$ by air in suspension at pH ≈ 7 [2]. When it is prepared in the presence of citric acid, the crystallization of lepidocrocite can be controlled by its concentration, which decreases with its increasing concentration [162]. Related to photocatalytic activity, visible light and the presence of ethylenediaminetetraacetic acid (EDTA) could accelerate the reaction rate of lepidocrocite [163].

Experimental laboratory results indicate that the morphology of catalysts influences the exposed facets and interaction with OH groups of iron oxides, having a direct effect on their reactivity. For example, the adsorption and degradation of Orange G (OG) on lath-like particles were mediated by the interaction with the μ-OH groups of the (010) facet, while in rod morphology, it was adsorbed laterally by the interactions with the μ-OH and μ_3-OH groups of the (010) and (001) facets [161].

When lepidocrocite VOOH hollow nanospheres were prepared by the hydrothermal method, a variation in the reaction temperature allowed the tuning of the surface area, modifying the catalytic activity. Also, it exhibited stability in alkaline media and achieved good performance in water splitting at a potential of 270 mV for OER [164].

The findings from the evaluation of the adsorption of arsenic suggest that pH affects it remarkably and the co-occurrence of aqueous Fe^{2+} and Fe^{3+} solids is beneficial for arsenic removal because Fe^{2+} can induce As^{3+} oxidation to As^{5+}. Above pH = 8, As^{5+} can be more mobile than As^{3+}; at low pH, the radical scavengers can compete with As^{3+} for the oxidants [165].

Theoretical results comparing goethite and lepidocrocite in the elimination of 5-bromosalicylic acid (BSA) via the Fenton process show that electrons are more easily transferred from the H_2O_2 molecule to the Fe atoms of α-FeOOH (goethite), enabling it to perform better in catalyzing the decomposition of H_2O_2. However, free radicals are more likely to desorb from γ-FeOOH (lepidocrocite), making the γ-FeOOH/H_2O_2 system more efficient in degrading BSA [168]. Lepidocrocite's catalytic applications are versatile, but its metastability, crystal morphology, and interaction with functional groups affect its reactivity and performance.

3.3. Ferrihydrite ($Fe_{8.2}O_{8.5}(OH)_{7.4}$ + $3H_2O$)

Ferrihydrite is a hydrated Fe(III) nano-oxide commonly found in various surface environments on Earth. Its presence has a significant impact on soil properties. Ferrihydrite was officially recognized as a mineral by the International Mineralogical Association (IMA) in 1975. In nature, it is commonly associated with silicates (with an atomic ratio of Si/Fe ranging from 0.17 to 0.37), and these silicates seem to directly influence its low crystallinity. Due to its small particle size and poor crystallinity, ferrihydrite exhibits crystal and compositional properties that are challenging to characterize [169]. The complexity of ferrihydrite's structural analysis is also influenced by the presence of various other species in its composition, including 0.5–31.5 wt% SiO_2, 0.5–9.6 wt% Al_2O_3, and up to 7.2 wt% PO_4, along with As(V), which ranges from 0.7 to 28.3 wt% [170]. Initially, its formula was identified as $5Fe_2O_3 + 9H_2O$, featuring a hexagonal unit cell with parameters of a = 0.508 nm and c = 0.94 nm and containing FeO_6 octahedra [171]. Eggleton and Fitzpatrick [172] proposed a trigonal cell structure for ferrihydrite, with two-thirds of the iron in octahedral coordination and one-third in tetrahedral coordination. However, other studies have questioned the presence of tetrahedral Fe [173]. Additional formulas such as $Fe_5HO_8 + 4H_2O$ [174] and $Fe_2O_3 + 2FeOOH + 6H_2O$ [175] have also been suggested. The number of X-ray diffraction peaks varies from two to five in natural ferrihydrites, while synthetic forms can show up to seven well-defined peaks, reflecting the degree of Fe atom ordering in the lattice.

Currently, one of the most accepted crystalline structures for two-line ferrihydrite is $Fe_{8.2}O_{8.5}(OH)_{7.4}$ + $3H_2O$ [176], which consists of 20% tetrahedral iron and approximately 80% octahedral iron, featuring a local δ-Keggin structure. Additionally, the structural depletion (SD) model proposed by Hiemstra in 2013 suggests a distinction between the nanoparticle's core and surface. While the core is largely free of defects, the surface is depleted of tetrahedral structures. The ideal structure of ferrihydrite consists of three types of Fe: two hexacoordinated and one tetracoordinated, arranged in a δ-Keggin framework with proportions of 60%, 20%, and 20%, respectively. The general structure includes three layers of polyhedra, with the first hexacoordinated Fe located in the outer layers, the second hexacoordinated Fe in the middle (surrounding the tetracoordinated Fe), and the third octahedral Fe, which is highly distorted, located in the inner layer [177].

Ferrihydrite, owing to its widespread presence, common occurrence as a coating on particles, and reactive surface with a small particle size (1–7 nm), is widely recognized for its role in influencing the mobility of both inorganic and organic pollutants via sorption processes. It typically forms under near-neutral pH conditions in a variety of redox-active environments as well as in large sedimentary deposits where there is a significant supply of ferrous iron from groundwater and high oxidation rates [170]. Its chemical composition and crystal structure have been debated for over five decades, largely due to its extremely small particle size and the high degree of structural defects it incorporates [178]. Ab initio calculation results suggest that its reaction sites are mainly disposed along rows at the edges of sheets of iron octahedra. Molecular dynamics studies on nanoparticles of up to 10 nm show that highly reactive hydroxo groups on the surface are typically unbound, but they participate as hydrogen bond acceptors in a network with less reactive groups [179]. Magnetic properties reveal that ferrihydrite exhibits antiferromagnetism with a ferromagnetic-like moment at lower temperatures (100 K and 10 K), but it is paramagnetic at room temperature. As the crystallite size of ferrihydrite increases, both magnetization and coercivity decrease, a trend attributed to the pronounced surface effects found in fine-grained materials [180].

The main catalytic applications of ferrihydrite are related to the Fenton reaction [181], primarily conducted with synthetic ferrihydrite. In their review, Zhu et al. [181] highlighted that ferrihydrite consistently exhibits higher reactivity with H_2O_2 compared to other iron minerals. Specifically, they noted performance hierarchies, such as ferrihydrite > hematite > goethite, ferrihydrite > feroxyhyte ≈ magnetite > maghemite > goethite > hematite, and ferrihydrite > hematite > goethite. These comparisons underscore ferrihydrite's superior catalytic potential. The use of ferrihydrite in the Fischer–Tropsch (FT) reaction has been studied by Kyu Seomoon [182]. By incorporating Al, Cu, and K into the synthesis of the catalyst and analyzing the FT reaction using gas chromatography–mass spectrometry (GC-MS) over 100 h, Seomoon observed a 25% increase in paraffin selectivity, attributed to reduced catalyst basicity. Increasing the H_2/CO ratio from 1 to 3 reduced CO_2 emissions by 40% and enhanced hydrocarbon formation. Similarly, Sumit Bali et al. [183] doped ferrihydrite with Al and Cu, achieving slightly better FT performance than a commercial fixed-bed catalyst, securing a patent for their synthesis method. However, there are some reports on the use of natural minerals as catalysts. The decomposition of hydrogen peroxide was tested using ferrihydrite obtained directly from acid mine drainage in the Zlaté Hory underground mine district (Czech Republic). In this case, the only further treatment was vacuum drying of the collected solid. The dried sample was gently powdered to minimize mechanical disruption (Figure 7B) [150]. Physicochemical characterization revealed a composition of mainly iron oxyhydroxides (approximately 96 wt%), with ferrihydrite as the principal phase, along with traces of goethite and possibly schwertmannite. The constant rate (6.59 min^{-1}) was slightly higher than that of commercial goethite, with the advantage of using extremely low-cost precursors or even industrial waste.

Ferrihydrite has been extensively used to adsorb heavy metals and other ions. The adsorption mechanisms for metals such as lead [184], copper [185], molybdenum [186], chromium [187,188], and zinc [189] occur via inner-sphere complexation, typically forming bidentate or bidentate–monodentate complexes depending on the pH and concentration. For radium and barium, tetradentate bonds with a single surface functional group are feasible [190]. Actinides like thorium [191], plutonium [192], and uranium [193] also prefer bidentate adsorption, with the latter two being extensively studied for removal from industrial systems through ferrihydrite nanoparticle precipitation.

Sorption capacity is strongly influenced by temperature and pH. Anion adsorption can significantly influence cation sorption by competing for available binding sites on the surface, thereby reducing the capacity for cation uptake. Carbonates alter ferrihydrite solubility [194], while phosphate addition enhances Cd^{2+} sorption by shifting from bidentate to ternary complex formation [195]. Zhu et al. [196] observed that ligands' ability to inhibit arsenic (As^{3+} and As^{5+}) sorption follows the sequence $SeO_4 \approx SO_4$ < oxalic acid $\approx$ malic acid < citric acid < SeO_3 << PO_4. Additionally, the content of other metals in ferrihydrite's lattice, such as aluminum, affects the sorption of chromates, selenates, and sulfates, as demonstrated by Johnston and Chrysochoou [187].

Maghemite, lepidocrocite, and ferrihydrite are considered less abundant oxides. Maghemite can only be distinguished from magnetite through characterizations that identify the oxidation states of iron. However, in the absence of such results, its catalytic activity may be mistakenly attributed to magnetite. Additionally, it is common to find traces of maghemite during hematite synthesis, making the evaluation of its performance as a catalyst a challenge.

Lepidocrocite, on the other hand, is rarely found in ore deposits as it is a metastable phase that serves as a precursor to other magnetic phases. Nevertheless, its synthesis can be achieved under controlled laboratory conditions by adjusting factors such as pH and the presence of additives. The morphology of the obtained particles is highly related to their catalytic activity.

As for ferrihydrite, it is found in nanometric sizes and, compared to other oxides and hydroxides, its recognition as a mineral is relatively recent. Its nanometric size and lack of crystallinity make it difficult to even determine its chemical formula. However, its capacity to adsorb heavy metals has been well established.

4. Rare Iron Oxides and Hydroxides

This section explores the catalytic applications of several rare iron oxides, including wüstite (FeO), akaganéite (β-FeOOH, feroxyhyte (δ′-FeOOH), and bernalite ($Fe(OH)_3$). The unique properties of these iron oxides and their potential catalytic applications will be discussed.

4.1. Wüstite (FeO)

Under ambient conditions, wüstite (FeO) crystallizes in a highly defective form of the ideal NaCl-type structure, with a nonstoichiometric formula of $Fe_{1-x}O$ [197]. It is metastable and tends to decay into a two-phase mixture of α-Fe and magnetite Fe_3O_4. In addition, cubic $Fe_{1-x}O$ can be synthesized at ambient pressure for iron deficiencies ranging from $0.05 < x < 0.15$ at 560 °C [198,199]. The number of vacant octahedral sites is twice the amount of iron deficiency (x), according to the neutron diffraction data, implying a fraction approximately equal to x of iron located at interstitial tetrahedral sites. Further analysis of several samples reveals a direct linear relationship between the stoichiometry of $Fe_{1-x}O$ and its lattice parameter, given by the equation a (Å) = 3.856 + 0.478(1 − x).

Wüstite exhibits antiferromagnetic ordering as temperatures descend below approximately 195 K. The nature of the magnetic order–disorder transition undergoes modulation in response to variations in the composition of $Fe_{1-x}O$, displaying heightened cooperativity in the presence of nearly stoichiometric FeO. The Fe^{2+} ion spins align antiferromagnetically along the (111) plane, with the predominant magnetic vector aligned parallel to the [111] direction. However, this axial component (3.8 μB for $Fe_{0.99}O$) is accompanied by a less perpendicular component (1.3 μB for $Fe_{0.99}O$). The magnitude of the ordered magnetic moment experiences a marked reduction in samples with decreased iron content. The paramagnetic phase assumes the structural attributes of a NaCl-type arrangement (space group $Fm\bar{3}m$), while the lower temperature phase exhibits a rhombohedral distortion (space group R3) [200].

Liu and colleagues [201] have conducted a review of the applications of wüstite in ammonia synthesis. The most successful ammonia synthesis method is the Haber–Bosch process, in which atmospheric nitrogen is converted into ammonia in the presence of a catalyst, high pressures, and high temperatures [202]. According to their findings, wüstite demonstrates ideal performance in terms of stability, activity, and cost. Its two industrial-level competitors are catalysts based on magnetite and ruthenium. The Ru/C catalyst can be up to 20 times more active than the magnetite-based catalyst (Haber catalyst), but in comparison to wüstite, it lacks competitiveness. Pernicone et al. [203] compared the performance of wüstite A301 to that of the best Ru/C catalyst [204], obtaining equivalent results. This positions wüstite as the most attractive option due to its abundance and cost-effectiveness. Its main challenges still lie in its production as wüstite does not exist naturally. In the laboratory, it is obtained by the decomposition of ferrous oxalate, resulting in a mixture with Fe_3O_4 and Fe that lacks the high-purity properties of wüstite. Industrially, it is obtained by reducing natural magnetite, using Fe as a reducing agent, directly in the smelting furnace at T > 834.5 K, incurring a significant energy cost.

Wüstite, with a bandgap of 2.1 eV (Figure 7C), has been utilized as a photocatalyst for removing fluoroquinolones from water [151]. When combined with H_2O_2, it functions as a photo-Fenton heterogeneous catalytic system, facilitating the generation of reactive oxygen species (ROS), including $O_2^{\bullet-}$, $^1O_2$, and $^\bullet OH$. The complete elimination of microbial activity is achieved within 3 h, accompanied by the mineralization of 21% of enrofloxacin. The catalyst's performance proves reproducible for at least four treatment cycles following the recovery and drying of wüstite in an oven at 105 °C. The reduction of nitrates (NO_3^-) is another application in which wüstite has been implicated [205]. Nearly stoichiometric conversion from NO_3^- to NH_4^+ is attained, with NO_2^- identified as an intermediate. In this case, it is not a catalysis but rather a heterogeneous redox reaction as FeO serves as a reactant, forming Fe_3O_4 as a product. Advanced oxidation processes represent a field where iron oxides and hydroxides have been extensively applied [206]. In a study by Expósito and colleagues [207], the performance of wüstite was compared to that of other iron oxides in the electro-Fenton removal of aniline. Wüstite and magnetite, along with hematite, goethite, and $FeSO_4$, exhibited approximately 80% total organic carbon (TOC) removal at a rate similar to $FeSO_4$ (homogeneous phase). Furthermore, the decay pattern and final concentrations of H_2O_2 and Fe^{2+} also mirrored the homogeneous phase process. This suggests that the mechanism of the degradation of wüstite aligns with the homogeneous phase, initiated by dissolved iron ions. The advantage lies in the minerals' ability to self-regulate the supply of iron ions to the solution based on the process's requirements. Also, FeO catalysis based on the catalytic degradation of chloramphenicol (CAP) by water falling film dielectric barrier discharge (WFFDBD) enhanced the degradation from 69.5% to 97.7%

in 8 min by just using WFFDBD. The reaction conditions are shown in Table 6. DFT studies of the catalytic mechanism showed that the key step in the catalytic reactions was the electron transfer from FeO to the antibonding orbitals of dissolved oxygen, H_2O_2, and O_3, which facilitated their transformation and decomposition. This process converted dissolved oxygen into $O_2^{\bullet-}$, which was then adsorbed onto FeO through electron transfer, crucial to the catalytic reaction. Additionally, oxygen vacancies increased the charge density of Fe, enhancing electron transfer [208].

The metastable nature of wüstite, its tendency to decompose into magnetite and hematite, and the difficulty in obtaining high-purity forms are the main limitations to its widespread use. Here, wüstite is effective for ammonia synthesis. However, its effectiveness in AOPs is dependent on acidic conditions (pH~3).

4.2. Akaganéite (β-FeOOH)

Akaganeite (β-FeOOH) derives its name from the Akagané mine in Japan, where its initial discovery occurred in 1962 [209]. Widely acknowledged as a corrosion byproduct of steel [210], it stands as the predominant iron oxide in both soils and geothermal brines [211]. Showing a hollandite-like structure, its empirical formula is $Fe_{7.6}Ni_{0.4}O_{6.35}(OH)_{9.65}Cl_{1.25}$, discerned from the corrosion crust formed on the Camp del Cielo meteorite. Exhibiting monoclinic symmetry (space group *I2/m*), its lattice parameters are characterized by a = 10.600(2) Å, b = 3.0339(5) Å, c = 10.513(2) Å, and β = 90.24(2)°. Within akaganeite, Fe^{3+} ions occupy octahedral sites, forming a double-chain structure that shares corners, thereby allowing ample space for the inclusion of large tunnels featuring square cross-sections, suitable for hosting Cl and H atoms [212]. This compound is recognized as an antiferromagnetic material, manifesting a Neel temperature of 259 K, coupled with ferromagnetic-type hysteresis [213].

Akaganeite is particularly characterized by its ability to remove contaminants through adsorption. According to Zhao et al.'s review [214], the successful adsorption of As^{3+}, As^{5+}, Cr^{6+}, Zn^{2+}, U^{6+}, Cd^{2+}, Cs^{+}, and PO_4^{3-} has been explored previously (for further details on the removal of these contaminants, please refer to that work and the references cited therein). Briefly, the morphology (size and shape) of akaganeite largely determines its properties. Nanocrystals with various shapes, such as rods, bundles, needles, fibrils, and cigars, have been obtained. These depend on the preparation method, usually involving hydrolysis and the use of different surfactant templates such as polymers, zeolites, activated carbon, and clays. Using ATR-FTIR and liquid chromatography, Marsac et al. [215] demonstrated that in the case of oxolinic acid removal, the removal occurs via the adsorption of the contaminant and not by oxidation on akaganeite. This adsorption occurred via metal complexes under slightly acidic pH and via hydrogen-bonded complexes under alkaline pH. They even determined that the most reactive crystal plane is (010).

Additionally, degradation applications have been found. In the laboratory, starting from $FeCl_3$, in the presence of glucose or the surfactant Brij30 at 60 °C for 3 days, Xiong et al. [216] determined that it can remove 100% of methyl orange in the presence of H_2O_2 and when pH = 4.5 using photo-Fenton reactions. This capacity is reduced to 86% after four working cycles. Polyaniline degrades acidic orange II even better than akaganeite alone [217] in Fenton processes. The β-FeOOH@GO catalyst, evaluated in a photo-Fenton-like system using UV irradiation, achieved 99.7% decolorization of MB after 60 min. The pseudo-first-order rate constant was 0.632 min^{-1}. The proposed degradation pathway of MB predominantly proceeded with the rupture of phenothiazine ring oxides with $^{\bullet}OH$, $HO^{2\bullet-}$, and singlet oxygen ($^1O_2$) radicals [218].

4.3. Feroxyhyte (δ′-FeOOH)

It is a relatively uncommon iron oxide mineral and one of the few phases in the Fe_2O_3-H_2O system [219]. The feroxyhyte structure is composed of O^{2-} and OH^- ions arranged in the closest packing, with the Fe^{3+} ions statistically distributed in half of the octahedral positions. It has a hexagonal cell, with parameters a = 2.93 Å and c = 4.6 Å. It is completely converted to goethite in 6 h at a temperature of 60 °C and is partially transformed into hematite at 80 °C. Feroxyhyte is ferrimagnetic with a low Curie temperature (155 °C). The widely varying degree of the order leads to the formation of both magnetic and nonmagnetic varieties. The specific magnetic susceptibility reached 5000×10^{-6} $cm^3 \cdot g^{-1}$ and only 400×10^{-6} $cm^3 \cdot g^{-1}$ in the studied coarse crystalline and fine crystalline samples of synthetic feroxyhyte, respectively. Magnetic susceptibility was even lower in the finer crystals. In natural occurrences, it is always fine-grained, although samples with larger particle sizes and better crystallinity can be prepared in the laboratory [220].

Feroxyhite has demonstrated strong catalytic performance in various applications, although reported uses primarily involve laboratory-synthesized materials, often as nanoparticles. Among FeOOH polymorphs, feroxyhite stands out for its sufficient magnetic properties, enabling easy recovery in aqueous media. It can be synthesized through a single-step reaction—a critical factor for scalability in laboratory chemical processes [80]—and features a significantly higher proportion of surface OH groups compared to common magnetic iron oxides [221]. Recent studies highlight its use in photo-Fenton reactions, water splitting, and metal adsorption [222–224].

Weiping Du [225] studied the photodegradation of orange II using synthetic hematite, maghemite, magnetite, goethite, lepidocrocite, and feroxyhite. According to their data, hydroxides outperformed iron oxides, contrary to the authors' expectations. Among hydroxides, feroxyhite showed the lowest performance. The mechanism involved dye adsorption, significantly enhanced by H_2O_2, $AgNO_3$, and NaF through different pathways. Pinto [226] utilized synthetic feroxyhite to decompose H_2O_2 into radicals via the Haber–Weiss mechanism, effectively degrading MB and IC. They demonstrated that the adsorption of anionic dyes is more effective than that of cationic dyes. Here, the degradation reaction occurred almost entirely on the surface, primarily through the generation of $^{\bullet}OH$ radicals. Li et al. [227] developed δ-FeOOH/$BiOBr_xI_{1-x}$ for the visible-light photodegradation of tetracycline, demonstrating improved charge transfer, synergy with H_2O_2, and enhanced recyclability. Radical species $O_2^{\bullet-}$ and $^{\bullet}OH$ were identified as key players.

Chen Wang [228], in a review of FeOOH applications, highlighted that δ-FeOOH exhibits the smallest band gap (1.94 eV) and the largest specific surface area among FeOOH crystalline phases, making its band edge positions ideal for oxidizing H_2O to $^{\bullet}OH$ radicals. Pereira [229] first applied synthetic feroxyhite for water splitting and H_2 production, citing its 2.2 eV band gap as being sufficient for water dissociation while absorbing visible light. The photocurrent generated was tenfold that of P25 (TiO_2). Da Silva Rocha et al. [230] demonstrated that Ni^{2+} doping in synthetic δ-FeOOH enhanced photocatalytic activity, conductivity, and charge transfer, while surface $Ni(OH)_2$ improved charge separation.

When modified with other materials, one of the most common applications of δ-FeOOH is the removal of heavy metals from water. Jing Hu [231] used δ-FeOOH-coated γ-Fe_2O_3 for Cr^{6+} adsorption via an outer-sphere complexation mechanism. In a Fe/Mn oxy-hydroxide (δ-$Fe_{0.76}Mn_{0.24}OOH$), Tresintsi et al. [232] found it to be highly efficient for As^{3+} and As^{5+} adsorption. Pinakidou [233] employed amorphous tetravalent manganese feroxyhyte nanoparticles (TMFx) to remove As^{5+}. Later, Kokkinos et al. [234] reported TMFx as being effective in eliminating Cd, Hg, and Ni.

The versatility of these materials is also evident in their ability to immobilize phosphates with the aid of Fe^{2+}, achieving up to 94% removal under acidic pH [221]. Additionally, they have biomedical applications, such as controlled heat release under an AC magnetic field [235]. Recently, Lacerda and coworkers proposed, using ab initio calculations, that feroxyhyte doped with niobium can function as a bifunctional catalyst. Their results suggest that doping δ-FeOOH layers leads to stronger van der Waals interactions, enhancing the catalyst's thermal stability. Additionally, the incorporation of Nb enhances Brönsted acid characteristics and introduces Lewis acidic sites on the catalyst's surface [236].

4.4. Bernalite ($Fe(OH)_3$)

Bernalite, $Fe(OH)_3$, is one of the more recently identified iron oxyhydroxide minerals, recognized in 1992. It is the scarcest iron mineral and has been documented in only a handful of locations [4], including New South Wales, Australia, and the central Black Forest, Germany [237]. McCammon et al. deduced [238] its empirical formula to be $[(H_2O)_{0.04}(CO_2)_{0.03}Pb_{0.01}]Fe^{3+}{}_{0.93}Si_{0.06}Zn_{0.01})(OH)_{2.95}O_{0.04}$. This mineral exhibits a protonated octahedral framework (POF), adopting a distorted perovskite-type structure (ReO_3) within the *Pmmn* space group [239]. Its tilt system of a + b + c$^-$ arises from the topology of hydrogen bonds, resulting in alternating layers of Fe(1) and Fe(2) octahedra with an inverted tilt along the [001] direction, n-glides passing through the Fe atoms, and notable anisotropy in hydrogen bond topology [239]. With a slight deviation from the collinear antiferromagnet, bernalite manifests a weak ferromagnetic moment [240]. To date, there exists scant specific information regarding its application as a catalyst, neither within Scopus nor Google Scholar. Scopus records a mere seven references with "bernalite" in the title, with only two in the last two decades, the remainder hailing from the 1990s. Of these, four discuss structural or magnetic properties.

Recently, a potential use for this mineral was serendipitously uncovered. Under conditions of pH = 10, in the presence of arsenates, goethite undergoes partial transformation into bernalite. Remarkably, bernalite exhibits 2.18 times greater arsenate sorption than goethite, achieved through the formation of monodentate mononuclear (MM) complexes. The goethite used in these studies was sourced from rod-shaped nano-goethite (US3162) crystals obtained from US Research Nanomaterials (USA) [241] or Sigma-Aldrich (St. Louis, MO, USA) [242]. In another possible application, a composite comprising bernalite, Fe_2O_3, and Ru was synthesized through microwave-assisted heating in deionized water. This resultant material demonstrates the potential for oxygen reduction and hydrogen oxidation reactions. In spite of that, discerning the catalytic role of each component remains challenging [243].

Wüstite, akaganéite, feroxyhyte, and bernalite are among the least abundant iron oxyhydroxides, limiting their use in large-scale applications due to the rarity of natural sources. Their catalytic applications are less explored compared to more common iron oxides. Nevertheless, these materials exhibit properties suitable for advanced oxidation processes (AOPs), particularly in acidic conditions and systems like Fenton, photo-Fenton, and electro-Fenton reactions. Their optical band gaps, i.e., 1–2.1 eV for wüstite [151,198], 1.94 eV for feroxyhyte, and 2.1 eV for akaganéite [244], make them suitable for solar light-driven photocatalysis. Excluding wüstite, their crystal structure and chemical composition provide a high affinity for adsorbing heavy metals and organic pollutants. Notable differences include akaganéite's superior chemical stability and wüstite's potential in ammonia synthesis. Bernalite remains underexplored due to its rarity, although recent studies hint at promising environmental and electrochemical applications.

Production challenges include synthesis complexity and stability, especially for large-scale use. With magnetic properties, feroxyhyte is easier to handle in aqueous environments, while wüstite's instability and bernalite's scarcity hinder broader adoption. These contrasts highlight the need for innovative applications that optimize cost-effectiveness.

5. Future Directions and Perspectives on Catalytic Systems

The reaction mechanism of Fe-OH catalysts from natural sources depends on the evaluated reaction, conditions, and impurities. For example, in photocatalytic processes with Fe_2O_3 over gelcasting porous ceramic [78] (cf. Figure 8A), UV-induced electron excitation produces ROS that attack phenolic compounds. In contrast, persulfate (PS) activation (cf. Figure 8D), used for the catalytic degradation of 2,4-D and MCPA [73], occurs via Fe^{2+}-generating $SO_4^{\bullet-}$ radicals both in homogeneous and heterogeneous phases. This dual-phase reaction provides enhanced degradation efficiency, with higher pH values compared to the more pH-dependent process photocatalytic, Fenton, and photo-Fenton processes.

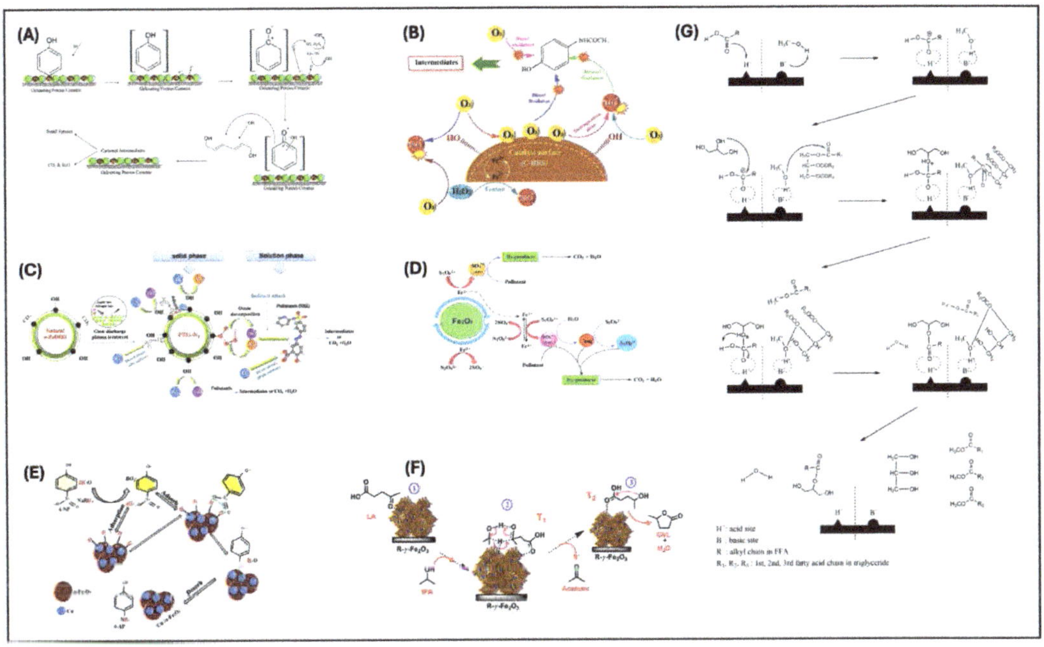

Figure 8. Schemes of proposed mechanism for different processes catalyzed by iron oxides and hydroxides from natural sources: (**A**) Photodegradation of phenol by α-Fe_2O_3 catalyst impregnated on gelcasting porous ceramic, (**B**) ozonation of ACT over HRS, (**C**) ozonation of SSZ over goethite, (**D**) Fenton-like oxidation of 2,4-D and MCPA over hematite, (**E**) photocatalytic reduction of 4 NP to 4 AP over hematite, (**F**) transfer hydrogenation of LA to form GVL over maghemite, and (**G**) waste cooking oil-based biodiesel production over α-Fe_2O_3/CaO_2. Reproduced with permission from References (**A**) [78], (**B**) [81], (**C**) [129], (**D**) [73], (**E**) [77], (**F**) [149], and (**G**) [87]. Copyright© 2022 IIETA/CC BY 4.0, 2022, 2020, 2018, and 2020 Elsevier; 2020 American Chemical Society; and 2022 Springer Nature, respectively.

The mechanism of the catalytic ozonation process mediated by HRS [81] (Figure 8B) involves three main pathways for the degradation of acetaminophen (ACT): (i) the direct oxidation of ACT by dissolved ozone (O_3) molecules in the bulk solution; (ii) the radical-based degradation of ACT by hydroxyl radicals ($^{\bullet}OH$) generated in the solution, and (iii) the indirect oxidation through the interaction of O_3 with the catalyst surface, which produces $^{\bullet}OH$ and superoxide radicals ($O_2^{\bullet-}$). The presence of ions like phosphate and carbonate can hinder the process by occupying active sites on the catalyst or reacting with $^{\bullet}OH$. In turn, two key mechanisms drive SSZ ozonation using goethite as a catalyst [129] (cf. Figure 8C): (i) the reductive degradation on the catalyst surface and (ii) the direct oxidation by non-adsorbed $O_2^{\bullet -}$ radicals. Additionally, plasma treatment on α-FeOOH enhances the production of $^{\bullet}OH$ and $O_2^{\bullet -}$, improving the ozonation process.

Figure 8E shows the mechanism of the photocatalytic reduction of 4-NP over Cu/α-Fe_2O_3 [77]. Here, we can observe that this follows three main steps: (i) the diffusion and adsorption of 4-NP onto the metal catalyst surface, (ii) electron transfer from BH_4^- to 4-NP, mediated by Fe and Cu metal particles, and (iii) the reduction of the nitro group by hydride transfer from a metal–hydride complex. This leads to the formation of 4-hydroxylaminophenol as a stable intermediate, which undergoes three hydrodeoxygenation reactions to produce 4-AP. Finally, 4-AP desorbs from the catalyst surface.

In transfer hydrogenation (cf. Figure 8F), the γ-Fe_2O_3 catalyst facilitates the initial activation of levulinic acid (LA) by interacting with its carbonyl carbon [149]. This interaction enhances the reactivity of LA, allowing it to interact with 2-propanol, which donates hydrogen to form 4-hydroxypentanoic acid as an intermediate. The iron catalyst promotes hydrogen transfer. For example, 4-hydroxypentanoic acid undergoes dehydration, which leads to the formation of Gamma-valerolactone (GVL). The Fe-OH minerals from natural sources can act as catalysts for biodiesel production. The mechanism proposed in Figure 8G [87] initiates with the adsorption of the carbonyl group in free fatty acids (FFAs) onto the acidic site and the adsorption of methanol onto the basic site of the catalyst. These interactions generate a carbocation and an oxygen anion, respectively. A nucleophilic attack occurs; the carbocation reacts with the hydroxyl group in glycerol on the acidic site, while the oxygen anion in methanol reacts with the carbonyl group in triglycerides on the basic site. This leads to the transesterification reaction, breaking down the three fatty acid chains of the triglyceride, yielding biodiesel molecules and glycerol as a byproduct. Simultaneously, the esterification of glycerol with FFAs on the acidic site produces water and diglycerides. Finally, the reaction products desorb, and the catalytic cycle restarts.

The separation of Fe-OH from other minerals and their purification can be difficult to achieve due to their textural intergrowths. However, the impurities present in Fe-OH catalysts can either inhibit or enhance catalytic reactions. There are few studies comparing synthetic catalysts to those prepared from natural sources. Based on existing comparative results [1,149], Fe-OHs from natural sources are more effective than their synthetic counterparts, but their catalytic activity is much more complex. For example, catalysts from different sources of mineral Fe_3O_4 with similar impurity content showed marked differences in the degradation of p-NP, indicating that the purity of the mineral might not be the determining factor for its catalytic activity. Here, the key factor in the performance is related to the surface site density of the parameter.

Maghemite prepared from river sand exhibited higher biomass conversion by transfer hydrogenation [149]. This was attributed to its crystallinity and higher pore size, allowing more interaction with the reactants. The presence of impurities of silica and metal carbonates also promotes the Lewis acidic sites. Therefore, one of the challenges for the use of these types of catalysts is understanding the effect of impurities on catalytic activity. A greater number of comparative studies between synthetic catalysts and those prepared from natural sources are needed, focusing on both their physicochemical characteristics and catalytic performances.

On the other hand, catalytic evaluation requires an appropriate experimental design as magnetic stirring could cause poor dispersion of particles in the solution, while mechanical stirring could hinder the scaling up of reactors to pilot or large-scale levels. Figure 9 shows different catalytic systems where Fe-OH prepared from natural sources were used as catalysts. At the laboratory scale, magnetic stirring (Figure 9C,F) can be used when there is a decrease in the magnetic response of catalysts. However, the use of pumps (Figure 9A,B,F) or gas carriers (Figure 9D,E) allows the continuous flow of solutions passing through the catalyst bed. To facilitate real-world scaling, we propose adapting and optimizing proven experimental reactor designs from analogous catalytic processes [80,245,246].

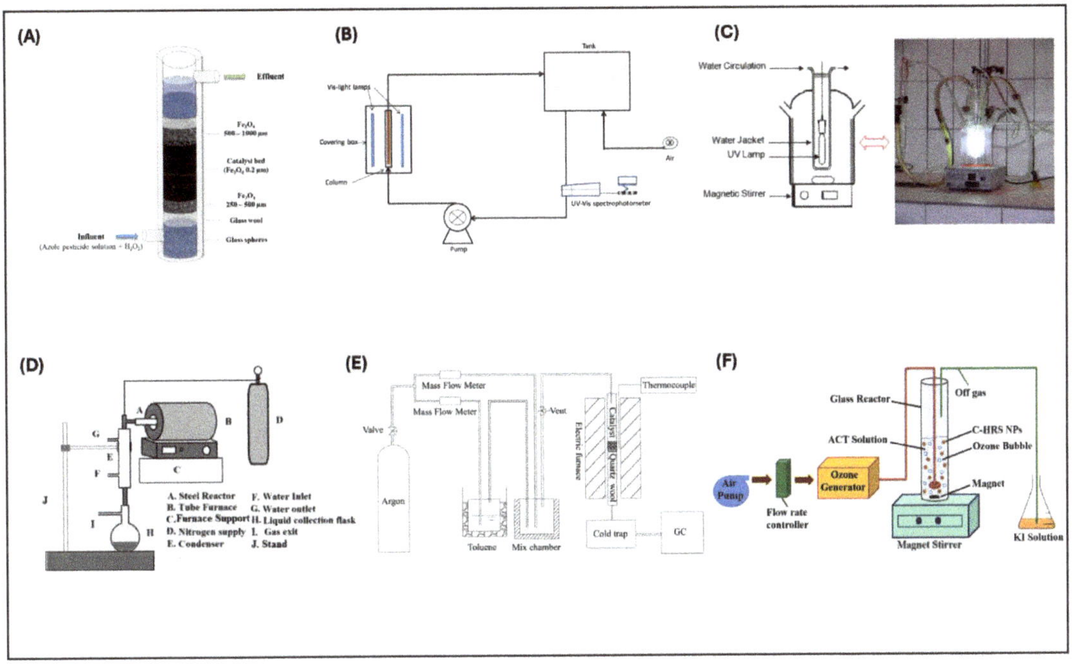

Figure 9. Schemes of reactors used in different processes catalyzed by Fe-OH from natural sources: (**A**) FBR packed with powdered magnetite catalyst used in the continuous CWPO experiments, (**B**) photocatalytic reactor continuous flow system for MB degradation, (**C**) photocatalytic reactor system for oxidation and mineralization of MO, (**D**) pyrolysis reactor to produce diesel-like fuel, (E) catalytic cracking of toluene, and (F) ozonation and peroxone-mediated removal of ACT. Reproduced with permission from References (**A**) [50], (**B**) [122], (**C**) [72], (**D**) [45], (**E**) [89], and (**F**) [81]. Copyright© 2024 Springer Nature, 2020 MDPI/CC BY, and 2020, 2016, 2016, and 2022 Elsevier, respectively.

The scalability of catalysts largely depends on their stability, lifespan, and resistance to poisoning. One of the most remarkable characteristics of Fe-OH is their ability to be recovered using magnetic fields, enabling their reuse. Figure 10 depicts conversion versus the number of cycles reached by natural Fe-OH. There is more information available on the reuse cycles of the most abundant iron oxide (magnetite (Figure 10A) and hematite (Figure 10B)) and hydroxide catalysts (goethite, Figure 10C). In contrast, information about less abundant and rare Fe-OH (maghemite and wüstite, Figure 10D) is rarely found.

According to reported results, magnetite prepared from natural sources is a stable catalyst in the photo-Fenton degradation of cefotaxime [46], CWPO degradation of phenol [62], and Fenton-like degradation of MeP reactions [52], showing less than 10% of deactivation until after five reused cycles. The primary cause of deactivation appears to be the partial oxidation of the catalyst surface, leading to a progressive decrease in the Fe^{2+}/Fe^{3+} ratio [57]. Nevertheless, to understand the decrease in reaction performance, it is necessary to characterize the catalysts after being recovered from the reaction. Comparisons between the physicochemical properties before and after the catalytic performance would help elucidate the mechanism of catalyst deactivation. The reuse results of hematite indicate that its conversion capacity decreases only slightly in the photocatalytic reduction of 4-NP to 4-AP [77], the ozonation of ACT [81], and the photo-Fenton degradation of phenol [74]. It also demonstrates stability in the cracking of toluene [89], where the reaction conditions are more aggressive than in AOPs. In this case, the catalysts' deactivation is attributed to carbon deposits, the aggregation of active metal particles, and poisonous impurities. This decline could be due to (i) the loss of Fe mass and/or iron leaching from the surface after each usage cycle, (ii) the reduction in the activation rate of PS molecules by the catalyst, and, as a result, (iii) the reduction in the efficiency of the heterogeneous catalytic degradation process. However, the iron loss was minimal, which confirmed the stability of the catalyst. Further characterization is needed to better understand the deactivation mechanism.

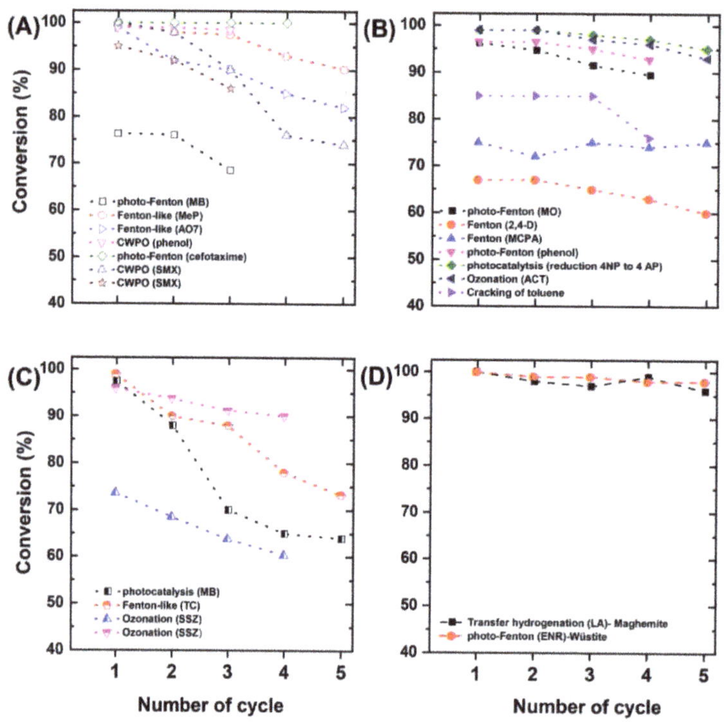

Figure 10. Conversion vs. recycled use of iron oxides and hydroxides: (**A**) magnetite, (**B**) hematite, (**C**) goethite, and (**D**) maghemite and wüstite. Elaborated with data from References [46,49,51,52,55,57,62], [72–74,77,81,89], [122,123,129], and [149,151], respectively. Dash lines are guides for the eye.

The decrease in catalytic conversion mediated by goethite is more evident (cf. Figure 10C) compared to magnetite and hematite, being more stable on the ozonation of SSZ using plasma-treated goethite (shown by pink triangles) [129]. Natural goethite (shown by blue triangles) interestingly exhibited a higher drop in conversion. Using plasma treatment, the surface enhancement of α-FeOOH can increase the density of hydroxyl groups at the surface.

Related to less abundant and rare iron oxides, the scant available results about reuse indicated that they are highly stable during transfer hydrogenation [149] and the photo-Fenton degradation of ENR [151]. The recovered catalyst (maghemite) from transfer hydrogenation remained stable and did not lose crystallinity, phase purity, and activity, even after extensive usage. Meanwhile, the surface area of the goethite decreased compared to that of the fresh catalyst [149]. This is consistent with the discussion throughout this review, which suggests that surface area is not the determining factor in the activity of these catalysts. In the photo-Fenton degradation of ENR, the reuse of wüstite resulted in the release of iron, but after the first cycle, iron leaching decreases, and the macroscopic solid undergoes disaggregation. Additionally, the first cycle shows a reduction in catalyst particle size. Furthermore, the observed disaggregation and particle size reduction suggest alterations to the solid surface throughout reuse cycles, potentially impacting the catalyst's interaction with light and H_2O_2 [151].

Generally, the reuse of Fe-OH catalysts does not require thermal treatments to reactivate the recovered catalysts, which reduces energy consumption and preparation costs. Additionally, the fact that they can be pretreated (or used directly without the need for thermal or chemical treatments) from inexpensive, accessible, and widely available sources may contribute to their scalability to real-world applications. One of the challenges to overcome in AOPs is the need for acidic pH conditions for these catalysts to remain active. One potential approach is exploring the use of oxidizing agents other than H_2O_2, such as PS, which has been shown to require nearly neutral pH conditions for effective catalytic reactions. In catalytic cracking, high temperatures are required, which can lead to the excessive consumption of network oxygen, causing catalyst deactivation. A thermal treatment in an oxidizing atmosphere could help reactivate it. However, this would increase energy consumption, potentially hindering its large-scale application.

Natural Fe-OH catalysts present challenges that must be addressed to enable their practical application. Their composition and mineralogical structure can vary depending on the source, affecting catalytic activity and reproducibility. Also, impurities may either inhibit or enhance catalytic performance, making it crucial to understand their role. In addition, AOP processes usually require acidic pH conditions, which limit natural Fe-OH applications in environmental samples or industrial waters. Another key challenge is the stability and reusability of these catalysts as iron leaching, surface fouling, or structural changes can lead to deactivation over multiple cycles. Furthermore, natural catalysts are widely available and cost-effective, but their efficiency must be comparable to synthetic catalysts to justify large-scale implementation without excessive energy-consuming treatments. Proper experimental design is also essential since factors such as stirring methods can impact catalyst dispersion and overall performance. To overcome these challenges, more comparative studies between natural and synthetic catalysts are needed, along with strategies to enhance stability, efficiency, and operational conditions.

6. Conclusions

While iron oxides are abundantly available in the Earth's crust, their utilization as catalysts has been largely overshadowed by the prevalence of laboratory-synthesized materials. A key advantage of iron oxide catalysts lies in their magnetic properties, facilitating easy recovery and potential for reuse. However, crucial knowledge gaps remain. First, the impact of repeated reuse cycles on the magnetic properties of these catalysts and their subsequent influence on catalytic performance requires further investigation. Second, a comprehensive comparative cost analysis between the synthesis of iron oxide catalysts and their extraction and preparation from natural sources is currently lacking. The main applications of iron oxides are AOPs, electrochemical applications (water splitting and OER), FTS, and ammonia synthesis. Hematite (α-Fe_2O_3) is the most widely used naturally sourced iron oxide, probably because it usually represents the final stage in the transformation of other iron oxides. However, AOPs are operative only in a narrow interval of pH compared to other iron oxides. The next most used iron oxides are magnetite (Fe_3O_4) and goethite

(α-FeOOH). The catalytic activity of magnetite has been related to the presence of Fe^{3+} and Fe^{2+} in its structure, where the oxidizing conditions can transform it into hematite. On the other hand, goethite can participate in redox reactions through the interconversion of Fe^{3+}; and Fe^{2+}, facilitating electron transfer in catalytic processes. This results in reactions where reactive species such as hydroxyl radicals ($^{\bullet}OH$) are necessary. Additionally, goethite exhibits the highest specific surface area among the three most used iron oxides.

Although conventional catalyst preparation often involves acid purification and thermal treatments, simpler approaches utilizing magnetic separation, grinding, sieving, and milling have demonstrated promising catalytic activity in some evaluations. Despite that, these methods may have limitations when precise control over particle size and morphology is critical as achieving consistent shape and size distribution through these techniques can be challenging.

The preparation and evaluation of catalytic applications of less abundant natural iron oxides, such as maghemite (γ-Fe_2O_3), lepidocrocite (γ-FeOOH), and ferrihydrite ($5Fe_2O_3 \cdot 9H_2O$), face diverse challenges, as follows: natural maghemite is not readily available in a pure form, and lepidocrocite is rarely found in natural ore deposits, limiting its accessibility for catalytic applications. On the one hand, ferrihydrite exhibits poor crystallinity, a small particle size, and structural defects that make its physicochemical characterization and reproducibility in catalytic processes difficult. On the other hand, the so-called rare iron oxides, i.e., wüstite (FeO), akaganeite (β-FeOOH), feroxyhyte (δ'-FeOOH), and bernalite ($Fe(OH)_3$), have been employed in their synthetic forms in few catalytic reactions. This may be related to their metastability, synthesis complexity, and dependence on specific operational conditions.

Despite the potential complexities associated with the preparation and evaluation of catalysts derived from natural sources, their abundance on the Earth's surface, coupled with their unique physicochemical properties and low-cost availability, presents a compelling opportunity to address environmental and energy-related challenges. Regardless, optimizing their catalytic performance necessitates a thorough understanding of the specific characteristics of each iron oxide to enable the development of effective strategies for scaling up these processes to pilot and industrial levels.

Author Contributions: Writing—original draft preparation and conceptualization, supervision, investigation, and methodology, A.J.-V.; Writing—original draft preparation, investigation, and conceptualization, R.J.-L.; Funding acquisition and methodology, S.P.-R.; investigation and methodology, C.M.M.-B. and Y.G.-P.; writing—review and editing, writing—original draft preparation and conceptualization, and funding acquisition, L.A.E.-W. All authors have read and agreed to the published version of the manuscript.

Funding: This work was partially funded by Instituto Politécnico Nacional through the SIP projects (Reference No. SIP-20240967, SIP-20242804, and SIP-20240049).

Data Availability Statement: No new data were created or analyzed in this study.

Acknowledgments: R. Jaimes-López acknowledges the postdoctoral scholarship from SECIHTI "Estancias Posdoctorales por México para la Formación y Consolidación de las y los Investigadores por México 2022".

Conflicts of Interest: The authors declare no conflicts of interest. The funders had no role in the design of this study, the collection, analysis, or interpretation of the data, the writing of the manuscript, or the decision to publish the results.

Abbreviations

The following abbreviations are used in this manuscript:

2,4-D	2,4-dichlorophenoxyacetic acid
4AP	4-aminophenol
4-NP	4-nitrophenol
ACT	Acetaminophen
AO7	Acid orange 7
AOP	Advanced oxidation process

ATR-FTIR	Attenuated total reflection–Fourier transform infrared spectroscopy
BIFs	Banded iron formations
BSA	5-bromosalicylic acid
CAP	Chloramphenicol
CGW	Contaminated groundwater
COD	Chemical oxygen demand
CWPO	Catalytic wet peroxide oxidation
DCF	Diclofenac
DFT	Density functional theory
DTA	Differential thermal analysis
EDTA	Ethylenediaminetetracetic acid
ENR	Enrofloxacin
FBR	Fixed-bed reactor
Fe-OH	Iron oxides and iron hydroxides
GC/MS	Gas chromatography–mass spectrometry
GHSV	Gas hourly space velocity
GVL	Gamma-valerolactone
Hc	Coercivity
HER	Hydrogen evolution reaction
HRS	Hormuz red soil
HTCT	High-temperature catalytic treatment
IC	Indigo carmine
IMA	International Mineralogical Association
IR	Infrared spectroscopy
LA	Levulinic acid
LEV	Levofloxacin
MB	Methylene blue
MC-LR	Microcystin-LR
MCPA	2-methyl-4-chlorophenoxyacetic acid
MeP	Methylparaben
MFCs	Microbial fuel cells
MM	Monodentate mononuclear
MO	Methyl orange
Mr	Remanent magnetization
NPs	Nanoparticles
OCs	Oxygen carriers
OER	Oxygen evolution reaction
ORR	Oxygen reduction reaction
OG	Orange G
PANI	Polyaniline
p-NP	p-nitrophenol
PS	Persulfate
ROS	Reactive oxygen species
SEM	Scanning electron microscopy
SMX	Sulfamethoxazole
SPS	Persulfate
SSZ	Sulfasalazine antibiotic
TC	Tetracycline
TEM	Transmission electron microscopy
TGA	Thermogravimetric analysis
T_M	Temperature of Morin transition
WFFDBD	Water falling film dielectric barrier discharge
WWTP	Wastewater treatment plant
XANES	X-ray absorption near-edge structure

XPS	X-ray photoelectron spectroscopy
XRD	X-ray diffraction
XRF	X-ray fluorescence

References

1. He, H.; Zhong, Y.; Liang, X.; Tan, W.; Zhu, J.; Wang, C.Y. Natural Magnetite: An Efficient Catalyst for the Degradation of Organic Contaminant. *Sci. Rep.* **2015**, *5*, 10139. [CrossRef] [PubMed]
2. Cornell, R.M.; Schwertmann, U. *The Iron Oxides*; Wiley: Hoboken, NJ, USA, 2003; ISBN 9783527302741.
3. Lagroix, F.; Banerjee, S.K.; Jackson, M.J. Geological Occurrences and Relevance of Iron Oxides. In *Iron Oxides*; Faivre, D., Ed.; Wiley: Hoboken, NJ, USA, 2016; pp. 7–30.
4. Jambor, J.L.; Dutrizac, J.E. Occurrence and Constitution of Natural and Synthetic Ferrihydrite, a Widespread Iron Oxyhydroxide. *Chem. Rev.* **1998**, *98*, 2549–2586. [CrossRef] [PubMed]
5. Misra, K.C. *Understanding Mineral Deposits*; Springer: Dordrecht, The Netherlands, 2000; ISBN 978-94-010-5752-3.
6. Schwertmann, U.; Fitzpatrick, R. Iron Minerals in Surface Environments. *Catena Suppl.* **1992**, *21*, 7–30.
7. Dowlath, M.J.H.; Musthafa, S.A.; Mohamed Khalith, S.B.; Varjani, S.; Karuppannan, S.K.; Ramanujam, G.M.; Arunachalam, A.M.; Arunachalam, K.D.; Chandrasekaran, M.; Chang, S.W.; et al. Comparison of Characteristics and Biocompatibility of Green Synthesized Iron Oxide Nanoparticles with Chemical Synthesized Nanoparticles. *Environ. Res.* **2021**, *201*, 111585. [CrossRef]
8. Patil, R.M.; Shete, P.B.; Thorat, N.D.; Otari, S.V.; Barick, K.C.; Prasad, A.; Ningthoujam, R.S.; Tiwale, B.M.; Pawar, S.H. Superparamagnetic Iron Oxide/Chitosan Core/Shells for Hyperthermia Application: Improved Colloidal Stability and Biocompatibility. *J. Magn. Magn. Mater.* **2014**, *355*, 22–30. [CrossRef]
9. Demirer, G.S.; Okur, A.C.; Kizilel, S. Synthesis and Design of Biologically Inspired Biocompatible Iron Oxide Nanoparticles for Biomedical Applications. *J. Mater. Chem. B* **2015**, *3*, 7831–7849. [CrossRef]
10. Pereira, M.C.; Oliveira, L.C.A.; Murad, E. Iron Oxide Catalysts: Fenton and Fentonlike Reactions—A Review. *Clay Min.* **2012**, *47*, 285–302. [CrossRef]
11. Rahim Pouran, S.; Abdul Raman, A.A.; Wan Daud, W.M.A. Review on the Application of Modified Iron Oxides as Heterogeneous Catalysts in Fenton Reactions. *J. Clean. Prod.* **2014**, *64*, 24–35. [CrossRef]
12. Wu, Q.; Siddique, M.S.; Wang, H.; Cui, L.; Wang, H.; Pan, M.; Yan, J. Visible-Light-Driven Iron-Based Heterogeneous Photo-Fenton Catalysts for Wastewater Decontamination: A Review of Recent Advances. *Chemosphere* **2023**, *313*, 137509. [CrossRef]
13. Gholami, Z.; Gholami, F.; Tišler, Z.; Hubáček, J.; Tomas, M.; Bačiak, M.; Vakili, M. Production of Light Olefins via Fischer-Tropsch Process Using Iron-Based Catalysts: A Review. *Catalysts* **2022**, *12*, 174. [CrossRef]
14. Okoye-Chine, C.G.; Mubenesha, S. The Use of Iron Ore as a Catalyst in Fischer–Tropsch Synthesis—A Review. *Crystals* **2022**, *12*, 1349. [CrossRef]
15. Ramadhani, B.; Kivevele, T.; Kihedu, J.H.; Jande, Y.A.C. Catalytic Tar Conversion and the Prospective Use of Iron-Based Catalyst in the Future Development of Biomass Gasification: A Review. *Biomass Convers. Biorefinery* **2022**, *12*, 1369–1392. [CrossRef]
16. Zhu, M.; Wachs, I.E. Iron-Based Catalysts for the High-Temperature Water–Gas Shift (HT-WGS) Reaction: A Review. *ACS Catal.* **2016**, *6*, 722–732. [CrossRef]
17. Cao, Y.; Yuan, X.; Zhao, Y.; Wang, H. In-Situ Soil Remediation via Heterogeneous Iron-Based Catalysts Activated Persulfate Process: A Review. *Chem. Eng. J.* **2022**, *431*, 133833. [CrossRef]
18. Cui, B.; Tian, T.; Duan, L.; Rong, H.; Chen, Z.; Luo, S.; Guo, D.; Naidu, R. Towards Advanced Removal of Organics in Persulfate Solution by Heterogeneous Iron-Based Catalyst: A Review. *J. Environ. Sci.* **2024**, *146*, 163–175. [CrossRef]
19. Sang, W.; Xu, X.; Zhan, C.; Lu, W.; Jia, D.; Wang, C.; Zhang, Q.; Gan, F.; Li, M. Recent Advances of Antibiotics Degradation in Different Environment by Iron-Based Catalysts Activated Persulfate: A Review. *J. Water Process Eng.* **2022**, *49*, 103075. [CrossRef]
20. Lai, L.; He, Y.; Zhou, H.; Huang, B.; Yao, G.; Lai, B. Critical Review of Natural Iron-Based Minerals Used as Heterogeneous Catalysts in Peroxide Activation Processes: Characteristics, Applications and Mechanisms. *J. Hazard. Mater.* **2021**, *416*, 125809. [CrossRef]
21. Roy, S.D.; Das, K.C.; Dhar, S.S. Conventional to Green Synthesis of Magnetic Iron Oxide Nanoparticles; Its Application as Catalyst, Photocatalyst and Toxicity: A Short Review. *Inorg. Chem. Commun.* **2021**, *134*, 109050. [CrossRef]
22. Ruan, Y.; Kong, L.; Zhong, Y.; Diao, Z.; Shih, K.; Hou, L.; Wang, S.; Chen, D. Review on the Synthesis and Activity of Iron-Based Catalyst in Catalytic Oxidation of Refractory Organic Pollutants in Wastewater. *J. Clean. Prod.* **2021**, *321*, 128924. [CrossRef]
23. Azfar Shaida, M.; Verma, S.; Talukdar, S.; Kumar, N.; Salim Mahtab, M.; Naushad, M.; Haq Farooqi, I. Critical Analysis of the Role of Various Iron-Based Heterogeneous Catalysts for Advanced Oxidation Processes: A State of the Art Review. *J. Mol. Liq.* **2023**, *374*, 121259. [CrossRef]
24. Munoz, M.; de Pedro, Z.M.; Casas, J.A.; Rodriguez, J.J. Preparation of Magnetite-Based Catalysts and Their Application in Heterogeneous Fenton Oxidation—A Review. *Appl. Catal. B* **2015**, *176–177*, 249–265. [CrossRef]

25. Buthelezi, A.S.; Tucker, C.L.; Heeres, H.J.; Shozi, M.L.; van de Bovenkamp, H.H.; Ntola, P. Fischer-Tropsch Synthesis Using Promoted, Unsupported, Supported, Bimetallic and Spray-Dried Iron Catalysts: A Review. *Results Chem.* **2024**, *9*, 101623. [CrossRef]
26. Jia, J.-Y.; Shan, Y.-L.; Tuo, Y.-X.; Yan, H.; Feng, X.; Chen, D. Review of Iron-Based Catalysts for Carbon Dioxide Fischer–Tropsch Synthesis. *Trans. Tianjin Univ.* **2024**, *30*, 178–197. [CrossRef]
27. Zhang, M.; Li, Y.; Tian, X.; Dai, L.; Wang, G.; Lei, Z.; Ma, G.; Zuo, Q.; Li, M.; Zhao, M.; et al. A Review of Iron-Based Catalysts for Persulfate Activation to Remove PFAS in Water: Catalytic Effects of Various Iron Species, Influencing Factors and Reaction Pathways. *Water Air Soil. Pollut.* **2025**, *236*, 42. [CrossRef]
28. Wu, H.; Li, J.; Gao, M.; Chen, Y.; Ren, S.; Yang, J.; Liu, Q. Deactivation Mechanisms and Strategies to Mitigate Deactivation of Iron-Based Catalysts in NH3-SCR for NO Reduction: A Comprehensive Review. *Sep. Purif. Technol.* **2025**, *358*, 130268. [CrossRef]
29. Al-Anazi, A.; Newair, E.F.; Han, C. The Development of Iron-Based Catalysts for the Reduction of Nitrogen and per- and Polyfluoroalkyl Substances in Wastewater Treatment: A Review. *Curr. Opin. Chem. Eng.* **2024**, *44*, 101024. [CrossRef]
30. Xu, H.; Ma, Z.; Wan, Z.; An, Z.; Wang, X. Recent Advances and Prospects of Iron-Based Noble Metal-Free Catalysts for Oxygen Reduction Reaction in Acidic Environment: A Mini Review. *Int. J. Hydrogen Energy* **2024**, *59*, 697–714. [CrossRef]
31. Tarek, M.; Santos, J.S.; Márquez, V.; Fereidooni, M.; Yazdanpanah, M.; Praserthdam, S.; Praserthdam, P. A Critical Review towards the Causes of the Iron-Based Catalysts Deactivation Mechanisms in the Selective Oxidation of Hydrogen Sulfide to Elemental Sulfur from Biogas. *J. Energy Chem.* **2024**, *90*, 388–411. [CrossRef]
32. Peng, H.; Hou, L.; Sun, C.; Zou, H.; Wang, T.-R.; Ma, Z.-Z. Geochemistry of Magnetite from the Hongniu–Hongshan Cu Skarn Deposit in Yunnan Province, SW China. *Ore Geol. Rev.* **2021**, *136*, 104237. [CrossRef]
33. Deditius, A.P.; Reich, M.; Simon, A.C.; Suvorova, A.; Knipping, J.; Roberts, M.P.; Rubanov, S.; Dodd, A.; Saunders, M. Nanogeochemistry of Hydrothermal Magnetite. *Contrib. Mineral. Petrol.* **2018**, *173*, 46. [CrossRef]
34. Usman, M.; Byrne, J.M.; Chaudhary, A.; Orsetti, S.; Hanna, K.; Ruby, C.; Kappler, A.; Haderlein, S.B. Magnetite and Green Rust: Synthesis, Properties, and Environmental Applications of Mixed-Valent Iron Minerals. *Chem. Rev.* **2018**, *118*, 3251–3304. [CrossRef] [PubMed]
35. Kozlenko, D.P.; Dubrovinsky, L.S.; Kichanov, S.E.; Lukin, E.V.; Cerantola, V.; Chumakov, A.I.; Savenko, B.N. Magnetic and Electronic Properties of Magnetite across the High Pressure Anomaly. *Sci. Rep.* **2019**, *9*, 4464. [CrossRef]
36. Liu, M.; Ye, Y.; Ye, J.; Gao, T.; Wang, D.; Chen, G.; Song, Z. Recent Advances of Magnetite (Fe_3O_4)-Based Magnetic Materials in Catalytic Applications. *Magnetochemistry* **2023**, *9*, 110. [CrossRef]
37. Gawande, M.B.; Branco, P.S.; Varma, R.S. Nano-Magnetite (Fe_3O_4) as a Support for Recyclable Catalysts in the Development of Sustainable Methodologies. *Chem. Soc. Rev.* **2013**, *42*, 3371. [CrossRef] [PubMed]
38. Niculescu, A.-G.; Chircov, C.; Grumezescu, A.M. Magnetite Nanoparticles: Synthesis Methods—A Comparative Review. *Methods* **2022**, *199*, 16–27. [CrossRef]
39. Nengsih, S.; Madjid, S.N.; Mursal; Jalil, Z. Synthesis and Characterization of Magnetite Particles from Syiah Kuala Iron Sand Prepared by Co-Precipitation Method. *J. Phys. Conf. Ser.* **2023**, *2582*, 012005. [CrossRef]
40. Safitri, I.; Wibowo, Y.G.; Rosarina, D.; Sudibyo. Synthesis and Characterization of Magnetite (Fe_3O_4) Nanoparticles from Iron Sand in Batanghari Beach. *IOP Conf. Ser. Mater. Sci. Eng.* **2021**, *1011*, 012020. [CrossRef]
41. Heryanto, H.; Tahir, D. The Correlations between Structural and Optical Properties of Magnetite Nanoparticles Synthesised from Natural Iron Sand. *Ceram. Int.* **2021**, *47*, 16820–16827. [CrossRef]
42. Darvina, Y.; Yulfriska, N.; Rifai, H.; Dwiridal, L.; Ramli, R. Synthesis of Magnetite Nanoparticles from Iron Sand by Ball-Milling. *J. Phys. Conf. Ser.* **2019**, *1185*, 012017. [CrossRef]
43. Morel, M.; Martínez, F.; Mosquera, E. Synthesis and Characterization of Magnetite Nanoparticles from Mineral Magnetite. *J. Magn. Magn. Mater.* **2013**, *343*, 76–81. [CrossRef]
44. Amiruddin, E.; Awaluddin, A.; Hariani, I.; Sihombing, R.; Angraini, R. The Influence of Milling Ball Size on the Structural, Morphological and Catalytic Properties of Magnetite (Fe_3O_4) Nanoparticles toward Methylene Blue Degradation. *J. Phys. Conf. Ser.* **2020**, *1655*, 012006. [CrossRef]
45. Khan, R.; Ahmad, I.; Khan, H.; Ismail, M.; Gul, K.; Yasin, A.; Ahmad, W. Production of Diesel-like Fuel from Spent Engine Oil by Catalytic Pyrolysis over Natural Magnetite. *J. Anal. Appl. Pyrolysis* **2016**, *120*, 493–500. [CrossRef]
46. Jiang, Q.; Zhu, R. Facile Synthesis of Highly Efficient and Cost-Effective Photo-Fenton Catalyst by Ball Milling Commercial TiO_2 and Natural Magnetite. *J. Alloys Compd.* **2021**, *862*, 158670. [CrossRef]
47. Priyadarshini, M.; Das, I.; Ghangrekar, M.M.; Blaney, L. Advanced Oxidation Processes: Performance, Advantages, and Scale-up of Emerging Technologies. *J. Environ. Manag.* **2022**, *316*, 115295. [CrossRef] [PubMed]
48. Vafaei Molamahmood, H.; Geng, W.; Wei, Y.; Miao, J.; Yu, S.; Shahi, A.; Chen, C.; Long, M. Catalyzed H_2O_2 Decomposition over Iron Oxides and Oxyhydroxides: Insights from Oxygen Production and Organic Degradation. *Chemosphere* **2022**, *291*, 133037. [CrossRef]

49. Aghdasinia, H.; Arehjani, P.; Vahid, B.; Khataee, A. Fluidized-Bed Fenton-like Oxidation of a Textile Dye Using Natural Magnetite. *Res. Chem. Intermed.* **2016**, *42*, 8083–8095. [CrossRef]
50. Lopez-Arago, N.; Munoz, M.; de Pedro, Z.M.; Casas, J.A. Natural Magnetite as an Effective and Long-Lasting Catalyst for CWPO of Azole Pesticides in a Continuous up-Flow Fixed-Bed Reactor. *Environ. Sci. Pollut. Res.* **2024**, *31*, 29148–29161. [CrossRef]
51. Mufti, N.; Maryam, S.; Setiyanto, H.; Taufiq, A.; Sunaryono, S. Recyclable Natural Magnetite Nanoparticles for Effective Degradation of Methylene Blue in Water under UV Light Irradiation. *Key Eng. Mater.* **2020**, *855*, 315–321. [CrossRef]
52. Rostamifasih, Z.; Pasalari, H.; Mohammadi, F.; Esrafili, A. Heterogeneous Catalytic Degradation of Methylparaben Using Persulfate Activated by Natural Magnetite; Optimization and Modeling by Response Surface Methodology. *J. Chem. Technol. Biotechnol.* **2019**, *94*, 1880–1892. [CrossRef]
53. Guo, R.; Xi, B.; Guo, C.; Cheng, X.; Lv, N.; Liu, W.; Borthwick, A.G.L.; Xu, J. Persulfate-Based Advanced Oxidation Processes: The New Hope Brought by Nanocatalyst Immobilization. *Environ. Funct. Mater.* **2022**, *1*, 67–91. [CrossRef]
54. Hajiahmadi, M.; Zarei, M.; Khataee, A. Introducing an Effective Iron-Based Catalyst for Heterogeneous Electro-Fenton Removal of Gemcitabine Using Three-Dimensional Graphene as Cathode. *J. Ind. Eng. Chem.* **2021**, *96*, 254–268. [CrossRef]
55. Munoz, M.; Conde, J.; de Pedro, Z.M.; Casas, J.A. Antibiotics Abatement in Synthetic and Real Aqueous Matrices by H_2O_2/Natural Magnetite. *Catal. Today* **2018**, *313*, 142–147. [CrossRef]
56. Rueda Márquez, J.J.; Levchuk, I.; Sillanpää, M. Application of Catalytic Wet Peroxide Oxidation for Industrial and Urban Wastewater Treatment: A Review. *Catalysts* **2018**, *8*, 673. [CrossRef]
57. Álvarez-Torrellas, S.; Munoz, M.; Mondejar, V.; de Pedro, Z.M.; Casas, J.A. Boosting the Catalytic Activity of Natural Magnetite for Wet Peroxide Oxidation. *Environ. Sci. Pollut. Res.* **2020**, *27*, 1176–1185. [CrossRef]
58. Lyu, L.; Bagchi, M.; Markoglou, N.; An, C.; Peng, H.; Bi, H.; Yang, X.; Sun, H. Towards Environmentally Sustainable Management: A Review on the Generation, Degradation, and Recycling of Polypropylene Face Mask Waste. *J. Hazard. Mater.* **2024**, *461*, 132566. [CrossRef]
59. He, L.; Hui, H.; Li, S.; Lin, W. Production of Light Aromatic Hydrocarbons by Catalytic Cracking of Coal Pyrolysis Vapors over Natural Iron Ores. *Fuel* **2018**, *216*, 227–232. [CrossRef]
60. Schulz, H. Short History and Present Trends of Fischer–Tropsch Synthesis. *Appl. Catal. A Gen.* **1999**, *186*, 3–12. [CrossRef]
61. Zhang, Q.; Gu, J.; Chen, J.; Qiu, S.; Wang, T. Facile Fabrication of Porous Fe@C Nanohybrids from Natural Magnetite as Excellent Fischer–Tropsch Catalysts. *Chem. Commun.* **2020**, *56*, 4523–4526. [CrossRef]
62. Munoz, M.; Domínguez, P.; de Pedro, Z.M.; Casas, J.A.; Rodriguez, J.J. Naturally-Occurring Iron Minerals as Inexpensive Catalysts for CWPO. *Appl. Catal. B* **2017**, *203*, 166–173. [CrossRef]
63. Widayat; Satriadi, H.; Setyojati, P.W.; Shihab, D.; Buchori, L.; Hadiyanto, H.; Nurushofa, F.A. Preparation $CaO/MgO/Fe_3O_4$ Magnetite Catalyst and Catalytic Test for Biodiesel Production. *Results Eng.* **2024**, *22*, 102202. [CrossRef]
64. Roberts, A.P.; Zhao, X.; Heslop, D.; Abrajevitch, A.; Chen, Y.-H.; Hu, P.; Jiang, Z.; Liu, Q.; Pillans, B.J. Hematite (α-Fe_2O_3) Quantification in Sedimentary Magnetism: Limitations of Existing Proxies and Ways Forward. *Geosci. Lett.* **2020**, *7*, 8. [CrossRef]
65. Manjunatha, M.; Kumar, R.; Anupama, A.V.; Khopkar, V.B.; Damle, R.; Ramesh, K.P.; Sahoo, B. XRD, Internal Field-NMR and Mössbauer Spectroscopy Study of Composition, Structure and Magnetic Properties of Iron Oxide Phases in Iron Ores. *J. Mater. Res. Technol.* **2019**, *8*, 2192–2200. [CrossRef]
66. Amiruddin, E.; Awaluddin, A.; Sinuraya, S.; Hadianto, H.; Noferdi, M.D.; Fitri, A.S. Study of Iron Oxide Nanoparticles Doped with Manganese for Catalytic Degradation of Methylene Blue. *J. Phys. Conf. Ser.* **2021**, *2049*, 012021. [CrossRef]
67. Jalil, Z.; Rahwanto, A.; Mulana, F.; Ismail, I.; Anggraini Akbar, I.; Budi Aulia, T. Effet of Hematite (Fe_2O_3) from Natural Iron Ore as a Catalyst in MgH_2 as Hydrogen Storage Materials. *J. Phys. Conf. Ser.* **2023**, *2498*, 012048. [CrossRef]
68. Mandal, S.; Müller, A.H.E. Facile Route to the Synthesis of Porous α-Fe_2O_3 Nanorods. *Mater. Chem. Phys.* **2008**, *111*, 438–443. [CrossRef]
69. Zhang, Y.C.; Tang, J.Y.; Hu, X.Y. Controllable Synthesis and Magnetic Properties of Pure Hematite and Maghemite Nanocrystals from a Molecular Precursor. *J. Alloys Compd.* **2008**, *462*, 24–28. [CrossRef]
70. Malaidurai, M.; Thangavel, R. Study of Structural and Magnetic Properties of Co-Precipitated α-Fe_2O_3 Nanocrystals. *Superlattices Microstruct.* **2018**, *120*, 553–560. [CrossRef]
71. Lohaus, C.; Klein, A.; Jaegermann, W. Limitation of Fermi Level Shifts by Polaron Defect States in Hematite Photoelectrodes. *Nat. Commun.* **2018**, *9*, 4309. [CrossRef]
72. Omri, A.; Hamza, W.; Benzina, M. Photo-Fenton Oxidation and Mineralization of Methyl Orange Using Fe-Sand as Effective Heterogeneous Catalyst. *J. Photochem. Photobiol. A Chem.* **2020**, *393*, 112444. [CrossRef]
73. Kermani, M.; Mohammadi, F.; Kakavandi, B.; Esrafili, A.; Rostamifasih, Z. Simultaneous Catalytic Degradation of 2,4-D and MCPA Herbicides Using Sulfate Radical-Based Heterogeneous Oxidation over Persulfate Activated by Natural Hematite (α-Fe_2O_3/PS). *J. Phys. Chem. Solids* **2018**, *117*, 49–59. [CrossRef]

74. Hollanda, L.R.; de Souza, J.A.B.; Dotto, G.L.; Foletto, E.L.; Chiavone-Filho, O. Iron-Bearing Mining Reject as an Alternative and Effective Catalyst for Photo-Fenton Oxidation of Phenol in Water. *Environ. Sci. Pollut. Res.* **2024**, *31*, 21291–21301. [CrossRef] [PubMed]
75. Mohammadi, S.; Moussavi, G.; Shekoohiyan, S.; Marín, M.L.; Boscá, F.; Giannakis, S. A Continuous-Flow Catalytic Process with Natural Hematite-Alginate Beads for Effective Water Decontamination and Disinfection: Peroxymonosulfate Activation Leading to Dominant Sulfate Radical and Minor Non-Radical Pathways. *Chem. Eng. J.* **2021**, *411*, 127738. [CrossRef]
76. Lubis, S.; Sheilatina; Murisna. Synthesis, Characterization and Photocatalytic Activity of α-Fe_2O_3/Bentonite Composite Prepared by Mechanical Milling. *J. Phys. Conf. Ser.* **2018**, *1116*, 042016. [CrossRef]
77. Elfiad, A.; Galli, F.; Boukhobza, L.M.; Djadoun, A.; Boffito, D.C. Low-Cost Synthesis of Cu/α-Fe_2O_3 from Natural $HFeO_2$: Application in 4-Nitrophenol Reduction. *J. Environ. Chem. Eng.* **2020**, *8*, 104214. [CrossRef]
78. Putri, S.E.; Pratiwi, D.E.; Tjahjanto, R.T.; Hasri; Andi, I.; Rahman, A.; Ramadani, A.I.W.S.; Ramadhani, A.N.; Subaer; Fudholi, A. The Renewable of Low Toxicity Gelcasting Porous Ceramic as Fe_2O_3 Catalyst Support on Phenol Photodegradation. *Int. J. Des. Nat. Ecodynamics* **2022**, *17*, 503–511. [CrossRef]
79. Elfiad, A.; Galli, F.; Djadoun, A.; Sennour, M.; Chegrouche, S.; Meddour-Boukhobza, L.; Boffito, D.C. Natural α-Fe_2O_3 as an Efficient Catalyst for the p-Nitrophenol Reduction. *Mater. Sci. Eng. B* **2018**, *229*, 126–134. [CrossRef]
80. Jaimes-López, R.; Jiménez-Vázquez, A.; Pérez-Rodríguez, S.; Estudillo-Wong, L.A.; Alonso-Vante, N. Catalyst for the Generation of OH Radicals in Advanced Electrochemical Oxidation Processes: Present and Future Perspectives. *Catalysts* **2024**, *14*, 703. [CrossRef]
81. Kohantorabi, M.; Moussavi, G.; Oulego, P.; Giannakis, S. Heterogeneous Catalytic Ozonation and Peroxone-Mediated Removal of Acetaminophen Using Natural and Modified Hematite-Rich Soil, as Efficient and Environmentally Friendly Catalysts. *Appl. Catal. B* **2022**, *301*, 120786. [CrossRef]
82. Yue, Y.; Niu, P.; Jiang, L.; Cao, Y.; Bao, X. Acid-Modified Natural Bauxite Mineral as a Cost-Effective and High-Efficient Catalyst Support for Slurry-Phase Hydrocracking of High-Temperature Coal Tar. *Energy Fuels* **2016**, *30*, 9203–9209. [CrossRef]
83. Robinson, P.R. Hydroconversion Processes and Technology for Clean Fuel and Chemical Production. In *Advances in Clean Hydrocarbon Fuel Processing*; Elsevier: Amsterdam, The Netherlands, 2011; pp. 287–325.
84. Kairbekov, Z.K.; Maloletnev, A.S.; Dzheldybaeva, I.M.; Sabitova, A.N.; Ermoldina, E.T. Application of Modified Iron-Containing Catalysts and Preliminary Ozonization of Coal from the Shubarkol Deposit to the Hydrogenation of This Coal. *Solid. Fuel Chem.* **2017**, *51*, 365–369. [CrossRef]
85. Kretzschmar, N.; Busse, O.; Seifert, M. Impact of Geometric and Electronic Factors on Selective Hydro-Deoxygenation of Guaiacol by Surface-Rich Metal/Silica Catalysts. *Catalysts* **2023**, *13*, 425. [CrossRef]
86. Widayat; Putra, D.A.; Nursafitri, I. Synthesis and Catalytic Evaluation of Hematite (α-Fe_2O_3) Magnetic Nanoparticles from Iron Sand for Waste Cooking Oil Conversion to Produce Biodiesel through Esterification-Transesterification Method. *IOP Conf. Ser. Mater. Sci. Eng.* **2019**, *509*, 012035. [CrossRef]
87. Prameswari, J.; Widayat, W.; Buchori, L.; Hadiyanto, H. Novel Iron Sand-Derived α-Fe_2O_3/CaO_2 Bifunctional Catalyst for Waste Cooking Oil-Based Biodiesel Production. *Environ. Sci. Pollut. Res.* **2022**, *30*, 98832–98847. [CrossRef]
88. Abdrafikova, I.M.; Kayukova, G.P.; Petrov, S.M.; Ramazanova, A.I.; Musin, R.Z.; Morozov, V.I. Conversion of Extra-Heavy Ashal'chinskoe Oil in Hydrothermal Catalytic System. *Pet. Chem.* **2015**, *55*, 104–111. [CrossRef]
89. Zou, X.; Chen, T.; Liu, H.; Zhang, P.; Chen, D.; Zhu, C. Catalytic Cracking of Toluene over Hematite Derived from Thermally Treated Natural Limonite. *Fuel* **2016**, *177*, 180–189. [CrossRef]
90. Li, Y.; Gong, J.; Huang, F.; Bai, H.; Wang, F.; Wang, C. Carbon Deposition and Sintering Characteristics on Iron-Based Oxygen Carriers in the Catalytic Cracking Process of Coal Tar. *Energy Fuels* **2017**, *31*, 6501–6506. [CrossRef]
91. Sayeed, M.A.; Fernando, J.F.S.; O'Mullane, A.P. Conversion of Iron Ore into an Active Catalyst for the Oxygen Evolution Reaction. *Adv. Sustain. Syst.* **2018**, *2*, 1800019. [CrossRef]
92. Fabbri, E.; Schmidt, T.J. Oxygen Evolution Reaction—The Enigma in Water Electrolysis. *ACS Catal.* **2018**, *8*, 9765–9774. [CrossRef]
93. Alduhaish, O.; Ubaidullah, M.; Al-Enizi, A.M.; Alhokbany, N.; Alshehri, S.M.; Ahmed, J. Facile Synthesis of Mesoporous α-Fe_2O_3@g-C_3N_4-NCs for Efficient Bifunctional Electro-Catalytic Activity (OER/ORR). *Sci. Rep.* **2019**, *9*, 14139. [CrossRef]
94. Farooq, A.; Khalil, S.; Basha, B.; Habib, A.; Al-Buriahi, M.S.; Warsi, M.F.; Yousaf, S.; Shahid, M. Electrochemical Investigation of C-Doped $CoFe_2O_4$/Fe_2O_3Nanostructures for Efficient Electrochemical Water Splitting. *Int. J. Hydrogen Energy* **2024**, *51*, 1318–1332. [CrossRef]
95. Ng, K.-L.; Kok, K.-Y.; Ong, B.-H. Facile Synthesis of Self-Assembled Cobalt Oxide Supported on Iron Oxide as the Novel Electrocatalyst for Enhanced Electrochemical Water Electrolysis. *ACS Appl. Nano Mater.* **2018**, *1*, 401–409. [CrossRef]
96. Zang, Y.; Zhang, H.; Zhang, X.; Liu, R.; Liu, S.; Wang, G.; Zhang, Y.; Zhao, H. Fe/Fe_2O_3 Nanoparticles Anchored on Fe-N-Doped Carbon Nanosheets as Bifunctional Oxygen Electrocatalysts for Rechargeable Zinc-Air Batteries. *Nano Res.* **2016**, *9*, 2123–2137. [CrossRef]

97. Guo, W.H.; Zhang, Q.; Fu, H.C.; Yang, Y.X.; Chen, X.H.; Luo, H.Q.; Li, N.B. Semi-Sacrificial Template Growth of FeS/Fe_2O_3 Heterogeneous Nanosheets on Iron Foam for Efficient Oxygen Evolution Reaction. *J. Alloys Compd.* **2022**, *909*, 164670. [CrossRef]
98. Khan, N.A.; Rashid, N.; Ahmad, I.; Zahidullah; Zairov, R.; Rehman, H.U.; Nazar, M.F.; Jabeen, U. An Efficient Fe_2O_3/FeS Heterostructures Water Oxidation Catalyst. *Int. J. Hydrogen Energy* **2022**, *47*, 22340–22347. [CrossRef]
99. Bandal, H.A.; Jadhav, A.R.; Chaugule, A.A.; Chung, W.-J.; Kim, H. Fe_2O_3 Hollow Nanorods/CNT Composites as an Efficient Electrocatalyst for Oxygen Evolution Reaction. *Electrochim. Acta* **2016**, *222*, 1316–1325. [CrossRef]
100. Ahmad, I.; Ahmed, J.; Batool, S.; Zafar, M.N.; Hanif, A.; Zahidullah; Nazar, M.F.; Ul-Hamid, A.; Jabeen, U.; Dahshan, A.; et al. Design and Fabrication of Fe_2O_3/FeP Heterostructure for Oxygen Evolution Reaction Electrocatalysis. *J. Alloys Compd.* **2022**, *894*, 162409. [CrossRef]
101. Cui, T.; Zhai, X.; Guo, L.; Chi, J.-Q.; Zhang, Y.; Zhu, J.; Sun, X.; Wang, L. Controllable Synthesis of a Self-Assembled Ultralow Ru, Ni-Doped Fe_2O_3 Lily as a Bifunctional Electrocatalyst for Large-Current-Density Alkaline Seawater Electrolysis. *Chin. J. Catal.* **2022**, *43*, 2202–2211. [CrossRef]
102. Xiao, X.; Ma, R.; Li, D.; Yang, M.; Tong, Y.; Qiu, Z.; Jing, B.; Yang, T.; Hu, R.; Yang, Y.; et al. Tungsten Oxide Decorating-Regulated Iron-Nickel Oxide Heterostructure as Electrocatalyst for Oxygen Evolution Reaction. *J. Alloys Compd.* **2023**, *968*, 171910. [CrossRef]
103. Tong, H.; Zheng, X.; Qi, M.; Li, D.; Zhu, J.; Jiang, D. Synergistically Coupled CoMo/Fe_2O_3 Electrocatalyst for Highly Efficient and Stable Overall Water Splitting. *J. Colloid. Interface Sci.* **2024**, *676*, 837–846. [CrossRef]
104. Yang, X.; Li, Y.; Deng, L.; Li, W.; Ren, Z.; Yang, M.; Yang, X.; Zhu, Y. Synthesis and Characterization of an IrO_2 –Fe_2O_3 Electrocatalyst for the Hydrogen Evolution Reaction in Acidic Water Electrolysis. *RSC Adv.* **2017**, *7*, 20252–20258. [CrossRef]
105. Kim, J.; Heo, J.N.; Do, J.Y.; Chava, R.K.; Kang, M. Electrochemical Synergies of Heterostructured Fe_2O_3-MnO Catalyst for Oxygen Evolution Reaction in Alkaline Water Splitting. *Nanomaterials* **2019**, *9*, 1486. [CrossRef] [PubMed]
106. Fang, Y.; Cao, X.; Zhou, W.; Li, Y.; Johnson, D.M.; Huang, Y. Catalytic Hydrolysis of Microcystin-LR Peptides on the Surface of Naturally Occurring Minerals. *Res. Chem. Intermed.* **2020**, *46*, 1141–1152. [CrossRef]
107. Ramutsindela, F.K.; Okoye-Chine, C.G.; Mbuya, C.O.L.; Mubenesha, S.; Gorimbo, J.; Okonye, L.U.; Liu, X.; Hildebrandt, D. The Effect of Reducing Gases on Raw Iron Ore Catalyst for Fischer-Tropsch Synthesis. *J. Taiwan. Inst. Chem. Eng.* **2022**, *131*, 104163. [CrossRef]
108. Liu, H.; Chen, T.; Xie, Q.; Zou, X.; Qing, C.; Frost, R.L. Kinetic Study of Goethite Dehydration and the Effect of Aluminium Substitution on the Dehydrate. *Thermochim. Acta* **2012**, *545*, 20–25. [CrossRef]
109. Liu, H.; Chen, T.; Frost, R.L. An Overview of the Role of Goethite Surfaces in the Environment. *Chemosphere* **2014**, *103*, 1–11. [CrossRef]
110. Strangway, D.W.; Honea, R.M.; McMahon, B.E.; Larson, E.E. The Magnetic Properties of Naturally Occurring Goethite. *Geophys. J. Int.* **1968**, *15*, 345–359. [CrossRef]
111. Valezi, D.F.; Carneiro, C.E.A.; Costa, A.C.S.; Paesano, A.; Spadotto, J.C.; Solórzano, I.G.; Londoño, O.M.; Di Mauro, E. Weak Ferromagnetic Component in Goethite (α-FeOOH) and Its Relation with Microstructural Characteristics. *Mater. Chem. Phys.* **2020**, *246*, 122851. [CrossRef]
112. Han, S.K.; Hwang, T.-M.; Yoon, Y.; Kang, J.-W. Evidence of Singlet Oxygen and Hydroxyl Radical Formation in Aqueous Goethite Suspension Using Spin-Trapping Electron Paramagnetic Resonance (EPR). *Chemosphere* **2011**, *84*, 1095–1101. [CrossRef]
113. Shindo, H.; Huang, P.M. Catalytic Effects of Manganese (IV), Iron(III), Aluminum, and Silicon Oxides on the Formation of Phenolic Polymers. *Soil. Sci. Soc. Am. J.* **1984**, *48*, 927–934. [CrossRef]
114. Cunningham, K.M.; Goldberg, M.C.; Weiner, E.R. Mechanisms for Aqueous Photolysis of Adsorbed Benzoate, Oxalate, and Succinate on Iron Oxyhydroxide (Goethite) Surfaces. *Environ. Sci. Technol.* **1988**, *22*, 1090–1097. [CrossRef]
115. Lin, K.; Ding, J.; Wang, H.; Huang, X.; Gan, J. Goethite-Mediated Transformation of Bisphenol A. *Chemosphere* **2012**, *89*, 789–795. [CrossRef] [PubMed]
116. Lu, M.-C. Oxidation of Chlorophenols with Hydrogen Peroxide in the Presence of Goethite. *Chemosphere* **2000**, *40*, 125–130. [CrossRef] [PubMed]
117. Ortiz de la Plata, G.B.; Alfano, O.M.; Cassano, A.E. Optical Properties of Goethite Catalyst for Heterogeneous Photo-Fenton ReactionsComparison with a Titanium Dioxide Catalyst. *Chem. Eng. J.* **2008**, *137*, 396–410. [CrossRef]
118. He, J.; Ma, W.; Song, W.; Zhao, J.; Qian, X.; Zhang, S.; Yu, J.C. Photoreaction of Aromatic Compounds at α-FeOOH/H2O Interface in the Presence of H_2O_2: Evidence for Organic-Goethite Surface Complex Formation. *Water Res.* **2005**, *39*, 119–128. [CrossRef]
119. Yeh, C.K.-J.; Hsu, C.-Y.; Chiu, C.-H.; Huang, K.-L. Reaction Efficiencies and Rate Constants for the Goethite-Catalyzed Fenton-like Reaction of NAPL-Form Aromatic Hydrocarbons and Chloroethylenes. *J. Hazard. Mater.* **2008**, *151*, 562–569. [CrossRef]
120. Lin, Z.-R.; Zhao, L.; Dong, Y.-H. Quantitative Characterization of Hydroxyl Radical Generation in a Goethite-Catalyzed Fenton-like Reaction. *Chemosphere* **2015**, *141*, 7–12. [CrossRef]
121. Andersen, S.L.F.; Flores, R.G.; Madeira, V.S.; José, H.J.; Moreira, R.F.P.M. Synthesis and Characterization of Acicular Iron Oxide Particles Obtained from Acid Mine Drainage and Their Catalytic Properties in Toluene Oxidation. *Ind. Eng. Chem. Res.* **2012**, *51*, 767–774. [CrossRef]

122. Aziz, F.; El Achaby, M.; Aziz, K.; Ouazzani, N.; Mandi, L.; Ghazzal, M.N. Nanocomposite Fiber Based on Natural Material for Water Disinfection under Visible Light Irradiation. *Nanomaterials* **2020**, *10*, 1192. [CrossRef]
123. Kuang, J.; Guo, H.; Si, Q.; Guo, W.; Ma, F. Highly Dispersed Natural Goethite/Fe_3O_4 Heterogeneous Fenton-like Catalyst for Efficient Tetracycline Degradation: Rapid Accumulation of Hydroxyl Radicals by Fe–O–Si Bonds. *J. Clean. Prod.* **2023**, *420*, 138262. [CrossRef]
124. Muruganandham, M.; Wu, J.J. Granular α-FeOOH—A Stable and Efficient Catalyst for the Decomposition of Dissolved Ozone in Water. *Catal. Commun.* **2007**, *8*, 668–672. [CrossRef]
125. Zhang, T.; Li, C.; Ma, J.; Tian, H.; Qiang, Z. Surface Hydroxyl Groups of Synthetic α-FeOOH in Promoting OH Generation from Aqueous Ozone: Property and Activity Relationship. *Appl. Catal. B* **2008**, *82*, 131–137. [CrossRef]
126. Li, X.; Fu, L.; Chen, F.; Zhao, S.; Zhu, J.; Yin, C. Application of Heterogeneous Catalytic Ozonation in Wastewater Treatment: An Overview. *Catalysts* **2023**, *13*, 342. [CrossRef]
127. Zhang, Y.; Yang, J.; Du, J.; Xing, B. Goethite Catalyzed Cr(VI) Reduction by Tartaric Acid via Surface Adsorption. *Ecotoxicol. Environ. Saf.* **2019**, *171*, 594–599. [CrossRef]
128. Zhang, T.; Wang, T.; Wang, W.; Liu, B.; Li, W.; Liu, Y. Reduction and Stabilization of Cr(VI) in Soil by Using Calcium Polysulfide: Catalysis of Natural Iron Oxides. *Environ. Res.* **2020**, *190*, 109992. [CrossRef]
129. Pelalak, R.; Alizadeh, R.; Ghareshabani, E. Enhanced Heterogeneous Catalytic Ozonation of Pharmaceutical Pollutants Using a Novel Nanostructure of Iron-Based Mineral Prepared via Plasma Technology: A Comparative Study. *J. Hazard. Mater.* **2020**, *392*, 122269. [CrossRef]
130. Fernández-Marchante, C.M.; Raschitor, A.; Mena, I.F.; Rodrigo, M.A.; Lobato, J. Evaluation of Goethite as a Catalyst for the Thermal Stage of the Westinghouse Process for Hydrogen Production. *Catalysts* **2021**, *11*, 1145. [CrossRef]
131. Shehzad, M.A.; Yasmin, A.; Ge, X.; Ge, Z.; Zhang, K.; Liang, X.; Zhang, J.; Li, G.; Xiao, X.; Jiang, B.; et al. Shielded Goethite Catalyst That Enables Fast Water Dissociation in Bipolar Membranes. *Nat. Commun.* **2021**, *12*, 9. [CrossRef]
132. Hu, Z.; Ai, Y.; Liu, L.; Zhou, J.; Zhang, G.; Liu, H.; Liu, X.; Liu, Z.; Hu, J.; Sun, H.; et al. Hydroxyl Assisted Rhodium Catalyst Supported on Goethite Nanoflower for Chemoselective Catalytic Transfer Hydrogenation of Fully Converted Nitrostyrenes. *Adv. Synth. Catal.* **2019**, *361*, 3146–3154. [CrossRef]
133. Kuncser, A.C.; Vlaicu, I.D.; Pavel, O.D.; Zavoianu, R.; Badea, M.; Radu, D.; Culita, D.C.; Rostas, A.M.; Olar, R. Soft Synthesis and Characterization of Goethite-Based Nanocomposites as Promising Cyclooctene Oxidation Catalysts. *RSC Adv.* **2021**, *11*, 27589–27602. [CrossRef]
134. Jadhav, D.A.; Ghadge, A.N.; Ghangrekar, M.M. Enhancing the Power Generation in Microbial Fuel Cells with Effective Utilization of Goethite Recovered from Mining Mud as Anodic Catalyst. *Bioresour. Technol.* **2015**, *191*, 110–116. [CrossRef]
135. Zhao, H.; Li, Y.; Song, Q.; Liu, S.; Ma, Q.; Ma, L.; Shu, X. Catalytic Reforming of Volatiles from Co-Pyrolysis of Lignite Blended with Corn Straw over Three Different Structures of Iron Ores. *J. Anal. Appl. Pyrolysis* **2019**, *144*, 104714. [CrossRef]
136. Verma, S.; Baig, R.B.N.; Nadagouda, M.N.; Varma, R.S. Oxidative C-H Activation of Amines Using Protuberant Lychee-like Goethite. *Sci. Rep.* **2018**, *8*, 2024. [CrossRef] [PubMed]
137. Zheng, X.; Wu, B.; Zhou, C.; Liu, T.; Wang, Y.; Zhao, G.; Chen, B.; Chu, C. Sunlight-Driven Production of Reactive Oxygen Species from Natural Iron Minerals: Quantum Yield and Wavelength Dependence. *Environ. Sci. Technol.* **2023**, *57*, 1177–1185. [CrossRef]
138. Liu, X.; Shaw, J.; Jiang, J.; Bloemendal, J.; Hesse, P.; Rolph, T.; Mao, X. Analysis on Variety and Characteristics of Maghemite. *Sci. China Earth Sci.* **2010**, *53*, 1153–1162. [CrossRef]
139. Chen, Y.H. Thermal Properties of Nanocrystalline Goethite, Magnetite, and Maghemite. *J. Alloys Compd.* **2013**, *553*, 194–198. [CrossRef]
140. Kim, W.; Suh, C.-Y.; Cho, S.-W.; Roh, K.-M.; Kwon, H.; Song, K.; Shon, I.-J. A New Method for the Identification and Quantification of Magnetite–Maghemite Mixture Using Conventional X-Ray Diffraction Technique. *Talanta* **2012**, *94*, 348–352. [CrossRef]
141. Yamashita, T.; Hayes, P. Analysis of XPS Spectra of Fe^{2+} and Fe^{3+} Ions in Oxide Materials. *Appl. Surf. Sci.* **2008**, *254*, 2441–2449. [CrossRef]
142. Gnanaprakash, G.; Mahadevan, S.; Jayakumar, T.; Kalyanasundaram, P.; Philip, J.; Raj, B. Effect of Initial PH and Temperature of Iron Salt Solutions on Formation of Magnetite Nanoparticles. *Mater. Chem. Phys.* **2007**, *103*, 168–175. [CrossRef]
143. Murad, E. Properties and Behavior of Iron Oxides as Determined by Mössbauer Spectroscopy. In *Iron in Soils and Clay Minerals*; Springer: Dordrecht, The Netherlands, 1988; pp. 309–350.
144. Dunlop, D.J.; Özdemir, Ö. Magnetizations in Rocks and Minerals. In *Treatise on Geophysics*; Elsevier: Amsterdam, The Netherlands, 2007; pp. 277–336.
145. Nikiforov, V.N.; Goldt, A.E.; Gudilin, E.A.; Sredin, V.G.; Irhin, V.Y. Magnetic Properties of Maghemite Nanoparticles. *Bull. Russ. Acad. Sci. Phys.* **2014**, *78*, 1075–1080. [CrossRef]
146. Grau-Crespo, R.; Al-Baitai, A.Y.; Saadoune, I.; De Leeuw, N.H. Vacancy Ordering and Electronic Structure of γ-Fe_2O_3(Maghemite): A Theoretical Investigation. *J. Phys. Condens. Matter* **2010**, *22*, 255401–255408. [CrossRef]

147. Ramirez, A.; Ould-Chikh, S.; Gevers, L.; Chowdhury, A.D.; Abou-Hamad, E.; Aguilar-Tapia, A.; Hazemann, J.; Wehbe, N.; Al Abdulghani, A.J.; Kozlov, S.M.; et al. Tandem Conversion of CO_2 to Valuable Hydrocarbons in Highly Concentrated Potassium Iron Catalysts. *ChemCatChem* **2019**, *11*, 2879–2886. [CrossRef]
148. Xu, D.; Lai, X.; Guo, W.; Dai, P. Microwave-Assisted Catalytic Degradation of Methyl Orange in Aqueous Solution by Ferrihydrite/Maghemite Nanoparticles. *J. Water Process Eng.* **2017**, *16*, 270–276. [CrossRef]
149. Vasanthakumar, P.; Karvembu, R. Unmodified Maghemite from River Sand as a Selective Catalyst for Base-Free Transfer Hydrogenation of Furfural, Levulinic Acid, and *o*-Vanillin: A Pathway for Sustainable Biomass Conversions. *ACS Sustain. Chem. Eng.* **2020**, *8*, 17069–17078. [CrossRef]
150. Filip, J.; Zboril, R.; Schneeweiss, O.; Zeman, J.; Cernik, M.; Kvapil, P.; Otyepka, M. Environmental Applications of Chemically Pure Natural Ferrihydrite. *Environ. Sci. Technol.* **2007**, *41*, 4367–4374. [CrossRef]
151. Jojoa-Sierra, S.D.; Herrero-Albillos, J.; Ormad, M.P.; Serna-Galvis, E.A.; Torres-Palma, R.A.; Mosteo, R. Wüstite as a Catalyst Source for Water Remediation: Differentiated Antimicrobial Activity of by-Products, Action Routes of the Process, and Transformation of Fluoroquinolones. *Chem. Eng. J.* **2022**, *435*, 134850. [CrossRef]
152. Husnain, N.; Wang, E.; Fareed, S. Low-Temperature Selective Catalytic Reduction of NO with NH3 over Natural Iron Ore Catalyst. *Catalysts* **2019**, *9*, 956. [CrossRef]
153. Cabrera, A.F.; Rodríguez Torres, C.E.; Stewart, S.J. Nanostructured Al-doped maghemite: A low-cost and eco-friendly material tested for methylene blue removal from water and as an accelerator in ammonium nitrate decomposition. *J. Phys. Chem. Solids* **2024**, *185*, 111784. [CrossRef]
154. Shopska, M.; Paneva, D.; Kolev, H.; Kadinov, G.; Briančin, J.; Fabián, M.; Cherkezova-Zheleva, Z.; Mitov, I. Characterization and Catalytic Activity in CO Oxidation of Biogenic Lepidocrocite Layered on Anodic Alumina. *Catal. Today* **2020**, *357*, 436–441. [CrossRef]
155. Guyodo, Y.; Bonville, P.; Till, J.L.; Ona-Nguema, G.; Lagroix, F.; Menguy, N. Constraining the Origins of the Magnetism of Lepidocrocite (γ-FeOOH): A Mössbauer and Magnetization Study. *Front. Earth Sci.* **2016**, *4*, 28. [CrossRef]
156. Zhukhlistov, A.P. Crystal Structure of Lepidocrocite FeO(OH) from the Electron-Diffractometry Data. *Crystallogr. Rep.* **2001**, *46*, 730–733. [CrossRef]
157. Hirt, A.M.; Lanci, L.; Dobson, J.; Weidler, P.; Gehring, A.U. Low-temperature Magnetic Properties of Lepidocrocite. *J. Geophys. Res. Solid. Earth* **2002**, *107*, EPM-5. [CrossRef]
158. Zhong, D.; Feng, W.; Ma, W.; Liu, X.; Ma, J.; Zhou, Z.; Du, X.; He, F. Goethite and Lepidocrocite Catalyzing Different Double-Oxidant Systems to Degrade Chlorophenol. *Environ. Sci. Pollut. Res.* **2022**, *29*, 72764–72776. [CrossRef]
159. Matta, R.; Hanna, K.; Chiron, S. Fenton-like Oxidation of 2,4,6-Trinitrotoluene Using Different Iron Minerals. *Sci. Total Environ.* **2007**, *385*, 242–251. [CrossRef] [PubMed]
160. Chou, S.; Huang, C. Application of a Supported Iron Oxyhydroxide Catalyst in Oxidation of Benzoic Acid by Hydrogen Peroxide. *Chemosphere* **1999**, *38*, 2719–2731. [CrossRef]
161. Qin, M.; Lu, B.; Feng, S.; Zhen, Z.; Chen, R.; Liu, H. Role of Exposed Facets and Surface OH Groups in the Fenton-like Reactivity of Lepidocrocite Catalyst. *Chemosphere* **2019**, *230*, 286–293. [CrossRef]
162. Chen, R.; Zhao, S.; Liu, H.; Song, X.; Wei, Y. Preparation and Photocatalytic Activity of Lepidocrocites Obtained Byphotocatalytic Oxidation of Fe(II) in the Presence of Citric Acid. *J. Photochem. Photobiol. A Chem.* **2015**, *312*, 73–80. [CrossRef]
163. Lin, Y.; Wei, Y.; Sun, Y. Room-Temperature Synthesis and Photocatalytic Properties of Lepidocrocite by Monowavelength Visible Light Irradiation. *J. Mol. Catal. A Chem.* **2012**, *353–354*, 67–73. [CrossRef]
164. Shi, H.; Liang, H.; Ming, F.; Wang, Z. Efficient Overall Water-Splitting Electrocatalysis Using Lepidocrocite VOOH Hollow Nanospheres. *Angew. Chem.* **2017**, *129*, 588–592. [CrossRef]
165. Wang, L.; Giammar, D.E. Effects of PH, Dissolved Oxygen, and Aqueous Ferrous Iron on the Adsorption of Arsenic to Lepidocrocite. *J. Colloid. Interface Sci.* **2015**, *448*, 331–338. [CrossRef]
166. He, F.; Zhong, D.; Ma, W.; Yuan, Y.; Li, K.; Dai, C. Activation of the Combined Hydrogen Peroxide and Peroxymonosulphate by Lepidocrocite for Chloramphenicol Removal: Kinetics and Mechanisms. *Environ. Technol.* **2023**, *44*, 1936–1946. [CrossRef]
167. Liu, H.; Li, X.; Zhang, X.; Coulon, F.; Wang, C. Harnessing the Power of Natural Minerals: A Comprehensive Review of Their Application as Heterogeneous Catalysts in Advanced Oxidation Processes for Organic Pollutant Degradation. *Chemosphere* **2023**, *337*, 139404. [CrossRef] [PubMed]
168. Huang, W.; Yuan, Y.; Zhong, D.; Zhang, P.; Liangdy, A.; Lim, T.-T.; Ma, W.; Yuan, Y. Catalytic Activity of H_2O_2 by Goethite and Lepidocrocite: Insight from 5-Bromosalicylic Acid Removal Mechanism and Density Functional Theory Calculation (ID:CHEM114760). *Chemosphere* **2023**, *329*, 138551. [CrossRef] [PubMed]
169. Childs, C.W. Ferrihydrite: A Review of Structure, Properties and Occurrence in Relation to Soils. *Z. Pflanzenernährung Bodenkd.* **1992**, *155*, 441–448. [CrossRef]
170. Cismasu, A.C.; Michel, F.M.; Tcaciuc, A.P.; Tyliszczak, T.; Brown, J.G.E. Composition and Structural Aspects of Naturally Occurring Ferrihydrite. *Comptes Rendus. Géoscience* **2011**, *343*, 210–218. [CrossRef]

171. Fleischer, M.; Chao, G.Y.; Cabri, L.J. New Minerals Names. *Am. Mineral. J. Earth Planet. Mater.* **1975**, *60*, 736–739.
172. Eggleton, R.A.; Fitzpatrick, R.W. New Data and a Revised Structural Model for Ferrihydrite. *Clays Clay Min.* **1988**, *36*, 111–124. [CrossRef]
173. Cardile, C.M. Tetrahedral Fe^{3+} in Ferrihydrite: ^{57} Fe Mössbauer Spectroscopic Evidence. *Clays Clay Min.* **1988**, *36*, 537–539. [CrossRef]
174. Towe, K.M.; Bradley, W.F. Mineralogical Constitution of Colloidal "Hydrous Ferric Oxides". *J. Colloid. Interface Sci.* **1967**, *24*, 384–392. [CrossRef]
175. Russell, J.D. Infrared Spectroscopy of Ferrihydrite: Evidence for the Presence of Structural Hydroxyl Groups. *Clay Min.* **1979**, *14*, 109–114. [CrossRef]
176. Barrón, V.; Torrent, J. Diffuse Reflectance Spectroscopy of Iron Oxides. *Encycl. Surf. Colloid. Sci.* **2002**, *1*, 1438–1446.
177. Hiemstra, T. Surface and Mineral Structure of Ferrihydrite. *Geochim. Cosmochim. Acta* **2013**, *105*, 316–325. [CrossRef]
178. Funnell, N.P.; Fulford, M.F.; Inoué, S.; Kletetschka, K.; Michel, F.M.; Goodwin, A.L. Nanocomposite Structure of Two-Line Ferrihydrite Powder from Total Scattering. *Commun. Chem.* **2020**, *3*, 22. [CrossRef] [PubMed]
179. Boily, J.-F.; Song, X. Direct Identification of Reaction Sites on Ferrihydrite. *Commun. Chem.* **2020**, *3*, 79. [CrossRef] [PubMed]
180. Wang, X.; Zhu, M.; Koopal, L.K.; Li, W.; Xu, W.; Liu, F.; Zhang, J.; Liu, Q.; Feng, X.; Sparks, D.L. Effects of Crystallite Size on the Structure and Magnetism of Ferrihydrite. *Environ. Sci. Nano* **2016**, *3*, 190–202. [CrossRef]
181. Zhu, Y.; Xie, Q.; Deng, F.; Ni, Z.; Lin, Q.; Cheng, L.; Chen, X.; Qiu, R.; Zhu, R. The Differences in Heterogeneous Fenton Catalytic Performance and Mechanism of Various Iron Minerals and Their Influencing Factors: A Review. *Sep. Purif. Technol.* **2023**, *325*, 124702. [CrossRef]
182. Seomoon, K. On-Line GC and GC–MS Analyses of the Fischer–Tropsch Products Synthesized Using Ferrihydrite Catalyst. *J. Ind. Eng. Chem.* **2013**, *19*, 2108–2114. [CrossRef]
183. Bali, S.; Bali, G.; Huggins, F.E.; Seehra, M.S.; Singh, V.; Hancock, J.M.; Harrison, R.; Huffman, G.P.; Pugmire, R.J.; Ernst, R.D.; et al. Synthetic Doped Amorphous Ferrihydrite for the Fischer–Tropsch Synthesis of Alternative Fuels. *Ind. Eng. Chem. Res.* **2012**, *51*, 4515–4522. [CrossRef]
184. Trivedi, P.; Dyer, J.A.; Sparks, D.L. Lead Sorption onto Ferrihydrite. 1. A Macroscopic and Spectroscopic Assessment. *Environ. Sci. Technol.* **2003**, *37*, 908–914. [CrossRef]
185. Scheinost, A.C.; Abend, S.; Pandya, K.I.; Sparks, D.L. Kinetic Controls on Cu and Pb Sorption by Ferrihydrite. *Environ. Sci. Technol.* **2001**, *35*, 1090–1096. [CrossRef]
186. Brinza, L.; Vu, H.P.; Neamtu, M.; Benning, L.G. Experimental and Simulation Results of the Adsorption of Mo and V onto Ferrihydrite. *Sci. Rep.* **2019**, *9*, 1365. [CrossRef]
187. Johnston, C.P.; Chrysochoou, M. Mechanisms of Chromate, Selenate, and Sulfate Adsorption on Al-Substituted Ferrihydrite: Implications for Ferrihydrite Surface Structure and Reactivity. *Environ. Sci. Technol.* **2016**, *50*, 3589–3596. [CrossRef] [PubMed]
188. Mamun, A.A.; Morita, M.; Matsuoka, M.; Tokoro, C. Sorption Mechanisms of Chromate with Coprecipitated Ferrihydrite in Aqueous Solution. *J. Hazard. Mater.* **2017**, *334*, 142–149. [CrossRef] [PubMed]
189. Trivedi, P.; Dyer, J.A.; Sparks, D.L.; Pandya, K. Mechanistic and Thermodynamic Interpretations of Zinc Sorption onto Ferrihydrite. *J. Colloid. Interface Sci.* **2004**, *270*, 77–85. [CrossRef]
190. Sajih, M.; Bryan, N.D.; Livens, F.R.; Vaughan, D.J.; Descostes, M.; Phrommavanh, V.; Nos, J.; Morris, K. Adsorption of Radium and Barium on Goethite and Ferrihydrite: A Kinetic and Surface Complexation Modelling Study. *Geochim. Cosmochim. Acta* **2014**, *146*, 150–163. [CrossRef]
191. Rojo, I.; Seco, F.; Rovira, M.; Giménez, J.; Cervantes, G.; Martí, V.; de Pablo, J. Thorium Sorption onto Magnetite and Ferrihydrite in Acidic Conditions. *J. Nucl. Mater.* **2009**, *385*, 474–478. [CrossRef]
192. Smith, K.F.; Morris, K.; Law, G.T.W.; Winstanley, E.H.; Livens, F.R.; Weatherill, J.S.; Abrahamsen-Mills, L.G.; Bryan, N.D.; Mosselmans, J.F.W.; Cibin, G.; et al. Plutonium(IV) Sorption during Ferrihydrite Nanoparticle Formation. *ACS Earth Space Chem.* **2019**, *3*, 2437–2442. [CrossRef]
193. Winstanley, E.H.; Morris, K.; Abrahamsen-Mills, L.G.; Blackham, R.; Shaw, S. U(VI) Sorption during Ferrihydrite Formation: Underpinning Radioactive Effluent Treatment. *J. Hazard. Mater.* **2019**, *366*, 98–104. [CrossRef]
194. Mendez, J.C.; Hiemstra, T. Carbonate Adsorption to Ferrihydrite: Competitive Interaction with Phosphate for Use in Soil Systems. *ACS Earth Space Chem.* **2019**, *3*, 129–141. [CrossRef]
195. Tiberg, C.; Gustafsson, J.P. Phosphate Effects on Cadmium(II) Sorption to Ferrihydrite. *J. Colloid. Interface Sci.* **2016**, *471*, 103–111. [CrossRef]
196. Zhu, J.; Pigna, M.; Cozzolino, V.; Caporale, A.G.; Violante, A. Sorption of Arsenite and Arsenate on Ferrihydrite: Effect of Organic and Inorganic Ligands. *J. Hazard. Mater.* **2011**, *189*, 564–571. [CrossRef]
197. Mao, H.; Shu, J.; Fei, Y.; Hu, J.; Hemley, R.J. The Wüstite Enigma. *Phys. Earth Planet. Inter.* **1996**, *96*, 135–145. [CrossRef]
198. Schrettle, F.; Kant, C.; Lunkenheimer, P.; Mayr, F.; Deisenhofer, J.; Loidl, A. Wüstite: Electric, Thermodynamic and Optical Properties of FeO. *Eur. Phys. J. B* **2012**, *85*, 164. [CrossRef]

199. Yin, M.; Chen, Z.; Deegan, B.; O'Brien, S. Wüstite Nanocrystals: Synthesis, Structure and Superlattice Formation. *J. Mater. Res.* **2007**, *22*, 1987–1995. [CrossRef]
200. Fjellvag, H.; Hauback, B.; Vogt, T.; Stolen, S. Monoclinic Nearly Stoichiometric Wüstite at Low Temperatures. *Am. Mineral.* **2002**, *87*, 347–349. [CrossRef]
201. Liu, H.; Han, W.; Huo, C.; Cen, Y. Development and Application of Wüstite-Based Ammonia Synthesis Catalysts. *Catal. Today* **2020**, *355*, 110–127. [CrossRef]
202. Humphreys, J.; Lan, R.; Tao, S. Development and Recent Progress on Ammonia Synthesis Catalysts for Haber–Bosch Process. *Adv. Energy Sustain. Res.* **2021**, *2*, 2000043. [CrossRef]
203. Pernicone, N. Wustite as a New Precursor of Industrial Ammonia Synthesis Catalysts. *Appl. Catal. A Gen.* **2003**, *251*, 121–129. [CrossRef]
204. Rossetti, I.; Pernicone, N.; Forni, L. Promoters Effect in Ru/C Ammonia Synthesis Catalyst. *Appl. Catal. A Gen.* **2001**, *208*, 271–278. [CrossRef]
205. Rakshit, S.; Matocha, C.J.; Haszler, G.R. Nitrate Reduction in the Presence of Wüstite. *J. Environ. Qual.* **2005**, *34*, 1286–1292. [CrossRef]
206. Poza-Nogueiras, V.; Rosales, E.; Pazos, M.; Sanromán, M.Á. Current Advances and Trends in Electro-Fenton Process Using Heterogeneous Catalysts—A Review. *Chemosphere* **2018**, *201*, 399–416. [CrossRef]
207. Expoósito, E.; Saánchez-Saánchez, C.M.; Montiel, V. Mineral Iron Oxides as Iron Source in Electro-Fenton and Photoelectro-Fenton Mineralization Processes. *J. Electrochem. Soc.* **2007**, *154*, E116. [CrossRef]
208. Ren, J.; Yao, Z.; Wei, Q.; Wang, R.; Wang, L.; Liu, Y.; Ren, Z.; Guo, H.; Niu, Z.; Wang, J.; et al. Catalytic Degradation of Chloramphenicol by Water Falling Film Dielectric Barrier Discharge and FeO Catalyst. *Sep. Purif. Technol.* **2022**, *290*, 120826. [CrossRef]
209. Mackay, A.L. β-Ferric Oxyhydroxide—Akaganéite. *Mineral. Mag. J. Mineral. Soc.* **1962**, *33*, 270–280. [CrossRef]
210. Jegdić, B.V.; Ristić, S.S.; Polić-Radovanović, S.R.; Alil, A.B. Corrosion and Conservation of Weapons and Military Equipment. *Vojnoteh. Glas.* **2012**, *60*, 169–182. [CrossRef]
211. Holm, N.G.; Dowler, M.J.; Wadsten, T.; Arrhenius, G. β-FeOOH · Cln (Akaganéite) and Fe1-XO (Wüstite) in Hot Brine from the Atlantis II Deep (Red Sea) and the Uptake of Amino Acids by Synthetic β-FeOOH · Cln. *Geochim. Cosmochim. Acta* **1983**, *47*, 1465–1470. [CrossRef]
212. Post, J.F.; Buchwald, V.F. Crystal Structure Refinement of Akaganeite. *Am. Mineral.* **1991**, *76*, 272–277.
213. Tadic, M.; Milosevic, I.; Kralj, S.; Saboungi, M.-L.; Motte, L. Ferromagnetic Behavior and Exchange Bias Effect in Akaganeite Nanorods. *Appl. Phys. Lett.* **2015**, *106*, 183706. [CrossRef]
214. Zhao, J.; Lin, W.; Chang, Q.; Li, W.; Lai, Y. Adsorptive Characteristics of Akaganeite and Its Environmental Applications: A Review. *Environ. Technol. Rev.* **2012**, *1*, 114–126. [CrossRef]
215. Marsac, R.; Martin, S.; Boily, J.-F.; Hanna, K. Oxolinic Acid Binding at Goethite and Akaganéite Surfaces: Experimental Study and Modeling. *Environ. Sci. Technol.* **2016**, *50*, 660–668. [CrossRef]
216. Xiong, H.; Shen, X.; Cui, C.; Cheng, L.; Xu, Y. Synthesis of Nanospindle-Akaganéite and Its Photocatalytic Degradation for Methyl Orange. *Water Sci. Technol.* **2020**, *82*, 481–491. [CrossRef]
217. Xiong, H.; Shi, K.; Han, J.; Cui, C.; Liu, Y.; Zhang, B. Synthesis of β-FeOOH/Polyaniline Heterogeneous Catalyst for Efficient Photo-Fenton Degradation of AOII Dye. *Environ. Sci. Pollut. Res.* **2023**, *30*, 59366–59381. [CrossRef] [PubMed]
218. Su, S.; Liu, Y.; Liu, X.; Jin, W.; Zhao, Y. Transformation Pathway and Degradation Mechanism of Methylene Blue through β-FeOOH@GO Catalyzed Photo-Fenton-like System. *Chemosphere* **2019**, *218*, 83–92. [CrossRef] [PubMed]
219. Majzlan, J.; Koch, C.B.; Navrotsky, A. Thermodynamic Properties of Feroxyhyte (Δ′-FeOOH). *Clays Clay Min.* **2008**, *56*, 526–530. [CrossRef]
220. Vodyanitskii, Y.N. Iron Hydroxides in Soils: A Review of Publications. *Eurasian Soil. Sci.* **2010**, *43*, 1244–1254. [CrossRef]
221. Li, Y.; Fu, F.; Cai, W.; Tang, B. Synergistic Effect of Mesoporous Feroxyhyte Nanoparticles and Fe(II) on Phosphate Immobilization: Adsorption and Chemical Precipitation. *Powder Technol.* **2019**, *345*, 786–795. [CrossRef]
222. Xu, Z.; Yu, Y.; Fang, D.; Xu, J.; Liang, J.; Zhou, L. Microwave–ultrasound assisted synthesis of β-FeOOH and its catalytic property in a photo-Fenton-like process. *Ultrason. Sonochem.* **2015**, *27*, 287–295. [CrossRef]
223. Liu, B.; Wang, Y.; Peng, H.Q.; Yang, R.; Jiang, Z.; Zhou, X.; Lee, C.-S.; Zhao, H.; Zhang, W. Iron vacancies induced bifunctionality in ultrathin feroxyhyte nanosheets for overall water splitting. *Adv. Mater.* **2018**, *30*, 1803144. [CrossRef]
224. Kokkinos, E.; Soukakos, K.; Kostoglou, M.; Mitrakas, M. Cadmium, mercury, and nickel adsorption by tetravalent manganese feroxyhyte: Selectivity, kinetic modeling, and thermodynamic study. *Environ. Sci. Pollut. Res.* **2018**, *25*, 12263–12273. [CrossRef]
225. Du, W.; Xu, Y.; Wang, Y. Photoinduced Degradation of Orange II on Different Iron (Hydr)Oxides in Aqueous Suspension: Rate Enhancement on Addition of Hydrogen Peroxide, Silver Nitrate, and Sodium Fluoride. *Langmuir* **2008**, *24*, 175–181. [CrossRef]

226. Pinto, I.S.X.; Pacheco, P.H.V.V.; Coelho, J.V.; Lorençon, E.; Ardisson, J.D.; Fabris, J.D.; de Souza, P.P.; Krambrock, K.W.H.; Oliveira, L.C.A.; Pereira, M.C. Nanostructured δ-FeOOH: An Efficient Fenton-like Catalyst for the Oxidation of Organics in Water. *Appl. Catal. B* **2012**, *119–120*, 175–182. [CrossRef]
227. Li, J.; Ding, Y.; Chen, K.; Li, Z.; Yang, H.; Yue, S.; Tang, Y.; Wang, Q. δ-FeOOH Coupled BiOBr0.5I0.5 for Efficient Photocatalysis-Fenton Synergistic Degradation of Organic Pollutants. *J. Alloys Compd.* **2022**, *903*, 163795. [CrossRef]
228. Wang, C.; Shi, P.; Wang, Z.; Guo, R.; You, J.; Zhang, H. Efficient Wastewater Disinfection through FeOOH-Mediated Photo-Fenton Reaction: A Review. *J. Environ. Chem. Eng.* **2023**, *11*, 111269. [CrossRef]
229. Pereira, M.C.; Garcia, E.M.; Cândido Da Silva, A.; Lorenon, E.; Ardisson, J.D.; Murad, E.; Fabris, J.D.; Matencio, T.; De Castro Ramalho, T.; Rocha, M.V.J. Nanostructured δ-FeOOH: A Novel Photocatalyst for Water Splitting. *J. Mater. Chem.* **2011**, *21*, 10280–10282. [CrossRef]
230. da Silva Rocha, T.; Nascimento, E.S.; da Silva, A.C.; dos Santos Oliveira, H.; Garcia, E.M.; de Oliveira, L.C.A.; Monteiro, D.S.; Rodriguez, M.; Pereira, M.C. Enhanced Photocatalytic Hydrogen Generation from Water by $Ni(OH)_2$ Loaded on Ni-Doped δ-FeOOH Nanoparticles Obtained by One-Step Synthesis. *RSC Adv.* **2013**, *3*, 20308. [CrossRef]
231. Hu, J.; Lo, I.; Chen, G. Performance and Mechanism of Chromate (VI) Adsorption by δ-FeOOH-Coated Maghemite (γ-Fe_2O_3) Nanoparticles. *Sep. Purif. Technol.* **2007**, *58*, 76–82. [CrossRef]
232. Tresintsi, S.; Simeonidis, K.; Estradé, S.; Martinez-Boubeta, C.; Vourlias, G.; Pinakidou, F.; Katsikini, M.; Paloura, E.C.; Stavropoulos, G.; Mitrakas, M. Tetravalent Manganese Feroxyhyte: A Novel Nanoadsorbent Equally Selective for As(III) and As(V) Removal from Drinking Water. *Environ. Sci. Technol.* **2013**, *47*, 9699–9705. [CrossRef]
233. Pinakidou, F.; Katsikini, M.; Paloura, E.C.; Simeonidis, K.; Mitraka, E.; Mitrakas, M. Monitoring the Role of Mn and Fe in the As-Removal Efficiency of Tetravalent Manganese Feroxyhyte Nanoparticles from Drinking Water: An X-Ray Absorption Spectroscopy Study. *J. Colloid. Interface Sci.* **2016**, *477*, 148–155. [CrossRef]
234. Kokkinos, E.; Kellartzis, I.; Diamantopoulou, I.; Stavropoulos, G.; Vourlias, G.; Mitrakas, M. Study of Elemental Mercury Removal from Flue Gases Using Tetravalent Manganese Feroxyhyte. *Chem. Eng. J.* **2017**, *315*, 152–158. [CrossRef]
235. Chagas, P.; Da Silva, A.C.; Passamani, E.C.; Ardisson, J.D.; De Oliveira, L.C.A.; Fabris, J.D.; Paniago, R.M.; Monteiro, D.S.; Pereira, M.C. δ-FeOOH: A Superparamagnetic Material for Controlled Heat Release under AC Magnetic Field. *J. Nanoparticle Res.* **2013**, *15*, 1544. [CrossRef]
236. Lacerda, L.C.T.; Pires, M.S.; Oliveira, I.S.S.; Silva, T.C.; de Castro, A.A.; Corrêa, S.; Vaiss, V.S.; Ramalho, T.C. Bulk and Surface Theoretical Investigation of Nb-Doped δ-FeOOH as a Promising Bifunctional Catalyst. *J. Mol. Model.* **2021**, *27*, 249. [CrossRef]
237. Kolitsch, U. Bernalite from the Clara Mine, Germany, and the Incorporation of Tungsten in Minerals Containing Ferric Iron. *Can. Mineral.* **1998**, *36*, 1211–1216.
238. McCammon, C.A.; Pring, A.; Keppler, H.; Sharp, T. A Study of Bernalite, $Fe(OH)_3$, Using Mössbauer Spectroscopy, Optical Spectroscopy and Transmission Electron Microscopy. *Phys. Chem. Min.* **1995**, *22*, 11–20. [CrossRef]
239. Welch, M.D.; Crichton, W.A.; Ross, N.L. Compression of the Perovskite-Related Mineral Bernalite $Fe(OH)_3$ to 9 GPa and a Reappraisal of Its Structure. *Min. Mag.* **2005**, *69*, 309–315. [CrossRef]
240. McCammon, C.A.; De Grave, E.; Pring, A. The Magnetic Structure of Bernalite, $Fe(OH)_3$. *J. Magn. Magn. Mater.* **1996**, *152*, 33–39. [CrossRef]
241. Han, J.; Ro, H.-M. Identification of Bernalite Transformation and Tridentate Arsenate Complex at Nano-Goethite under Effects of Drying, PH and Surface Loading. *Sci. Rep.* **2018**, *8*, 8369. [CrossRef]
242. Han, J.; Ro, H.-M. Interpreting competitive adsorption of arsenate and phosphate on nanosized iron (hydr)oxides: Effects of pH and surface loading. *Environ. Sci. Pollut. Res.* **2018**, *25*, 28572–28582. [CrossRef]
243. Su-Gallegos, J.; Magallón-Cacho, L.; Borja-Arco, E.; García-Valdés, J.; Sebastian, P.J. Oxygen Reduction and Hydrogen Oxidation Reactions on Ru-Fe Electrocatalyst Synthesized by a Microwave-Assisted Synthesis. *Mater. Res. Express* **2020**, *7*, 035505. [CrossRef]
244. Nur'aeni; Chae, A.; Jo, S.; Choi, Y.; Park, B.; Park, S.Y.; In, I. Synthesis of β-FeOOH/Fe_3O_4 Hybrid Photocatalyst Using Catechol-Quaternized Poly(N-Vinyl Pyrrolidone) as a Double-Sided Molecular Tape. *J. Mater. Sci.* **2017**, *52*, 8493–8501. [CrossRef]
245. Sciscenko, I.; Mestre, S.; Climent, J.; Valero, F.; Escudero-Oñate, C.; Oller, I.; Arques, A. Magnetic Photocatalyst for Wastewater Tertiary Treatment at Pilot Plant Scale: Disinfection and Enrofloxacin Abatement. *Water* **2021**, *13*, 329. [CrossRef]
246. Mahy, J.G.; Wolfs, C.; Vreuls, C.; Drot, S.; Dircks, S.; Boergers, A.; Tuerk, J.; Hermans, S.; Lambert, S.D. Advanced Oxidation Processes for Waste Water Treatment: From Laboratory-Scale Model Water to on-Site Real Waste Water. *Environ. Technol.* **2021**, *42*, 3974–3986. [CrossRef]

catalysts

MDPI

Article

Enhanced Photocatalytic Degradation of Herbicide 2,4-Dichlorophenoxyacetic Acid Using Sulfated CeO_2

Carlos Rodríguez [1], Claudia Castañeda [1,*], Edwin Sosa [1], José J. Martínez [1], Sonia Mancipe [1], Hugo Rojas [1], Francisco Tzompantzi [2] and Ricardo Gómez [2]

1 Grupo de Catálisis, Escuela de Ciencias Químicas, Universidad Pedagógica y Tecnológica de Colombia UPTC, Avenida Central del Norte 39-115, Tunja 150003, Boyacá, Colombia; carlos.rodriguez21@uptc.edu.co (C.R.); edwin.sosa01@uptc.edu.co (E.S.); jose.martinez@uptc.edu.co (J.J.M.); sonia.mancipe@uptc.edu.co (S.M.); hugo.rojas@uptc.edu.co (H.R.)

2 Grupo de Ecocatálisis, Universidad Autónoma Metropolitana-Iztapalapa, Av. San Rafael Atlixco No. 186, Ciudad de México 09340, Mexico; fjtz@xanum.uam.mx (F.T.); gomr@xanum.uam.mx (R.G.)

* Correspondence: claudia.castaneda@uptc.edu.co

Abstract: The present study presents the results obtained from evaluating the photocatalytic behavior of a series of sulfated CeO_2 materials in the photocatalytic degradation of the herbicide 2,4-dichlorophenoxyacetic acid. The CeO_2 photocatalytic support was prepared using the precipitation synthesis method. Subsequently, the support was wetly impregnated with different contents of sulfate ions (0.5, 1.0, and 2.0 wt.%). The materials were characterized using X-ray diffraction, nitrogen physisorption, infrared spectroscopy, diffuse reflectance UV–Vis spectrophotometry, and thermal analysis. The characterization results showed that the sulfation of the material promoted an increase in the surface area and a decrease in the average size of the crystallites. Likewise, it was possible to demonstrate the surface sulfation of the support through bidentate coordination of the sulfate groups to the semiconductor metal. Concerning photoactivity, the convenience of the surface modification of CeO_2 was confirmed because the sulfate groups acted as capturers of the electrons generated during the photocatalytic process, reducing the frequency of recombination of the charge carriers and allowing the availability of the gaps to favor the degradation reaction of the contaminant. Finally, it was evident that a percentage of 1.0 wt.% of the sulfate anion is the optimal content to improve the photocatalytic properties of CeO_2.

Keywords: photocatalysis; degradation; 2,4-dichlorophenoxyacetic acid; sulfated CeO_2

1. Introduction

The frequent use of herbicides for weed control and to improve agricultural production processes represents an important environmental challenge, due to the resistance of these products to conventional degradation, associated with factors such as their chemical structure, limited biodegradability, microbial resistance, and adverse environmental conditions [1]. A herbicide highly recognized for its extensive use is 2,4-dichlorophenoxyacetic acid (2,4-D), classified as an auxin herbicide due to its antitumor, antimicrobial, and inhibitory properties towards the growth and development of weeds. Still, they are harmful to human health, because they can be absorbed through the skin, or inhaled, which can cause damage to the liver, kidneys, muscles, and brain tissue.

Recently, some techniques have been used, such as adsorption on activated carbon [2], flocculation [3], and ultrafiltration [4] with porous resins or biological treatments [5], to eliminate contaminants present in water sources. However, applying advanced oxidation processes (AOPs), such as heterogeneous photocatalysis, is considered an attractive, sustainable, and versatile strategy to remove organic contaminants at low concentrations, potentially leading to their total mineralization. The photocatalytic process requires a photocatalytic material, generally a semiconductor solid, which is activated by irradiation, generating highly reactive charge carriers (electron–hole pairs). The released electrons can participate in reduction reactions. At the same time, the holes can react with water to form highly oxidizing hydroxyl radicals (•OH) capable of decomposing organic contaminants present in water, converting them into less harmful products [6].

Thus, one of the main challenges for researchers in photocatalysis is related to the development of materials with adequate physicochemical characteristics that favor activity in reactions promoted via activation of the material with radiation. In this sense, studies on the textural, structural, and morphological modification of materials have experienced significant growth in recent years, given that these modifications can have a significant impact on photocatalytic behavior [7]. Recent research indicates that rare earth oxides are high-performance materials in catalytic and photocatalytic reactions. Among these solids, CeO_2 has been successfully tested in photocatalysis, due to its chemical stability, non-toxicity, biocompatibility [8,9], strong absorption in the ultraviolet region (typical absorption edge between 385–400 nm), and the strong redox potential of the Ce^{4+}/Ce^{3+} pair that promotes the separation of the photogenerated electron–hole pairs [10]. Regarding this last aspect, in the electron transfer process, the Ce^{4+}/Ce^{3+} pair acts as an electron trap; thus, the Ce^{4+} species can accept an electron and be reduced to Ce^{3+}. Subsequently, the Ce^{3+} species can return to the oxidized state by releasing the electron, which is transferred to the substrates adsorbed on the semiconductor [11,12].

However, recent studies confirm the convenience of the surface modification of CeO_2 with anionic species such as phosphates, reflected in an improvement in the physicochemical and photocatalytic properties of the semiconductor in degradation reactions of phenolic contaminants [6]. This behavior is associated with the phosphate species bound to the surface of the support in bidentate mode, which could favor a decrease in the recombination frequency of the charge carriers.

To improve the physicochemical properties and effectiveness of CeO_2 for photocatalytic applications, it is interesting to evaluate the effects produced by surface modification through sulfation processes. The use of the sulfation route to modify catalysts has been shown to improve photocatalytic properties, which results in a highly effective removal of organic compounds and dyes [13–15].

In this way, we intend to verify the potential use of the series of sulfated ceria materials in photocatalytic applications, specifically in environmental remediation processes, through the degradation of herbicide 2,4-dichlorophenoxyacetic acid, which is characterized by a long half-life, which allows enough time for it to be transferred to surface and groundwater and create kidney and liver problems in animals and humans.

According to the above, the main aim of this work is to present the results of the synthesis, physicochemical characterization, and evaluation of the photocatalytic behavior of a series of materials of sulfated CeO_2 with different contents of anion (0.5, 1.0, and 2.0 wt.%) during the degradation of 2,4-D under UV irradiation.

2. Results and Discussion

2.1. Characterization of the Photocatalysts

The crystallinity of the CeO_2 support and the sulfated materials was determined through XRD analysis, as depicted in Figure 1. Characteristic X-ray diffraction signals of cerium dioxide were detected at 2θ angles of 28.2°, 32.4°, 47.0°, 55.9°, 58.5°, 69.0°, 76.5°, and 78.8°. These signals correspond to the pure cubic fluorite structure characteristic of CeO_2 (JCPDS 34-0394).

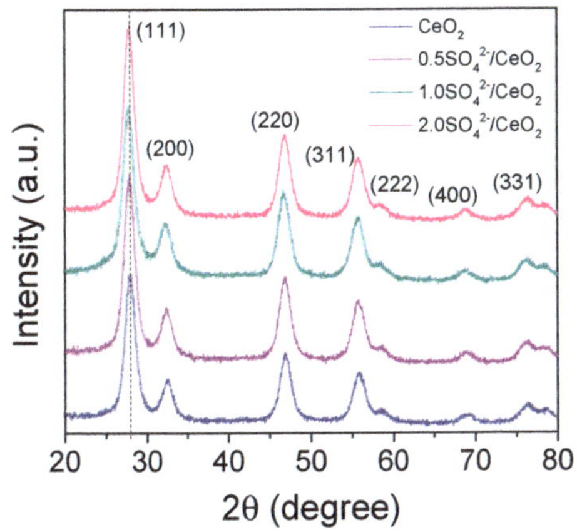

Figure 1. X-ray diffraction results of the materials studied.

Furthermore, as shown in Figure 1, no shifts in diffraction signals are observed in the sulfated materials compared to the photocatalytic support. This could suggest that the bulk structure of CeO_2 does not change during the sulfation procedure. It is crucial to note that substituting Ce^{4+} ions (with an ionic radius of 0.092 nm) with S^{6+} ions (with an ionic radius of 0.029 nm), which could be originated from the sulfuric acid used in the impregnation process of the support, would probably promote shifts in the diffraction signal towards higher angles, suggesting lattice contraction. Consequently, to corroborate the surface modification of the CeO_2 support, lattice parameter values were calculated and included in Table 1. As observed, there is no change in the edges, confirming that no substitutions of oxygen or cerium by sulfide ions have occurred.

Table 1. Physicochemical properties of the materials studied.

Photocatalyst	Crystallite Size (nm)	Lattice Parameter *a* (nm)	S_{BET} (m^2/g)	Pore Size (Å)	Eg (eV)
CeO_2	6.2	0.552	146	46	3.00
$0.5SO_4^{2-}/CeO_2$	5.6	0.554	149	46	3.08
$1.0SO_4^{2-}/CeO_2$	5.2	0.552	158	43	3.08
$2.0SO_4^{2-}/CeO_2$	5.0	0.552	161	45	3.08

Figure 2 presents nitrogen adsorption–desorption isotherms for the CeO_2 support and SO_4^{2-}/CeO_2 photocatalysts. According to the IUPAC classification, typical type IV(a) isotherms were observed in the samples associated with mesoporous solids [6]. A hysteresis loop of type H2 (b) was observed in the isotherms of the support and the materials sulfated with 0.5 and 1.0% anion content, characteristic of materials featuring pores with a narrow distribution in the body and a broad distribution in neck size [16]. However, a detailed observation of the desorption branch in the isotherm corresponding to the $2.0SO_4^{2-}/CeO_2$ material shows a H2 (a) hysteresis loop, indicating a narrow distribution of pore neck widths, resulting in a very pronounced desorption branch, possibly due to the amount of H_2SO_4 and the pH reached during the synthesis of the material [17].

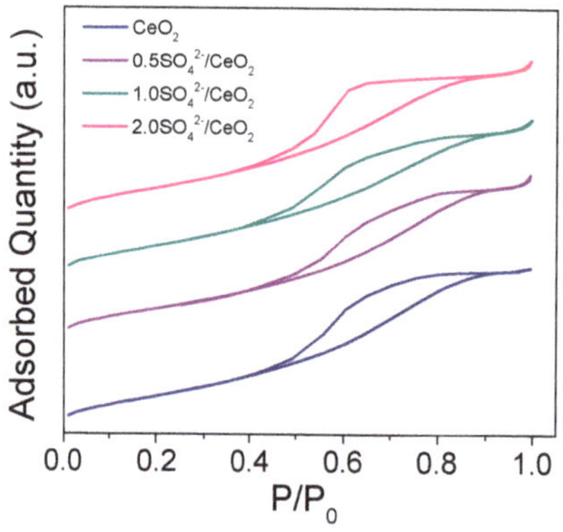

Figure 2. N_2 adsorption–desorption isotherms of the materials studied.

The surface areas and pore size values calculated for the materials studied are listed in Table 1. The modification of the CeO_2 support with sulfate ions promotes an increase in surface areas. Literature reports indicate that sulfated metal oxides, such as TiO_2 [18], ZrO_2 [19], and SnO_2 [20], present a higher surface area compared to the unmodified material. These results are associated with the dispersed sulfate species on the surface, which prevent the agglomeration of metal oxide particles, leading to a reduction in crystallite size and consequently an increase in the surface area [21].

Figure 3 presents the ATR-FTIR spectra of the CeO_2 support and the sulfated materials. ATR-FTIR analysis was conducted to confirm the presence of sulfate groups in the sulfated ceria catalyst and to assess the effect of sulfate ion concentration on the intensity of the band for each functional group of the synthesized catalyst. The CeO_2 support presents a small band around 3720 cm^{-1}, corresponding to isolated hydroxyl groups. At approximately 3360 cm^{-1}, a broadband corresponding to the stretching vibration of the O-H bond of adsorbed water molecules was identified [22]. Additionally, the band near 1634 cm^{-1} is attributed to the bending vibration of the hydroxyl group assigned as chemisorbed water; this band increases in intensity, which could be associated with a more hydroxylated surface as sulfate content increases, in line with the previously described surface area results [23]. The increase in the surface hydroxylation could be associated with the negative charges of the sulfate ions anchored on the surface of CeO_2, which attracts water molecules. It is important to consider that the presence of hydroxylated surfaces is desirable in photocatalytic processes because the surface hydroxyls could experience oxidation reactions with the photogenerated holes in the semiconductor and favor both the decrease in the recombination frequency of the charge carriers and the formation of hydroxyl radicals, which are mainly responsible for the degradation of organic contaminants [24,25].

The stretching vibration of Ce-O-Ce bonds becomes evident in the band at 1320 cm^{-1}, whose intensity decreases with the increase in the sulfate ion content impregnated in the material. The stretching vibration mode of the Ce-O bond of CeO_2 nanoparticles is shown in the band at 610 cm^{-1}. Regarding the sulfated photocatalysts, their spectra reveal a band near 1110 cm^{-1}, corresponding to the vibration of the S=O group of sulfate species, which exhibits a characteristic band between 1222 cm^{-1} and 1100 cm^{-1}. Furthermore, bands between 1057 cm^{-1} and 994 cm^{-1} are attributed to the stretching vibrations in symmetric and asymmetric modes of the S-O bonds [23]. The presence of these bands confirms the sulfation process and indicates the bidentate coordination of the sulfate groups to the metal atom of the support [26,27]. Besides, the decrease in the intensity of the band located at 1320 cm^{-1} mentioned previously could be a consequence of the surface sulfation of the semiconductor, which would inhibit the vibration of the Ce-O-Ce bonds.

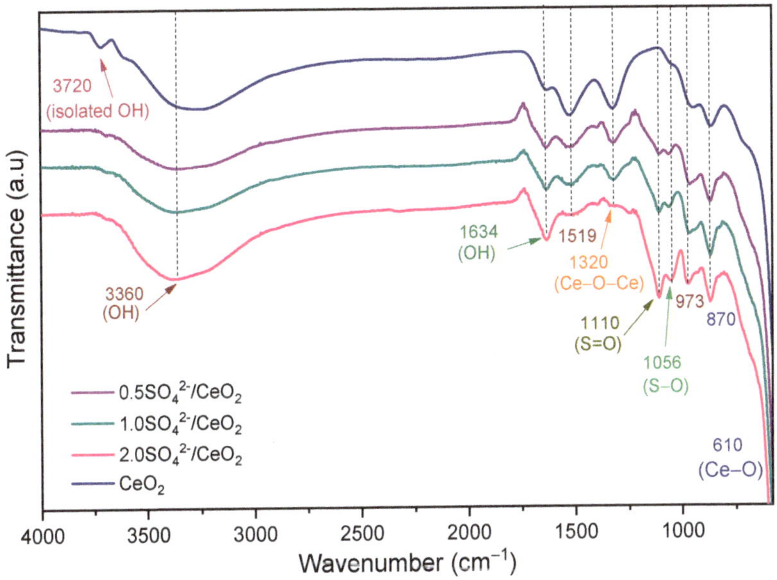

Figure 3. ATR-FTIR spectra of the materials studied.

The UV–Vis spectra of the CeO_2 support and the sulfated materials are shown in Figure 4a. The typical absorption region of the CeO_2 semiconductor was observed around 420 nm. Two bands at 228 and 280 nm, corresponding to the direct charge transition between the Ce^{4+} and O^{2-} states, are evident [28].

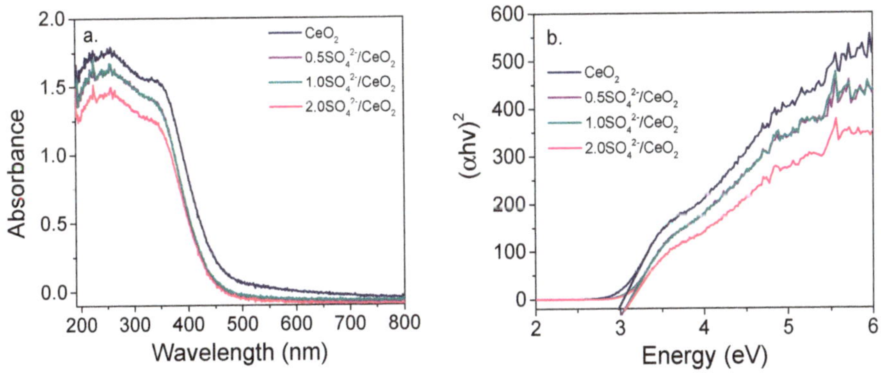

Figure 4. (**a**) UV–Vis spectra and (**b**) Tauc plot of the materials studied.

The Tauc plot as a function of absorbed light energy is depicted in Figure 4b. The values of the bandgap energy (Eg) were calculated through linear fitting with the energy axis and are presented in Table 1. As observed, the modification of the support with sulfate anions promoted a slight blue shift in the UV–Vis spectra, which suggests a slight increase in the value of Eg (CeO_2 3.00 eV and sulfated CeO_2 materials 3.08 eV), which is consistent with the decrease in crystallite size in the sulfated materials, as evidenced by X-ray diffraction analyses. It is widely recognized that the increase in the energy of the band gap with the decrease in crystallite size is associated with the quantum confinement effect (particle sizes less than 10 nm), which favors the presence of discrete energy levels [29,30].

Furthermore, based on these results, the insertion of sulfide ions into the CeO_2 lattice can be ruled out, as such a modification in the crystal structure would lead to a decrease in the energy of the band gap [31]. Thus, the surface sulfation of the semiconductor in the materials evaluated in this study is confirmed. On the other hand, the increase in sulfate ion content on the material surface does not produce significant variations in the values of Eg.

The results of the thermal analysis of the materials coupled with mass spectroscopy (TGA-MS) are shown in Figure 5. The TGA-MS of the uncalcined CeO_2 support is shown in Figure 5a; the mass loss between 50–200 °C is associated with the desorption of adsorbed water and ammonia remaining following the support synthesis [32,33]. CO_2 was released during the various mass-loss steps between 110–250 °C and 310–460 °C. In addition, at temperatures ranging from 200 to 460 °C, the dehydration of water strongly bound to the surface of the nanoparticles occurs, generating a network of water and dehydroxylation of surface OH groups [34]. The mass loss recorded at the end of the process was 19%.

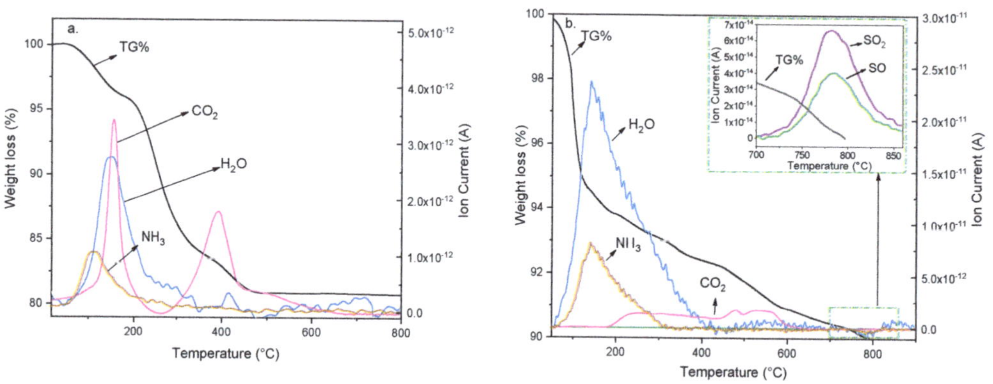

Figure 5. TGA-MS results of (**a**) uncalcined CeO_2 support and (**b**) $1.0SO_4^{2-}/CeO_2$ photocatalyst.

In Figure 5b, the thermogram of the photocatalyst $1.0SO_4^{2-}/CeO_2$ is shown. The removal of adsorbed water from the material is observed at a temperature close to 200 °C, resulting in a mass loss of 6.2%. In addition, the release of residual ammonia following the synthesis was seen. The CO_2 released in this material is significantly lower than in the support considering that the impregnation of the anions occurred on the calcined CeO_2 support. Additionally, another stage in the thermogram is evident, corresponding to the loss of sulfates from the material at higher temperatures (500–800 °C); the SO_4^{2-} ions from the sulfated sample are released in the form of SO_2 and SO [23], which were confirmed by MS analysis. The mass loss recorded at the end of the process was 10%. The difference in total mass loss between the uncalcined CeO_2 support and the sulfated material can be attributed to changes in the chemical composition of the materials, due to the surface modification of the support with sulfate anions carried out on the calcined CeO_2.

The recombination of the photogenerated electron–hole pairs is a critical factor to consider in photocatalysis and likely the most significant one, as it prevents the oxidation–reduction processes from taking place. To understand the behavior of the photogenerated charge carriers in the materials, photoluminescence spectra were obtained for the CeO_2 support and the sulfated solids (Figure 6).

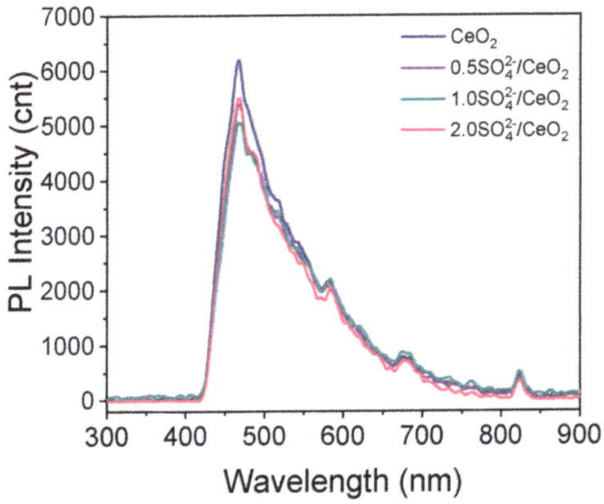

Figure 6. Photoluminescence spectra of the materials studied.

The photocatalytic materials exhibited a fluorescence emission signal at 468 nm when excited using a fixed wavelength of 254 nm. The emission band ranging from 360 nm to 530 nm is attributed to the presence of defects related to the Ce 4f and O 2p states [35,36]. As can be seen, the CeO_2 support showed the signal with the highest intensity, suggesting a greater charge recombination capacity. Conversely, the 1.0 wt.% sulfated material evidenced an emission signal with lower intensity compared to the other materials, which is associated with a lower capacity for electron–hole pair recombination in the semiconductor. This result suggests that the $1.0SO_4{}^{2-}/CeO_2$ material is likely to exhibit the best photocatalytic behavior in the degradation of the herbicide 2,4-D, as it favors a decrease in the charge carrier recombination rate [37].

2.2. Evaluation of the Photocatalytic Activity

The photocatalytic behavior of CeO_2 support and sulfated materials was evaluated in the degradation of the 2,4-dichlorophenoxyacetic acid (2,4-D) herbicide using UV radiation.

Photocatalytic Degradation of 2,4-Dichlorophenoxyacetic Acid

The ultraviolet absorption spectra corresponding to the degradation monitoring of 2,4-dichlorophenoxyacetic acid using the $1.0SO_4{}^{2-}/CeO_2$ photocatalyst are presented in Figure 7a. The absorption bands identified at 203 and 227 nm are associated with the $\pi\rightarrow\pi^*$ transition of the aromatic ring, while the band located at 282 nm is related to $n\rightarrow\pi^*$ transitions [38,39]. As the reaction progresses, a decrease in the intensity of the absorption band is observed, indicating herbicide degradation.

The variation of the relative concentration of 2,4-dichlorophenoxyacetic acid with reaction time is presented in Figure 7b. As observed, conducting the reaction in the absence of a photocatalytic material (photolysis) promotes a 37% degradation of the contaminant. The degradation of the molecule is contingent upon its ability to absorb incident radiation and reach an excited state, leading to bond cleavages. If the molecule does not degrade directly via photolysis, indirect degradation may occur by forming hydroxyl radicals from water [40,41]. In the presence of the photocatalysts, a reduction in the herbicide concentration is observed during the irradiation time. After 180 min of reaction, a degradation of the contaminant of approximately 87%, 87%, 95%, and 88% was achieved using CeO_2, $0.5SO_4{}^{2-}/CeO_2$, $1.0SO_4{}^{2-}/CeO_2$, and $2.0SO_4{}^{2-}/CeO_2$, respectively.

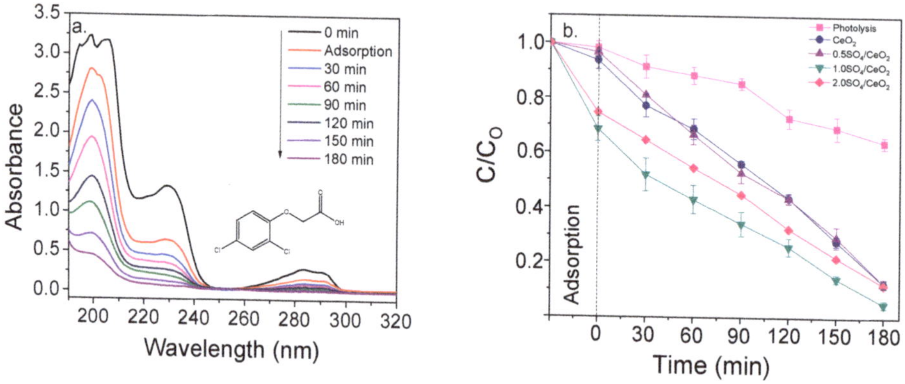

Figure 7. (**a**) UV spectra of the 2,4-D photodegradation using $1.0SO_4^{2-}/CeO_2$ photocatalyst, and (**b**) variation of the relative concentration of 2,4-D as a function of reaction time using materials studied.

As can be seen in Figure 7b, the $1.0SO_4^{2-}/CeO_2$ material shows significant adsorption at 30 min in the dark. To confirm the behavior of the sulfated material with an anion content of 1.0 wt.% in the adsorption process, the herbicide degradation reaction was conducted under the experimental conditions described in the methodology, in the absence of radiation. The results indicated that after 30 min of reaction, the pollutant was adsorbed by the material; however, no further decrease in concentration was observed over time. This suggests that the material has reached its adsorption equilibrium or maximum capacity, and extending the time in the dark may not lead to a significant reduction in herbicide concentration. UV spectra of the 2,4-D adsorption using the $1.0SO_4^{2-}/CeO_2$ photocatalyst and the comparison between the adsorption and the photocatalytic degradation of the herbicide are shown in Figure 8.

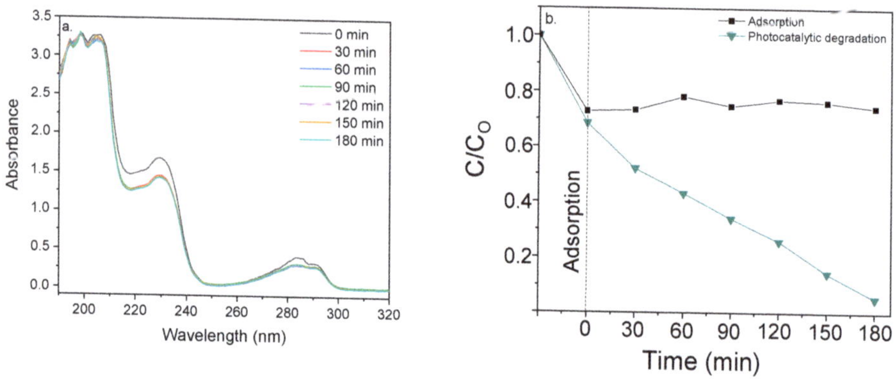

Figure 8. (**a**) UV spectra of the 2,4-D adsorption using $1.0SO_4^{2-}/CeO_2$ photocatalyst. (**b**) Comparison between the adsorption reaction and the photocatalytic degradation of the 2,4-D herbicide.

Based on the results obtained from the evaluation of the photocatalytic behavior of the studied materials, it was observed that the degradation of 2,4-D acid follows a first-order kinetic behavior. The values of the kinetic constants, as well as the half-life time ($t_{1/2}$) and the coefficient of linearization R^2, are detailed in Table 2.

Table 2. Kinetic parameters of the herbicide's photocatalytic degradation using the materials studied.

Photocatalyst	2,4-Dichlorophenoxyacetic Acid		
	k ($\times 10^{-3}$ min^{-1})	$t_{1/2}$ (min)	R^2
CeO_2	5.5	126	0.972
$0.5SO_4{}^{2-}/CeO_2$	5.8	119	0.970
$1.0SO_4{}^{2-}/CeO_2$	8.6	80	0.989
$2.0SO_4{}^{2-}/CeO_2$	6.9	100	0.982

As can be seen, a higher photodegradation rate and a shorter half-life time ($t_{1/2}$) were recorded when using the $1.0SO_4{}^{2-}/CeO_2$ photocatalyst. The reaction rate constant when employing the material with a 1.0 wt.% anion content is 1.6 times higher than that obtained using the CeO_2 support, indicating that sulfation enhances the degradation rate of the contaminant.

Mineralization of the contaminant was analyzed by measuring the percentage of residual total organic carbon (TOC) content after 180 min of reaction. Conducting the reaction in the absence of a photocatalyst results in approximately a 2% TOC removal; however, using CeO_2 as a support promotes a 28% removal. On the other hand, employing modified materials achieves a TOC removal of 25%, 50%, and 31% for $0.5SO_4{}^{2-}/CeO_2$, $1.0SO_4{}^{2-}/CeO_2$, and $2.0SO_4{}^{2-}/CeO_2$, respectively.

Based on the results obtained, the suitability of modifying the CeO_2 support via surface sulfation was confirmed, as it enhanced its photocatalytic properties, resulting in increased effectiveness in degrading the contaminant and the percentage of mineralization of this herbicide. Additionally, it was shown that a 1% sulfate anion content is optimal for promoting higher photodegradation of the compound. Through FTIR results, it was possible to confirm the bidentate coordination of the $SO_4{}^{2-}$ groups on the semiconductor surface, which could act as electron collectors during photocatalytic activation, thereby decreasing carrier velocity and increasing herbicide degradation.

In this regard, the photocatalytic degradation of the herbicide 2,4-D using the $1.0SO_4{}^{2-}/CeO_2$ material is depicted in Figure 9. Once the photocatalyst is irradiated, electron–hole pairs are generated; the photogenerated electrons are attracted to the sulfate groups anchored on the semiconductor surface, acting as trap centers for these charge carriers. Adsorbed oxygen (O_2) is then reduced through interaction with the electrons, leading to the formation of superoxide radicals ($O_2{}^{\bullet -}$). Consequently, the photogenerated holes remain available in the semiconductor to undergo reactions with hydroxyl groups from water, facilitating the formation of hydroxyl radicals ($^{\bullet}OH$), which are responsible for degrading the organic contaminant.

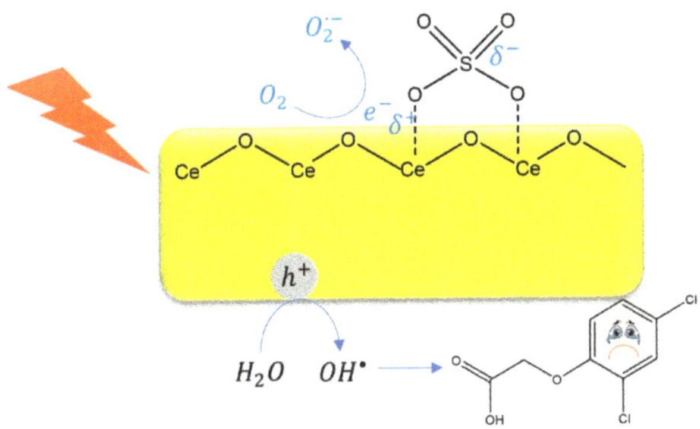

Figure 9. Representation of the photocatalytic process in the degradation of 2,4-D using the $1.0SO_4{}^{2-}/CeO_2$ material.

However, the results also demonstrated that sulfate anion concentrations exceeding 1.0 wt.% exert a negative effect on the photoactivity. This could be attributed to a significant coverage of the active sites on the semiconductor, hindering effective interaction between the contaminant and the photocatalyst. Similar findings have been observed by other researchers in sulfated metal oxide materials [42].

2.3. Response Surface Methodology (RSM)

The randomly chosen photocatalyst was $1.0SO_4^{2-}/CeO_2$, which was utilized in the central composite design with a constant reaction time of 180 min. Experiments were randomized using the following code in RStudio software (https://posit.co/download/rstudio-desktop/). Several models were explored, and among them, the one showing the best fit was obtained from the CCD matrix. This matrix was constructed with the 13 randomized experiments, and the results along with predicted values are detailed in Table 3.

Table 3. Experiment randomization for response surface design.

Test N°	Catalyst Mass (x_1)	Contaminant Concentration (x_2)	%Y Calculated	%Y Experimental	% Error
1	(−1) 50	(−1) 25	71.9	71	1.22
2	(1) 150	(−1) 25	80.2	79	1.48
3	(−1) 50	(1) 55	72.5	65	10.3
4	(1) 150	(1) 55	80.8	69	14.6
5	(−0.5) 10	(0) 40	52.1	73	40.2
6	(0.5) 190	(0) 40	67.0	75	11.9
7	(0) 100	(−0.5) 5	46.9	64	36.6
8	(0) 100	(0.5) 75	48.3	60	24.3
9	(0) 100	(0) 40	96.9	92	5.09
10	(0) 100	(0) 40	96.9	91	6.12
11	(0) 100	(0) 40	96.9	93	4.06
12	(0) 100	(0) 40	96.9	92	5.09
13	(0) 100	(0) 40	96.9	90	7.15

Values exceeding a 10% error may be attributed to significant changes in reaction conditions. These changes can induce effects external to the reaction itself. For example, an excess of material may lead to shielding phenomena in the reaction, resulting in not only photocatalytic but also catalytic processes, altering the degradation mechanisms of the molecule [43]. In situations where there is an excess of photons and a shortage of material capable of capturing them and generating charged species responsible for degrading the contaminant, the removal efficiency of the contaminant is limited [44]. Additionally, at low contaminant concentrations, the interaction between it and the photocatalyst may be restricted due to the limited availability of contaminant molecules, competition with other species, secondary reactions, and reduced quantum efficiency, resulting in limited removal efficiency of the contaminant through photocatalytic processes [43]. On the other hand, discrepancies between calculated and experimental values may be due to random laboratory errors. These errors, occurring unpredictably, may result from variations in environmental conditions, fluctuations in measurement instruments, and random human errors, among other factors. Although they can be minimized, it is difficult to eliminate them, and they cannot be fully controlled.

During the analysis of variance (ANOVA), the factors under study and their interactions were explored, and their influence on the variable of interest was examined. Table A1 presents the analysis results (Appendix A), providing a detailed view of how each variable contributes to the proposed model.

Based on the data analysis an empirical quadratic equation was proposed for the photocatalytic degradation of 2,4-D using the photocatalyst $1.0SO_4^{2-}/CeO_2$:

$$Y\% = -22.782 + (1.0057) \times M + (3.2438) \times C - (0.0046131) \times M^2 - (0.040299) \times C^2 \quad (1)$$

In the presented equation, Y% denotes the degradation percentage, while M and C represent the catalyst mass and the contaminant concentration, respectively. Upon examining the regression coefficients in Equation (1), it is observed that the value associated with C (3.2438) is considerably larger than the value associated with M (1.0057). This suggests that the contaminant concentration (C) has a more significant impact on Y% compared to the catalyst mass (M). The analysis of the variance of the data supports this conclusion, showing an F-value for the model of 35.5 and a P-value for the model fit (0.0001973 < 0.05), confirming consistency between the proposed equation and experimental data. Furthermore, the influence of variables M and C on the degradation percentage (Y%) is depicted in Figure 10a,b, where it can be observed that the maximum value for Y% is achieved when the concentration of 2,4-D and the mass of the photocatalyst are 38.5 ppm and 103.4 mg, respectively (optimal conditions).

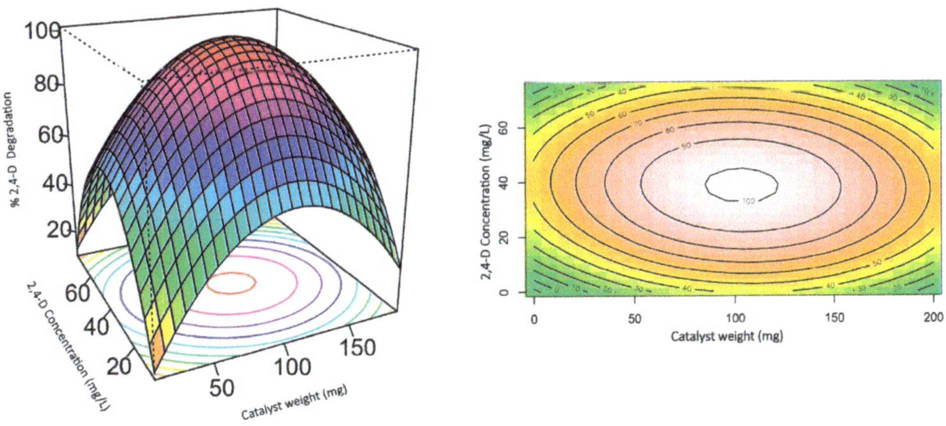

Figure 10. (**a**) Surface plots and (**b**) contour plots representing the effect of 2,4-D concentration and photocatalyst mass on the degradation percentage.

2.4. Reuse of the $1.0SO_4^{2-}/CeO_2$ Photocatalyst in the Degradation of 2,4-Dichlorophenoxyacetic Aucid

The reusability of the $1.0SO_4^{2-}/CeO_2$ photocatalyst was evaluated over three photocatalytic cycles under identical conditions, reintroducing the recovered solid into a new solution of 2,4-dichlorophenoxyacetic acid. The results, depicted in Figure 11, show that the photodegradation of 2,4-D experiences a decrease, with degradation percentages of 95%, 85%, and 71% for each cycle, respectively, after 180 min of reaction. In terms of the percentage of removal of the total organic carbon (TOC) content, values of 50%, 42%, and 32% were observed in the three cycles, respectively.

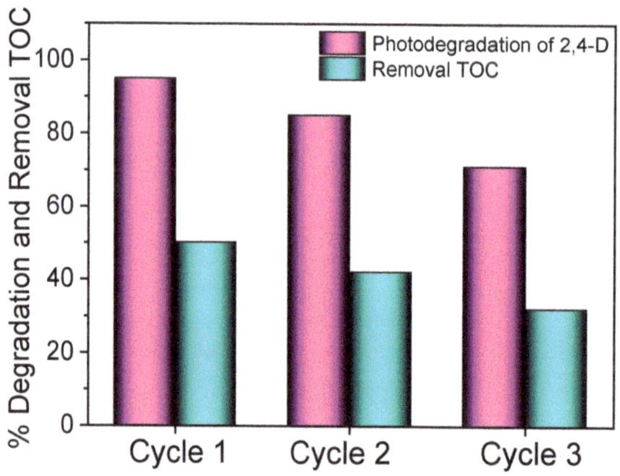

Figure 11. Reuse of the $1.0SO_4^{2-}/CeO_2$ photocatalyst in the photocatalytic degradation of 2,4-dichlorophenoxyacetic acid (2,4-D).

Figure 12 shows the ATR-FTIR spectra of the $1.0SO_4^{2-}/CeO_2$ material reused in the cycle experiments and the material used in the adsorption test. As can be seen, the intensity of the signal associated with the vibrational mode of the Ce-O-Ce bonds at 1320 cm^{-1} decreases as a function of the reuse cycles.

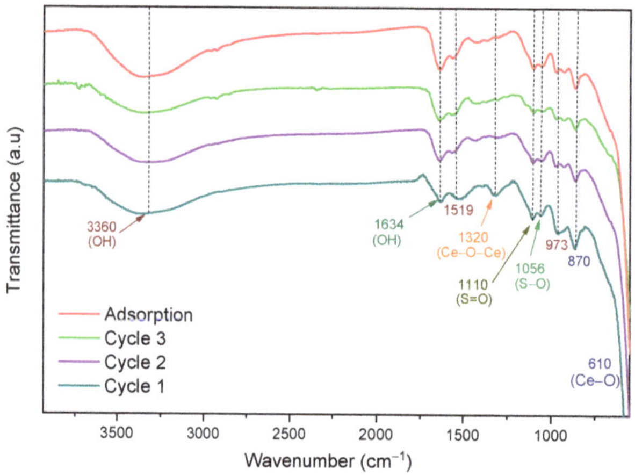

Figure 12. ATR-FTIR spectra of the $1.0SO_4^{2-}/CeO_2$ material reused in the cycle experiments and the material used in the adsorption test.

These results could be related to the permanent adsorption of the herbicide or the degradation intermediates formed on the material surface during each reaction cycle. To corroborate this information, the ATR-FTIR spectrum of the material collected after the adsorption study was analyzed and it is shown in Figure 12. The spectrum reveals the decrease in the intensity in the signal located at 1320 cm^{-1}, which suggests that the Ce-O-Ce sites are probably responsible for the pollutant adsorption. Thus, the progressive decrease in the percentage of degradation as the cycles increase would be related to the lower availability of the sites (Ce-O-Ce) for herbicide adsorption.

3. Materials and Methods

3.1. Synthesis of the Photocatalysts

The CeO_2 support was synthesized by slowly adding a 0.33 M sodium hydroxide (NaOH) solution to a solution prepared with 4.34 g of cerium (III) nitrate hexahydrate ($Ce(NO)_3 \cdot 6H_2O$) and 2.19 g of cetyltrimethylammonium bromide (CTAB) in 200 mL of distilled water. The mixture was continuously stirred at 300 rpm for 24 h. Subsequently, the solution was thermally treated at 90 °C for 3 h, forming a precipitate, which was filtered and washed using hot water. The material was dried at 100 °C for 6 h, ground, and then calcined at 450 °C for 4 h.

The sulfation procedure of the support was carried out by adding 1 g of CeO_2 into a beaker and 20 mL of distilled water was added. Then, the appropriate amount of 0.1 M sulfuric acid (H_2SO_4) solution was added to achieve sulfate ion concentrations of 0.5%, 1.0%, and 2.0 wt.%, respectively. The solution was stirred for 12 h at 400 rpm, and subsequently dried for 12 h at 85 °C. The material was then ground to obtain a fine powder and finally calcined at 450 °C for 2 h. The catalysts were labeled as $0.5SO_4^{2-}/CeO_2$, $1.0SO_4^{2-}/CeO_2$, and $2.0SO_4^{2-}/CeO_2$. The sulfate anion contents were selected based on recent studies with similar materials, where it was shown that the surface modification of the semiconductors such as TiO_2 and CeO_2 with sulfate, fluoride, and phosphate anions in contents higher than 2.0 or 3.0 wt.% does not cause photocatalytic activity, which is associated with a hindered effective interaction between the pollutant with the semiconductor, which leads to a decline in photoactivity [6,17,45,46].

3.2. Characterization of the Photocatalysts

The photocatalytic materials were characterized using various techniques to determine their physicochemical properties. The crystalline structure and crystallite size of the materials were obtained from X-ray diffraction (XRD) analyses performed using a Bruker 2D Phaser instrument irradiated with Cu Kα radiation (Bruker Corporation, Billerica, MA, USA), considering the range of 10 to 80°. The textural properties of the solids were obtained through nitrogen physisorption analysis at −196 °C. For this purpose, the materials were degassed at 200 °C for 6 h and analyzed using an ASAP 2020 porosimetry instrument (Micromeritics Instrument Corporation, Norcross, Georgia, EE. UU). The values of specific surface area were calculated using the BET method, and the average pore sizes were determined using the BJH method applied to the desorption branch of the isotherms. Fourier transform infrared (FTIR) spectroscopy analyses were performed using a Thermo Scientific Nicolet iS50 FTIR spectrometer equipped with an attenuated total reflectance (ATR) device (Thermo Scientific, Waltham, MA, USA), considering a range between 4000 to 200 cm^{-1}, a resolution of 8 cm^{-1}, and 32 scans. The band gap energies (Eg) of the materials were determined from UV-Vis spectra obtained using an Evolution 300 spectrophotometer with a Praying Mantis accessory (Thermo Scientific, Waltham, MA, USA), in the range of 200 to 700 nm. $BaSO_4$ was used as a reference material. Additionally, the materials were studied using thermogravimetric analysis (TGA) performed on thermal analyzer DTA DSC SETARAM 1600 (SETARAM Instrumentation, Lyon, France), in a temperature range between 20 to 800 °C, employing a heating rate of 10 °C/min. Photoluminescence analyses were conducted on the synthesized samples using a Scinco FS-2 fluorescence spectrometer (Scinco, Seoul, Republic of Korea). Excitation occurred at a wavelength of 254 nm, with readings taken within the 300 to 700 nm range.

3.3. Evaluation of the Photocatalytic Activity

The photocatalytic behavior of the series of sulfated CeO_2 materials was studied in the photocatalytic degradation of the herbicide 2,4-dichlorophenoxyacetic acid (2,4-D). For this purpose, 200 mL of an aqueous solution with a concentration of 40 ppm of the herbicide and 100 mg of the photocatalyst were added to a glass reactor. The suspension was continuously stirred at 700 rpm, bubbled with airflow, and irradiated with a high-pressure mercury lamp emitting at 254 nm with an intensity of 4.4 mW, protected with a quartz tube and immersed in the solution. To achieve an adsorption–desorption equilibrium of the corresponding acid on the surface of the photocatalyst, the suspension was kept in darkness for the first 30 min of the reaction. Then, it was irradiated, and samples were taken every 30 min for 3 h. The samples were filtered using a 0.45-micron membrane filter and analyzed by UV spectrophotometry using an Evolution 300 instrument. The mineralization process was monitored by determining the samples' residual total organic carbon (TOC) content.

3.4. Experimental Design-Data Analysis

A central composite design (CCD) and response surface methodology (RSM) were employed to model and optimize the photocatalytic degradation of 2,4-D. The variables under study were the catalyst mass (x_1) and the 2,4-D concentration (x_2). The percentage degradation of 2,4-dichlorophenoxyacetic acid (Y%) after 180 min of treatment was considered the response variable, according to Equation (2):

$$Y(\%) = \left[\frac{C_o - C_t}{C_o}\right] \times 100 \quad (2)$$

where C_o is the initial concentration of 2,4-D in ppm and C_t is the concentration after photocatalytic treatment (ppm). In the experimental design, low (−1) and high (+1) levels were considered, as well as central points (0), for the investigated independent variables. A total of 13 experiments were conducted using this dataset to optimize the response of the percentage of photocatalytic degradation of 2,4-D.

The statistical analysis was performed using RStudio software, employing a second-order polynomial response equation (Equation (3)) to correlate the response variable with the independent variables:

$$y = \beta_0 + \sum_{i=1}^{k} \beta_i x_i + \sum_{i=1}^{k} \beta_{ii} x_i^2 + \sum_{i=1}^{k} \sum_{<j=1}^{k} \beta_{ij} x_i x_j + \varepsilon \quad (3)$$

The response variable Y represents the degradation percentage, while β_0 denotes the constant term. The regression coefficients, β_i, β_{ii}, and β_{ij} indicate the linear, quadratic, and interaction effects, respectively, where x_i and x_j represent the independent variables. Additionally, ε represents the random error term in the model. The data were analyzed through an analysis of variance (ANOVA) based on the proposed model and subsequently evaluated. The precision and applicability of the polynomial model were determined using the coefficient of determination (R^2, adjusted R^2), while statistical significance was verified through F values (Fisher's variation coefficient) and P-values. All terms of the model were assessed at a significance level of 95%. Additionally, response surface and contour plots were generated to visualize the interaction effects between the independent variables on the degradation of 2,4-D.

4. Conclusions

The photocatalytic activity of a series of sulfated CeO_2 materials during the degradation of the herbicide 2,4-dichlorophenoxyacetic acid using ultraviolet radiation was explored. Characterization of the materials confirmed the effect of surface sulfation of the semiconductor on the textural and structural properties of the support. Specifically, the anchoring of sulfate groups on the surface of CeO_2 in a bidentate mode was shown, promoting an increase in the surface area of the materials, which enhances the adsorption and subsequent degradation of the contaminant. Additionally, the role of sulfate groups as traps for photogenerated electrons was confirmed, leading to a reduction in the recombination rate of charge carriers. Finally, the effect of the percentage of sulfate anion on the photocatalytic activity was observed, with an optimal content of the anion at 1.0 wt.%, suggesting that higher amounts could generate recombination centers, leading to a decrease in herbicide degradation.

Author Contributions: Conceptualization, C.C., J.J.M., S.M., H.R., F.T. and R.G.; methodology, H.R., C.C. and J.J.M.; validation, C.C., F.T., J.J.M. and H.R.; formal analysis, F.T., C.R., C.C., E.S., J.J.M., S.M., and H.R.; investigation, C.R., R.G., F.T., C.C. and J.J.M.; writing—original draft preparation, C.C., C.R., J.J.M., H.R. and F.T.; writing—review and editing, C.C., C.R., J.J.M., H.R. and F.T.; supervision, C.C., J.J.M. and H.R.; project administration, C.C. and H.R.; funding acquisition, H.R. and C.C. All authors have read and agreed to the published version of the manuscript.

Funding: The authors express their special gratitude to the Universidad Pedagógica y Tecnológica de Colombia for its financial support in the development of this research, within the framework of project SGI 3684, and to the "CONVOCATORIA 01 DE 2024—INVESTIGAR DA MÁS" of the Vicerrectoría de Investigación y Extensión (VIE).

Data Availability Statement: The original contributions presented in the study are included in the article. Further inquiries can be directed to the corresponding authors.

Acknowledgments: Special thanks to the Ecocatalysis Group of the Universidad Autónoma Metropolitana-Iztapalapa, and the Catalysis Group of the Universidad Pedagógica y Tecnológica de Colombia.

Conflicts of Interest: The authors declare no conflicts of interest.

Appendix A

Table A1. Analysis of Variance (ANOVA).

	Calculated	Standard Error	t-Value	Pr (>|t|)
(Intercept)	−22.782	12.182	−1.8701	0.1106605
Block 2	19.766	3.5074	5.6355	0.0013366
Mass	1.0057	1.2655×10^{-1}	7.9473	0.0002111
Concentration	3.2438	3.4174×10^{-1}	9.4919	7.791×10^{-5}
Mass:Concentration	-1.3333×10^{-3}	1.9027×10^{-3}	−0.7008	0.5097046
Mass^2	-4.6131×10^{-3}	4.9778×10^{-4}	−9.2673	8.923×10^{-5}
Concentration^2	-4.0299×10^{-2}	3.4945×10^{-3}	−11.5321	2.5560×10^{-5}

R^2 = 0.9726 **R^2 adjusted** = 0.9452 *p*-value: 0.0001973
F-Statistic: 35.5
The stationary point of the response surface:
OPTIMAL CONDITIONS

Mass (mg)	Concentration (ppm)
103.3840	38.53572

References

1. Mehdizadeh, M.; Mushtaq, W.; Siddiqui, S.A.; Ayadi, S.; Kaur, P.; Yeboah, S.; Mazraedoost, S.; K.A.Al Taey, D.; Tampubolon, K. Herbicide Residues in Agroecosystems: Fate, Detection, and Effect on Non-Target Plants. *Rev. Agric. Sci.* **2021**, *9*, 157–167. [CrossRef] [PubMed]
2. McGinley, J.; Healy, M.G.; Ryan, P.C.; Mellander, P.E.; Morrison, L.; O'Driscoll, J.H.; Siggins, A. Batch adsorption of herbi-cides from aqueous solution onto diverse reusable materials and granulated activated carbon. *J. Environ. Manag.* **2022**, *323*, 116102. [CrossRef]
3. Shabeer, T.A.; Saha, A.; Gajbhiye, V.; Gupta, S.; Manjaiah, K.; Varghese, E. Simultaneous removal of multiple pesticides from water: Effect of organically modified clays as coagulant aid and adsorbent in coagulation–flocculation process. *Environ. Technol.* **2014**, *35*, 2619–2627. [CrossRef] [PubMed]
4. Acosta Santoyo, G.; Llanos, J.; Raschitor, A.; Bustos, E.; Cañizares, P.; Rodrigo, M. Performance of ultrafiltration as a pre-concentration stage for the treatment of oxyfluorfen by electrochemical BDD oxidation. *Sep. Purif. Technol.* **2019**, *237*, 116366. [CrossRef]
5. Wang, Y.; Tian, Y.S.; Gao, J.J.; Xu, J.; Li, Z.J.; Fu, X.Y.; Han, H.J.; Wang, L.J.; Zhang, W.H.; Deng, Y.D.; et al. Complete biodegradation of the oldest organic herbicide 2,4-Dichlorophenoxyacetic acid by engineering Escherichia coli. *J. Hazard. Mater.* **2023**, *451*, 131099. [CrossRef] [PubMed]
6. Castañeda, C.; Gutiérrez, K.; Alvarado, I.; Martínez, J.J.; Rojas, H.; Tzompantzi, F.; Gómez, R. Effective phosphated CeO_2 materials in the photocatalytic degradation of phenol under UV irradiation. *J. Chem. Technol. Biotechnol.* **2020**, *95*, 3213–3220. [CrossRef]
7. Zhang, X.; Su, L.; Kong, Y.; Ma, D.; Ran, Y.; Peng, S.; Wang, L.; Wang, Y. CeO2 nanoparticles modified by CuO nanoparticles for low-temperature CO oxidation with high catalytic activity. *J. Phys. Chem. Solids* **2020**, *147*, 109651. [CrossRef]

8. Li, B.; Zhang, B.; Nie, S.; Shao, L.; Hu, L. Optimization of plasmon-induced photocatalysis in electrospun Au/CeO_2 hybrid nanofibers for selective oxidation of benzyl alcohol. *J. Catal.* **2017**, *348*, 256–264. [CrossRef]
9. Chang, J.; Ma, Q.; Ma, J.; Ma, H. Synthesis of Fe_3O_4 nanowire@CeO_2/Ag nanocomposites with enhanced photocatalytic activity under sunlight exposure. *Ceram. Int.* **2016**, *42*, 11827–11837. [CrossRef]
10. Valente, J.S.; Tzompantzi, F.; Prince, J. Highly efficient photocatalytic elimination of phenol and chlorinated phenols by CeO_2/MgAl layered double hydroxides. *Appl. Catal. B Environ* **2011**, *102*, 276–285. [CrossRef]
11. Vita, A. Catalytic Applications of CeO_2-Based Materials. *Catalysts* **2020**, *10*, 576. [CrossRef]
12. Montini, T.; Melchionna, M.; Monai, M.; Fornasiero, P. Fundamentals and Catalytic Applications of CeO_2-Based Materials. *Chem. Rev.* **2016**, *116*, 5987–6041. [CrossRef] [PubMed]
13. Patil, S.M.; Deshmukh, S.P.; More, K.V.; Shevale, V.B.; Mullani, S.B.; Dhodamani, A.G.; Delekar, S.D. Sulfated TiO_2/WO_3 nanocomposite: An efficient photocatalyst for degradation of Congo red and methyl red dyes under visible light irradiation. *Mater. Chem. Phys.* **2019**, *225*, 247–255. [CrossRef]
14. Zhao, Y.; Li, J.; Wang, L.; Hao, Y.; Yang, L.; He, P.; Xue, J. Preparation and Characterization of Sulfated TiO_2/Zeolite Composite Catalysts with Enhanced Photocatalytic Activity. *Nano* **2018**, *13*, 1850117. [CrossRef]
15. Svensson, F.G.; Österlund, L. Adsorption and Photo-Degradation of Organophosphates on Sulfate-Terminated Anatase TiO_2 Nanoparticles. *Catalysts* **2023**, *13*, 526. [CrossRef]
16. Cychosz, K.A.; Thommes, M. Progress in the Physisorption Characterization of Nanoporous Gas Storage Materials. *Engineering* **2018**, *4*, 559–566. [CrossRef]
17. Castañeda, C.; Tzompantzi, F.; Gómez, R.; Rojas, H. Enhanced photocatalytic degradation of 4-chlorophenol and 2,4-dichlorophenol on in situ phosphated sol-gel TiO_2. *J. Chem. Technol. Biotechnol.* **2016**, *91*, 2170–2178. [CrossRef]
18. Gómez, R.; López, T.; Ortiz-Islas, E.; Navarrete, J.; Sánchez, E.; Tzompanztzi, F.; Bokhimi, X. Effect of sulfation on the photo-activity of TiO_2 sol–gel derived catalysts. *J. Mol. Catal. A Chem.* **2003**, *193*, 217–226. [CrossRef]
19. Reddy, P.S.; Sudarsanam, P.; Mallesham, B.; Raju, G.; Reddy, B.M. Acetalisation of glycerol with acetone over zirconia and promoted zirconia catalysts under mild reaction conditions. *J. Ind. Eng. Chem.* **2011**, *17*, 377–381. [CrossRef]
20. Khder, A.S.; El Sharkawy, E.A.; El Hakam, S.A.; Ahmed, A.I. Surface characterization and catalytic activity of sulfated tin oxide catalyst. *Catal. Commun.* **2007**, *9*, 769–777. [CrossRef]
21. Marlowe, J.; Acharya, S.; Zuber, A.; Tsilomelekis, G. Characterization of Sulfated SnO_2-ZrO_2 Catalysts and Their Catalytic Performance on the Tert-Butylation of Phenol. *Catalysts* **2020**, *10*, 726. [CrossRef]
22. Vinothini, R.; Tamizhdurai, P.; Mangesh, V.L.; Vanjinathan, M.; Pazhanisamy, P.; Kumaran, R.; Kumar, N.S.; Al Fatesh, A.S.; Kasim, S. Synthesis of CeO_2/SO_4/Ni nanoparticle with a greener perspective for the selective oxidation of propylene glycol in continuous reactor. *Mol. Catal.* **2023**, *549*, 113475. [CrossRef]
23. Farisya, M.R.S.; Irmawati, R.; Shafizah, I.N.; Taufiq Yap, Y.H.; Muhamad, E.N.; Lee, S.L.; Salamun, N. Assessment on the effect of sulfuric acid concentration on physicochemical properties of Sulfated-Titania catalyst and glycerol acetylation performance. *Catalysts* **2021**, *11*, 1542. [CrossRef]
24. Ghorai, T.K.; Chakraborty, M.; Pramanik, P. Photocatalytic performance of nano-photocatalyst from TiO_2 and Fe_2O_3 by mechanochemical synthesis. *J. Alloys Compd.* **2011**, *509*, 8158–8164. [CrossRef]
25. Castañeda, C.; Santos, D.; Hernández, J.S.; Álvarez, A.; Rojas, H.; Gómez, R.; Rajabi, F.; Martínez, J.J.; Luque, R. Efficient NiO/F–TiO_2 nanocomposites for 4-chlorophenol photodegradation. *Chemosphere* **2023**, *315*, 137606. [CrossRef]
26. Humelnicu, A.C.; Samoila, P.; Asandulesa, M.; Cojocaru, C.; Bele, A.; Marinoiu, A.T.; Sacca, A.; Harabagiu, V. Chitosan-Sulfated Titania Composite Membranes with Potential Applications in Fuel Cell: Influence of Cross-Linker Nature. *Polymers* **2020**, *12*, 1125. [CrossRef]
27. Lefèvre, G. In situ Fourier-transform infrared spectroscopy studies of inorganic ions adsorption on metal oxides and hydroxides. *Adv. Colloid Interface Sci.* **2004**, *107*, 109–123. [CrossRef]
28. Sumalatha, V.; Anujya, C.; Balchander, V.; Dhanalaxmi, B.; Kumar, M.P.; Ayodhya, D. Hydrothermal fabrication of n-CeO_2/p-CuS heterojunction nanocomposite for enhanced photodegradation of pharmaceutical drugs in wastewater under visible-light and fluorometric sensor for detection of uric acid. *Inorg. Chem. Commun.* **2023**, *155*, 110962. [CrossRef]
29. Chen, P. Crystal Sizes and Energy Gaps of Cerium Oxide Using Co-Precipitation Method. *Mater. Sci. Appl.* **2022**, *13*, 213–231. [CrossRef]
30. Arora, N.; Joshi, D.P. Band gap dependence of semiconducting nano-wires on cross-sectional shape and size. *Indian J. Phys.* **2017**, *91*, 1493–1501. [CrossRef]
31. Chen, J.; Feng, W.; Zhao, W. Anion-doped CeO_2 for high-performance lithium-sulfur batteries. *Appl. Surf. Sci.* **2022**, *584*, 152613. [CrossRef]
32. Bourja, L.; Bakiz, B.; Benlhachemi, A.; Ezahri, M.; Villain, S.; Favotto, C.; Valmalette, J.C.; Gavarri, J.R. Structural modifications of nanostructured ceria CeO_2, xH_2O during dehydration process. *Powder Technol.* **2012**, *215–216*, 66–71. [CrossRef]
33. Heering, C.; Nateghi, B.; Janiak, C. Charge-Assisted Hydrogen-Bonded Networks of NH_4^+ and $[Co(NH_3)_6]^{3+}$ with the New Linker Anion of 4-Phosphono-Biphenyl-4′-Carboxylic Acid. *Crystals* **2016**, *6*, 22. [CrossRef]
34. Afshari, B.B.; Jamshidi, M.; Rostami, M.; Ghamarpoor, R. Improving the Mechanical/Anticorrosive Properties of a Nitrile Rubber-Based Adhesive Filled with Cerium Oxide Nanoparticles Using a Two-Step Surface Modification Method. *ACS Omega* **2022**, *7*, 44912–44927. [CrossRef] [PubMed]

35. Poornaprakash, B.; Subramanyam, K.; Kumar, M.; Kim, Y.; Reddy, M.S.P. Enhanced photoluminescence characteristics and intrinsic ferromagnetism in Co-substituted CeO_2 nanoparticles. *Mater. Sci. Semicond. Process.* **2020**, *123*, 105566. [CrossRef]
36. Malleshappa, J.; Nagabhushana, H.; Prasad, B.D.; Sharma, S.C.; Vidya, Y.S.; Anantharaju, K.S. Structural, photoluminescence and thermoluminescence properties of CeO_2 nanoparticles. *Optik* **2016**, *127*, 855–861. [CrossRef]
37. Tzompantzi Flores, C.; Castillo Rodríguez, J.C.; Gómez, R.; Hernández, R.P.; Santolalla Vargas, C.E.; Tzompantzi, F. Photocatalytic evaluation of the ZrO_2:$Zn_5(OH)_6(CO_3)_2$ composite for the H_2 production via water splitting. *Top. Catal.* **2020**, *63*, 575–585. [CrossRef]
38. Castañeda Martínez, C.P.; Alvarado Ortega, I.A.; Rojas Sarmiento, H.A.; Tzompantzi Morales, F.J.; Gómez Romero, J.R. Photocatalytic degradation of the 2,4-dichlorophenoxyacetic acid herbicide using supported iridium materials. *Ciencia Desarrollo* **2021**, *12*, 125–134. [CrossRef]
39. Carvalho, L.M.; Soares Filho, A.F.; Lima, M.S.; Cruz Filho, J.F.; Dantas, T.C.M.; Luz, G.E. 2,4-Dichlorophenoxyacetic acid (2,4-D) photodegradation on WO_3-TiO_2-SBA-15 nanostructured composite. *Environ. Sci. Pollut. Res.* **2020**, *28*, 7774–7785. [CrossRef] [PubMed]
40. Castañeda, C.; Martínez, J.J.; Santos, L.; Rojas, H.; Osman, S.M.; Gómez, R.; Luque, R. Caffeine photocatalytic degradation using composites of NiO/TiO_2–F and CuO/TiO_2–F under UV irradiation. *Chemosphere* **2021**, *288*, 132506. [CrossRef] [PubMed]
41. Nawaz, T.; Sengupta, S. Chapter 4-Contaminants of Emerging Concern: Occurrence, fate, and remediation. In *Advances in Water Purification Techniques*; Elsevier: Amsterdam, The Netherlands, 2019; pp. 67–114. [CrossRef]
42. Yan, X.M.; Mei, P.; Gao, L. Photocatalytic Degradation of Rhodamine B over Sulfated SiO_2-TiO_2 Composite by Sulphuric Acid Impregnation Method. *J. Adv. Oxid. Technol.* **2016**, *19*, 52–58. [CrossRef]
43. Hernández, J.S.; Guevara, D.; Shamshurin, M.; Benassi, E.; Sokolov, M.N.; Feliz, M. Octahedral Tantalum Bromide Clusters as Catalysts for Light-Driven Hydrogen Evolution. *Inorg. Chem.* **2023**, *62*, 19060–19069. [CrossRef] [PubMed]
44. Antonopoulou, M.; Konstantinou, I. Photocatalytic degradation of pentachlorophenol by visible light N–F–TiO_2 in the presence of oxalate ions: Optimization, modeling, and scavenging studies. *Environ. Sci. Pollut. Res.* **2015**, *22*, 9438–9448. [CrossRef] [PubMed]
45. Castañeda, C.; Tzompantzi, F.; Gómez, R. Photocatalytic reduction of 4-nitrophenol on in situ fluorinated sol–gel TiO_2 under UV irradiation using Na_2SO_3 as reducing agent. *J. Sol-Gel Sci. Technol.* **2016**, *80*, 426–435. [CrossRef]
46. Elghniji, K.; Hentati, O.; Mlaik, N.; Mahfoudh, A.; Ksibi, M. Photocatalytic degradation of 4-chlorophenol under P-modified TiO_2/UV system: Kinetics, intermediates, phytotoxicity and acute toxicity. *J. Environ. Sci.* **2012**, *24*, 479–487. [CrossRef]

catalysts

MDPI

Article

Exploring the Synthesis of Novel Sillenite $Bi_{12}SnO_{20}$: Effect of Calcination Temperature on the Phase Formation and Catalytic Performance

Oussama Baaloudj [1,2,*], Hamza Kenfoud [3], Monica Brienza [1], Atef El Jery [4], Moutaz Aldrdery [4] and Aymen Amin Assadi [5,6]

1 Department of Science, University of Basilicata, Viale dell'Ateneo Lucano, 10, 85100 Potenza, Italy; monica.brienza@unibas.it
2 Laboratory of Advanced Materials for Energy and Environment, Université du Québec à Trois-Rivières (UQTR), 3351 Boulevard des Forges, Trois-Rivieres, QC G8Z 4M3, Canada
3 Laboratory of Reaction Engineering, Faculty of Mechanical Engineering and Process Engineering, University of Science and Technology Houari Boumediene, BP 32, Algiers 16111, Algeria; hkenfoud@usthb.dz
4 Department of Chemical Engineering, College of Engineering, King Khalid University, Abha 61411, Saudi Arabia; ajery@kku.edu.sa (A.E.J.); maldrdery@kku.edu.sa (M.A.)
5 College of Engineering, Imam Mohammad Ibn Saud Islamic University, Riyadh 11432, Saudi Arabia; aaassadi@imamu.edu.sa
6 Univ Rennes, ENSCR-équipe Chimie et Ingénierie des Procédés, URM 6226 CNRS, ENSCR-11, allee de Beaulieu, CS 508307, 35708 Rennes, France
* Correspondence: oussama.baaloudj@unibas.it or oussama.baaloudj@uqtr.ca

Abstract: Sillenite materials have been the focus of intense research in recent years due to their unique properties and distinct structure with the I23 space group. This electronic structure has reflected high-quality applications and results for some environmental processes such as photocatalysis. This paper investigates the synthesis of a new sillenite, $Bi_{12}SnO_{20}$, and its characteristics, emphasizing its potential for photocatalytic applications. The sillenite $Bi_{12}SnO_{20}$ has been synthesized through the co-precipitation method by mixing the appropriate ratio of Bi and Sn ions. The obtained particles after precipitation and drying were characterized by thermogravimetric analysis (TGA) and then calcined at different temperatures based on this analysis. The phase has been identified by structural analysis using X-ray diffraction (XRD), and its morphology after identification was carried out by scanning electron microscopy (SEM). The calcination temperature has been found to have a critical role in obtaining the phase, where the phase was found to be formed at temperatures between 310 and 400 °C and changed to other phases within higher temperatures. The physicochemical properties of this sillenite were also studied by Fourier-transform infrared spectroscopy (FTIR) and UV Visible Spectrometer (UV-Vis). To study the obtained phases at different calcination temperatures, performance testing was performed under visible light to remove different contaminants, which are Tetracycline, Bisphenol A, and Rhodamine B. The phase $Bi_{12}SnO_{20}$ obtained at 350 °C with a catalyst dose of 1 g/L showed the highest performance for removing these pollutants with concentrations of 20 mg/L, with an efficiency of almost 100% within 2 h. This work will be useful as an important resource and strategy for the development of this sillenite material in its pure phase.

Keywords: sillenite; synthesis; calcination temperature; characterization; photocatalysis

1. Introduction

The catalysts' core metal ion is one of the most important components in the creation of effective photocatalysts [1,2]. As the metal core of catalysts, bismuth ions have attracted a lot of attention recently because of their many unique characteristics, including their high chemical stability, strong light absorption, low rate of pair recombination, and small band gap [3]. Currently, bismuth sillenites, with the chemical formula $Bi_x[M]O_y$ (M is the core metal), have been identified as materials with high photocatalytic potential [4]. Among them, ideal sillenite, $Bi_{12}[M]O_{20}$, has been studied the most [5,6]. Researchers have recently become interested in ideal sillenite materials because of their optical properties and photocatalytic potential [3,7], and they are considered desirable materials for a range of potentially beneficial applications, such as photocatalysis and other photonic domains [8,9].

Although numerous recent studies conducted by researchers have synthesized and examined bismuth sillenites as catalysts, the sillenite composed of bismuth (Bi) and tin (Sn), two of the most preferred ions for producing catalysts [10,11], has not yet been studied due to the challenges associated with its synthesis, including perceived difficulties in preparation and instability. In a study by Rafael et al., they found Bi_1 nO_{20} as an impurity during the synthesis of $Bi_2Sn_2O_7$ [12]. It was also claimed that $Bi_{12}SnO_{20}$ demonstrates instability and is difficult to synthesize, requiring quenching, which would limit its widespread investigation and application. In another study by Aming Lin et al. [13], it is stated that $Bi_{12}SnO_{20}$ has not been studied in terms of its properties, except for sporadic reports on its possible existence. By theoretical analysis, they explained the lack of comprehensive research on this sillenite can be attributed to its high formation energy, as indicated by its high ΔG value (Gibbs free energy). This formation energy can be influenced by the stability of the binary oxide SnO_2, since SnO_2 is highly stable, which makes the formation of $Bi_{12}SnO_{20}$ energetically unfavorable. Other studies have reported the presence of sillenite $Bi_{12}SnO_{20}$ as an unintended byproduct during the synthesis of other materials [14–16]. However, these studies did not delve into a dedicated investigation or understand its characteristics.

In the context of synthesizing the novel sillenite $Bi_{12}SnO_{20}$ and investigating its activity, the co-precipitation method was used. Thermogravimetric analysis (TGA) was utilized to analyze the powder that was obtained after the coprecipitation in order to determine the phases that had formed. X-ray diffraction analysis (XRD) was then employed to determine the phase. Further characterization techniques were employed to examine its optical characteristics, composition, and features. To comprehend the phase formation, the photocatalytic performance of this material was assessed using three distinct pollutants: Tetracycline, Bisphenol A, and Rhodamine B, with the sillenite at varying calcination temperatures. Finally, this investigation ended with perspectives that offered broad recommendations. To our knowledge, no study has been carried out yet concerning the synthesis of the sillenite $Bi_{12}SnO_{20}$ and its use for photocatalysis applications.

2. Materials and Methods

2.1. Chemicals

The chemicals used in the synthesis of sillenite were $Bi(NO_3)_3{\cdot}5H_2O$ (Chem-Lab, Zedelgem, Belgium, 98%), $SnCl_4$ (Sigma Aldrich, Saint Louis, MO, USA, 98%), nitric acid HNO_3 (Sigma Aldrich, 60%) and NaOH (Fluka, Buchs, Switzerland, $\geq$97%). Tetracycline, Bisphenol A, and Rhodamine B were sourced from a pharmaceutical company in Canada. Distilled water was used as a solvent.

2.2. Preparation of the Sillenite $Bi_{12}SnO_{20}$

In order to create the sillenite, bismuth nitrate $[Bi(NO_3)_3{\cdot}5H_2O]$ and Tin(IV) chloride $SnCl_4$ were combined in stoichiometric proportions and dissolved while being stirred in 10% nitric acid. The solutions were then heated to 50 °C for an hour on a hot plate. Then, the mixture was mixed with the NaOH (6 M) solution to bring the pH for precipitation to levels between 9 and 10. Once the pH was reached, white precipitation was observed. Following that, the precipitate was removed and repeatedly cleaned with distilled water before being dried in an autoclave for twenty-four hours at 120 °C. After this procedure, an amorphous powder was produced. After being ground in an agate mortar to a homogeneous powder, the mixture was calcined for 6 h at various temperatures. The final product after calcination, which had a whitish-yellow color, was passed on for optical and structural characterization and phase identification. The preparation process to obtain $Bi_{12}SnO_{20}$ sillenite is shown in Figure 1.

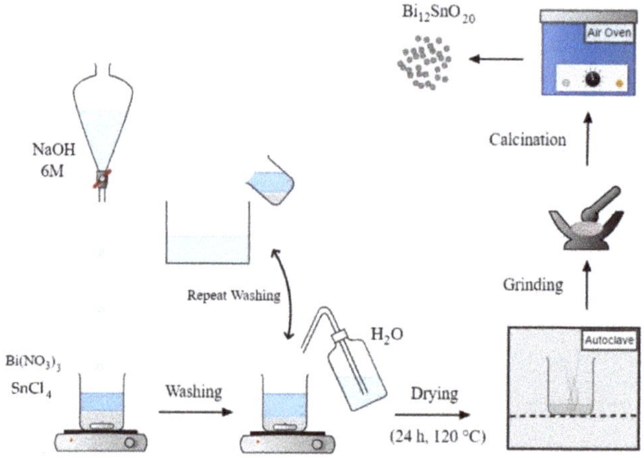

Figure 1. Co-precipitation synthesis of sillenite-type $Bi_{12}SnO_{20}$.

2.3. Characterization

Thermogravimetric analysis (TGA) was performed using a PerkinElmer Diamond TGA/DSC apparatus, the sample weight was 12.232 mg, and it was held at 50.00 °C for 1 min, then heated from 50.00 °C to 800.00 °C at a rate of 10.00 °C/min. After reaching 800.00 °C, the sample was cooled back to 50.00 °C at a rate of 20.00 °C/min. X-ray diffraction (XRD) analysis was performed using a Rigaku MiniFlexTM 600 diffractometer using filtered Cu Ka radiation (1.540593 Å) at an accelerating voltage of 40 kV. Fourier Transformed InfraRed Spectrometer (FTIR) analysis was performed using Cary 630 Agilent. The morphology of the samples, along with energy dispersive X-ray (EDX) analysis and mapping, was characterized using a Scanning Electron Microscope (SEM, Hitachi SU3500, Tokyo, Japan). UV-Vis diffuse reflectance spectra were recorded on a SHIMADZU (Tokyo, Japan) UV-2550 spectrophotometer with an integrating sphere.

2.4. Photocatalytic Application

To evaluate the photocatalytic performance of the materials at each calcination temperature, three hazardous and non-biodegradable pollutants, i.e., Rhodamine B, Bisphenol A, and Tetracycline, were examined. The photocatalytic removal tests were conducted using a photoreactor equipped with 60 W high-intensity 5000K LEDs that were purchased from a local store. An amount of 5.0 mg of the catalyst powder (1 g/L) was mixed with 5.0 mL of the pollutants (20 mg/L as inlet concentration) to investigate the photocatalytic efficiency of the materials. After that, the mixture was continuously stirred at a speed of 300 rpm. Before starting the photocatalytic process and irradiation, it was crucial to examine the catalyst's adsorption of the materials by mixing the pollutant samples with it in a dark environment for two hours to preserve the equilibrium between adsorption and desorption [17]. An air-cooling system was used to maintain a nearly constant temperature at 25 °C throughout the photocatalytic process. The photoreaction was carried out for 2 h. A Cary 60 UV-Vis Spectrophotometer was used to monitor the pollution concentrations. Using Equation (1), the removal efficiency both in the dark and under irradiation was assessed, and Equation (2) was used to study the kinetics using the pseudo-first-order kinetic model.

$$\text{Removal efficiency } (\%) = \frac{(C_0 - C_t)}{C_0} \times 100\% \quad (1)$$

$$ln\frac{C_0}{C_t} = k_{app}(t) \quad (2)$$

C_0 is the initial concentration of the pollutants, and C_t is the concentration on the time (t), k_{app} is the apparent pseudo-first-order rate constant represented in min^{-1}.

3. Results and Discussion

3.1. Characterization of the Materials

3.1.1. Phase Identification and Structural Investigation

The thermogravimetric analysis (TGA) and derivative curve (DTG) shown in Figure 2 have been used to study the thermal transitions and phase changes in the materials. The peak at 100.15 °C in the derivative curve is attributed to the evaporation of water in an endothermic reaction, which is accompanied by weight loss at this stage. The peaks at 212.83 °C and 307.55 °C can be associated with denitrification and nitrate decomposition as exothermic reactions [18,19], whereas the significant peak at 307.55 °C accompanied by the mass loss and significant exothermic effect can be due to the decomposition of hydroxide of bismuth tin mixture (Bi_1 n$(OH)_{20}$) into oxide ($Bi_{12}SnO_{20}$) where the material started to crystallize [20,21]. It can be observed that a phase has been reached after a temperature of 310 °C. A chemical reaction occurs with a significant downward peak at 481 °C accompanied by significant weight loss, which is associated with an endothermic melting reaction. Within this temperature, the obtained phase has been transformed into other phases stable across higher temperatures. Based on the derivative curve results, the $Bi_{12}SnO_{20}$ phase can be obtained at temperatures between 310 and 400 °C in air and this is in line with the literature [22]. To maintain a sufficient temperature and prevent phase loss, a calcination temperature of 350 °C was selected to be investigated.

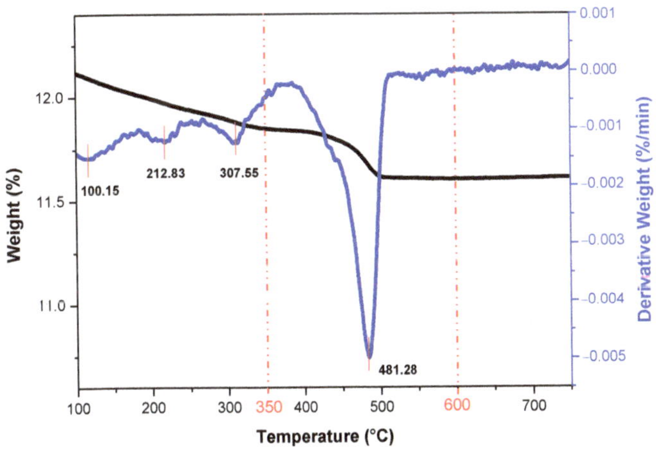

Figure 2. Thermogravimetric analysis and derivative curve of the amorphic powder before calcination.

X-ray diffractometry (XRD) was used to study the development of the sillenite phase. The nature of the phases generated after a calcination temperature of 310 °C has been further investigated. Figure 3 displays the usual XRD patterns of Bi-Sn powder that has been calcined at 350 °C and 600 °C. In the sample that calcined at 350 °C, crystalline sillenite nanoparticles of $Bi_{12}SnO_{20}$ were identified by comparing them to the calculated sillenite phase XRD generated by Vista software (ver. 3.5.8), and the results were in good agreement with the international card ASTM 054-1004 [14]. However, the phase was not pure, and some impurities such as $Bi_{64}SnO_{98}$ were also observed. In this calcination temperature, there is the possibility of the formation of $Bi_{12}Sn(OH)_{20}$ as another impurity due to the presence of the undecomposed hydroxyl group product –OH at low temperatures. The XRD pattern of the sample calcinated at 600 °C shows sharper and narrower diffraction lines, a mixed phase of SnO_2 (JCPDS No. 41-1445 [23]), tetragonal Bi_2O_3 (JCPDS No. 78-1793 [24]), Bi_2O_3 doped Sn, and other new phases such as $Bi_2Sn_3O_9$ and $Bi_2Sn_2O_7$ [12,15,16]. The material calcinated at this temperature was not identical to the calculated XRD of the sillenite $Bi_{12}SnO_{20}$. According to the TGA/DTG and XRD analysis, it can be concluded that the sillenite phase formed between the temperatures 310 and 400 °C, and it is unstable, whereby increasing the temperature causes the decomposition and melting of the phase and leads to the formation of other stable compounds such as SnO_2, which is stable at high calcination temperatures. Based on the strongest diffraction lines in the XRD at 350 °C, the average size of the nanocrystals in the obtained phase was determined to be D = 34.215 nm. This was calculated using the full width at half maximum (β, in radians) and the Debye–Scherrer equation.

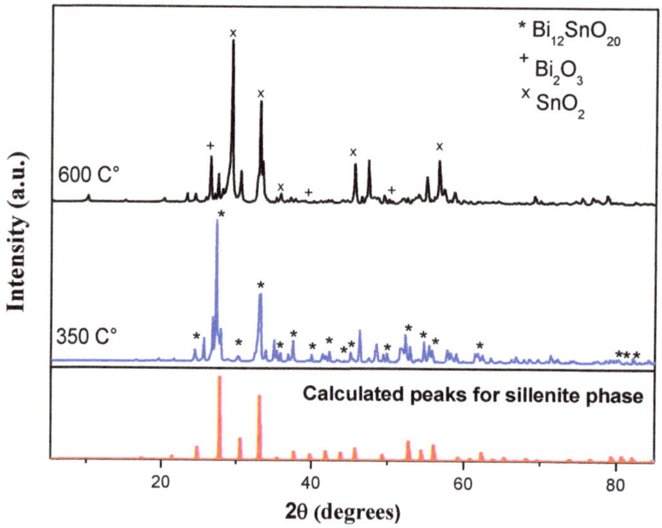

Figure 3. XRD analysis of Bi-Sn powder calcinated at 350 °C and 600 °C.

In order to investigate if the temperature of the calcination 350 °C is sufficient for the development of sillenite catalysts, Fourier Transform Infrared (FT-IR) has been integrated [23]. This temperature was selected based on XRD where the appropriate calcination temperature at which the sillenite crystalline nanoparticles were detected. The spectra illustrated in Figure 4 show the contents of the sample after 350 °C calcination over the range of 700–3600 cm^{-1}, and the focus was on this area in order to investigate the remaining impurities such as nitrates in the phase post-calcination. In general, the metal oxide bonds Sn–O and Bi–O are frequently detected by absorption peaks in the 500–700 cm^{-1} region, as already demonstrated in the literature [25,26]. The stretching vibration of hydroxyl groups and the deformation vibration of adsorbed water molecules, respectively, are responsible for the absorption peaks around 3400 cm^{-1} and 1600 cm^{-1}, respectively [27]. The remaining nitrates in the sample were responsible for the broadband absorption peaks at 1300–1400 cm^{-1} (N–O asymmetric stretching) [28], 1040 cm^{-1} (N–O symmetric stretching) [26], and 840 cm^{-1} (N–O bending vibrations) [29]. The presence of nitrates in the sample was expected since the remaining nitrogen after thermal treatment cannot be completely removed by calcination at 350 °C, and if higher heat is used during calcination, the sillenite phase may be lost, which is a limitation in obtaining this phase.

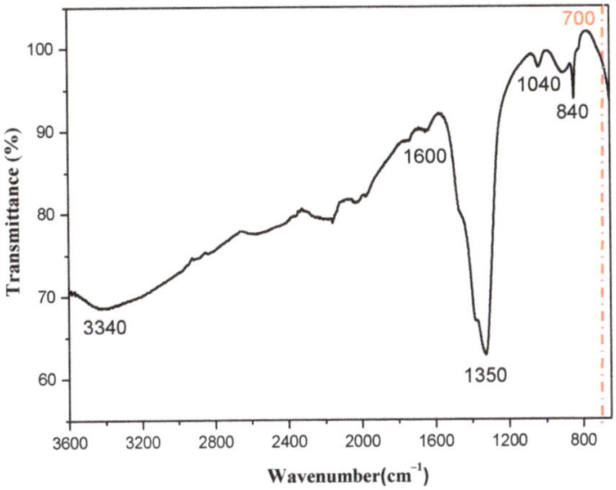

Figure 4. FTIR spectra of the sillenite $Bi_{12}SnO_{20}$ calcinated at 350 °C.

3.1.2. Morphology Investigation

The morphology of the particles calcined at 350 °C is depicted in the scanning electron microscope (SEM) image in Figure 5a. The figure shows that the powder consists of irregular sub-micron aggregate grains, dispersed throughout the agglomerate in a nearly orderly manner, often forming a sphere-like geometry. The particles forming the agglomeration are less than 1 μm and are confirmed to be nanoparticles, which aligns with the XRD-calculated crystallite size. It is also evident from the morphology that the sample is porous, which promotes both photocatalysis and adsorption. Additionally, the EDX spectra were used to examine the elemental composition. The representative EDX spectrum of the nanoparticles is shown in Figure 5b. Several distinct lines detected the elements O, Sn, and Bi. The measured average content ratios of O/Sn and Bi/Sn were 8.45 and 15.55, respectively, which is close to the theoretical value found in the $Bi_{12}SnO_{20}$ formulas.

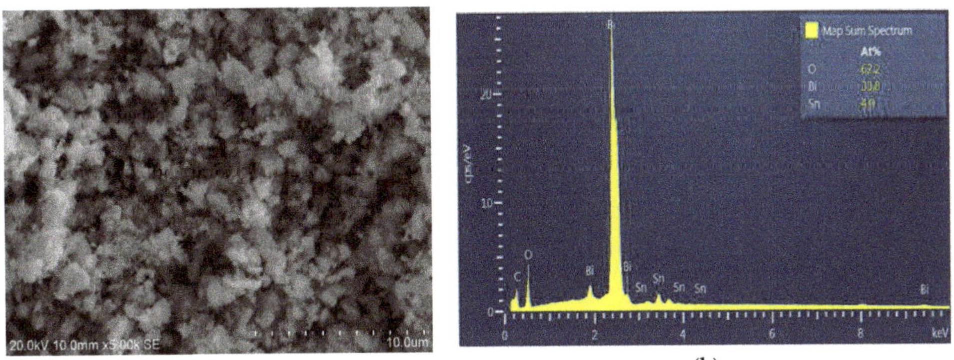

(a) (b)

Figure 5. (**a**) SEM images and (**b**) EDX analysis of $Bi_{12}SnO_{20}$ calcinated at 350 °C.

The mapping analysis depicted in the images in Figure 6 demonstrates a homogenous distribution of the elements throughout the sample. This finding confirms the successful formation of homogeneous phases within the material. Additionally, this homogeneous distribution demonstrates that the Sn and Bi oxides have not significantly existed in the sample; instead, they have formed the sillenite phase. This uniformity ensures predictable and reliable performance in the application of this material.

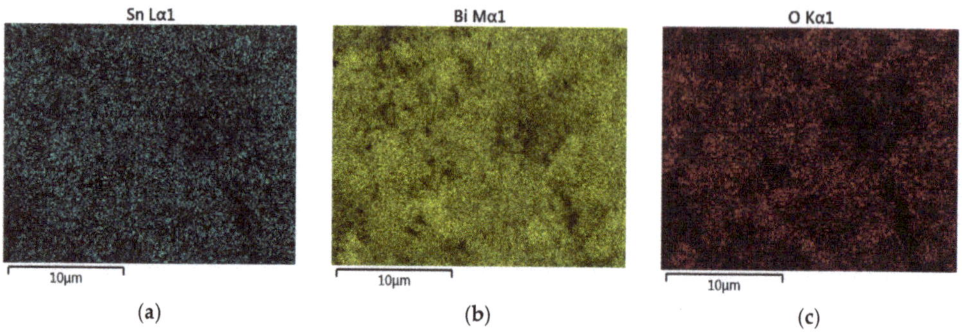

Figure 6. Mapping analysis of the elements (**a**) Sn, (**b**) Bi, and (**c**) O for the sillenite $Bi_{12}SnO_{20}$.

3.1.3. Optical Study

The optical characteristics of photocatalyst materials are important information for their application, and the band gap energy is one of the parameters controlling the photocatalytic activity [30]. UV–Vis diffuse reflectance spectra of $Bi_{12}SnO_{20}$ at room temperature are shown in Figure 7a. The reflectance spectrum of $Bi_{12}SnO_{20}$ exhibits a sharp increase before 470 nm. This falls higher than the ultraviolet (UV) region (>400 nm), which suggests that this catalyst can also absorb visible spectrum rather than only the UV region. Tauc's formula (Equation (3)) can be used to determine the energy bandgap (E_g)—a significant parameter that can be used to describe the optical properties of photocatalysts. The E_g value can be calculated by the intercept of the linear plots of direct transition $(\alpha h\nu)^2$ with the hν-axis (Figure 7b).

$$(\alpha h\nu)^2 = K(h\nu - E_g) \quad (3)$$

where $h\nu$ is the photon energy, α is the absorption coefficient and K is the constant relative to the material type.

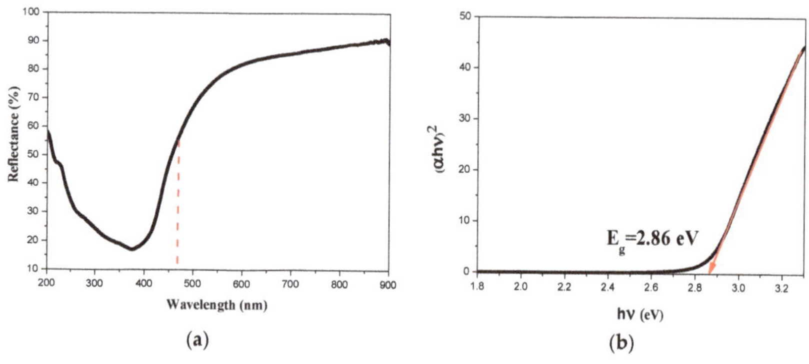

Figure 7. (**a**) Diffuse reflectance spectrum, (**b**) Tauc plot for measurement of the band gap of $Bi_{12}SnO_{20}$.

The slope of the linear part suggests the band gap energy of $Bi_{12}SnO_{20}$ powder is 2.86 eV, which matches previous reports [11,31]. This value of E_g can be favorable for photo generation upon photoexcitation, even with visible light ($E_g < 3$). Therefore, the $Bi_{12}SnO_{20}$ catalyst has the ability of photocatalytic degradation with both visible light and UV light irradiation.

3.2. Material Performance Tests

Following the examination of the characteristics of the Bi-Sn materials after their calcination at varying temperatures, a sequence of experiments was conducted to assess their effectiveness and potential applications for treating various contaminants. Before performing the photocatalytic degradation tests under irradiation, these experiments involved assessing the materials under dark adsorption tests at different temperatures. Figure 8 presents the efficiency of removal under irradiation as well as the adsorption efficiency after 1 h for the material Bi-Sn calcinated at 300, 350, 400, 500, and 600 °C for 6 h. The experiments were performed with an initial concentration of 20 mg/L with the materials added to the solution with a catalyst dose of 1 g/L and a reaction time of 1 h in the dark and under irradiation. These experiments showed that Tetracycline and Bisphenol A had slightly high adsorption efficiencies in dark conditions under different calcination temperatures, but Rhodamine B had low adsorption efficiency. In contrast, when exposed to radiation, the materials underwent considerable increases in the removal of all contaminants under certain conditions, albeit with variations across different temperatures. With all studied pollutants, almost the same behavior was recorded, where the photocatalytic activity of the Bi-Sn material increased with the calcination temperature, peaking at 350 °C, and then decreased with further temperature increases. The removal efficiency after 1 h of reaction for the Bi-Sn sample calcined at 350 °C for Rhodamine B, Bisphenol A, and Tetracycline were 98.92, 95.04 and 76.97% respectively. These results suggest that at 350 °C, there is a phase that can be activated by light to generate radicals, and that phase, as XRD showed, is the sillenite $Bi_{12}SnO_{20}$. Samples that are calcined above 350 °C exhibit lower photocatalytic activity as a result of the loss of this phase. The XRD analysis indicates that the phase has nearly completely disappeared after calcining at 600 °C, which likely contributed to the decreased photocatalytic efficiency. The enhanced efficiency of phase $Bi_{12}SnO_{20}$, which was detected at 350 °C, was higher than that of SnO_2 and Bi_2O_3, which were detected at 600 °C. This may be because $Bi_{12}SnO_{20}$ has a greater absorption of incident visible light, which results in the generation of more photogenerated electrons to improve the efficiency of charge separation. This, in turn, enhances the photocatalytic activity, as demonstrated by UV–Vis diffuse reflectance spectra. The improvement in removal performance can also be attributed to several factors such as particle size, morphology, porosity, crystallinity, and surface area [32–35]. However, in this material, the difference is mainly due to phase transformation as it is a total phase change, and for that, the calcination temperature played a crucial role in determining the efficiency of the samples. The results obtained from this application were in line with previous reports about similar materials [36–38]. From the data and results shown in Figure 8, we can conclude that the calcination at 350 °C is identified as the optimum temperature for sillenite $Bi_{12}SnO_{20}$, where the catalyst exhibited the highest catalytic activity, avoiding the phase transformation that reduces photocatalytic efficiency, and the enhanced photocatalytic activity could be attributed to the presence of the $Bi_{12}SnO_{20}$ phase within this temperature.

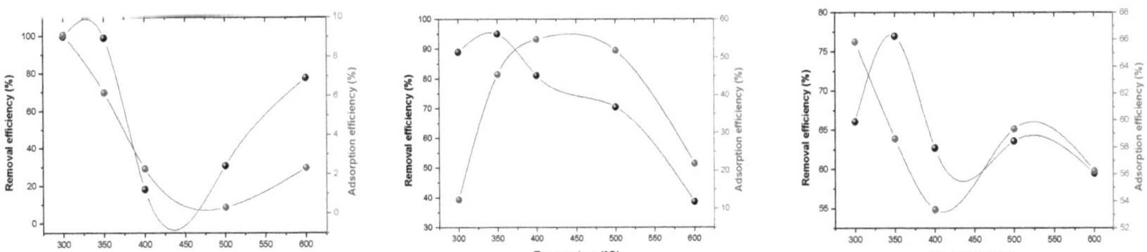

(a) (b) (c)

Figure 8. Adsorption and removal efficiency of the system Bi-Sn as a function of the calcination temperature for (**a**) Rhodamine B (**b**) Bisphenol A (**c**) Tetracycline, $[Pollutants]_0$ = 20 mg/L, catalyst dose 1 g/L, Reaction time 1 h.

Bi-Sn material calcined at 350 °C, which exhibits the dominant phase sillenite $Bi_{12}SnO_{20}$, has demonstrated an effective removal for the removal of Tetracycline, Bisphenol A, and Rhodamine B. Thus, a kinetic study has been conducted using this material for the removal of these pollutants. The kinetics of photodegradation of these pollutants on the sillenite can be described by a pseudo-first-order reaction model (PFO). Figure 9 shows the kinetic data of the removal of the three pollutants that were fitted with the first-order model, and as can be seen, the sillenite $Bi_{12}SnO_{20}$ calcined at 350 °C demonstrated a strong photodegradation rate and nearly a total removal of the three pollutants within two hours.

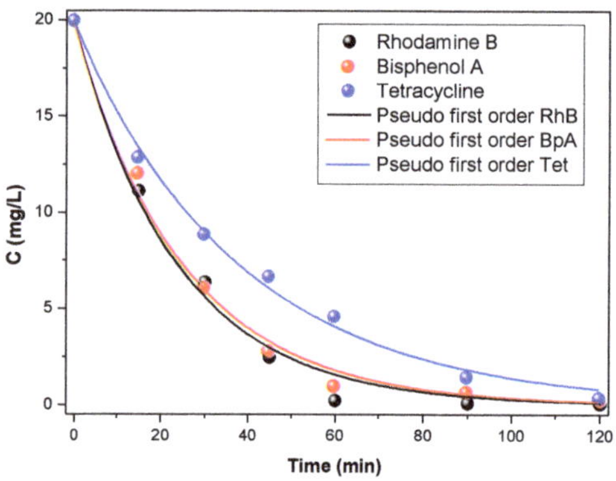

Figure 9. Removal kinetics of the pollutants using sillenite $Bi_{12}SnO_{20}$, $[concentration]_0$ = 20 mg/L and catalyst dose 1 g/L.

From the fitting with the PFO model, the reaction rate constant (k) and correlation coefficient (R^2) were calculated and are summarized in Table 1. The slight differences in reaction rate constants between the three pollutants may be attributed to factors such as chemical structure, stability, and adsorption affinity on the photocatalyst surface. Specifically, Tetracycline's complex structure and large molecular size likely contribute to its lower reaction rate constant compared to other pollutants.

Table 1. Kinetic parameters of the pseudo-first-order model for the removal of pollutants using the sillenite $Bi_{12}SnO_{20}$.

Pollutant	k_{app} (min^{-1})	R^2
Rhodamine B	0.042	0.99
Bisphenol A	0.04	0.99
Tetracycline	0.026	0.99

4. Conclusions

This study successfully obtained the challenging sillenite material $Bi_{12}SnO_{20}$ via a co-precipitation method. Thermogravimetric analysis (TGA) guided the selection of appropriate calcination temperatures, followed by X-ray diffraction (XRD) for phase identification and scanning electron microscopy (SEM) for morphological characterization. The results revealed that $Bi_{12}SnO_{20}$ formation occurs between 300 °C and 400 °C, transforming to other phases at higher temperatures. Fourier-transform infrared spectroscopy (FTIR) and UV-Vis spectroscopy provided further insights into the material's physicochemical properties, where the sillenite was found to have a bandgap of 2.8 eV. Performance testing under visible light for contaminant removal on the Bi-Sn material calcined in different temperatures demonstrated a high efficiency with the Bi-Sn material in which the $Bi_{12}SnO_{20}$ phase was identified. The efficiency has achieved a nearly total removal of Tetracycline, Bisphenol A, and Rhodamine B (20 mg/L each) within 2 h with a catalyst dose of 1 g/L. This research paves the way for further exploration of $Bi_{12}SnO_{20}$ as a novel sillenite photocatalyst, but more efforts are needed to refine the synthesis method to obtain a pure $Bi_{12}SnO_{20}$ phase. This sillenite has demonstrated exceptional performance, achieving nearly 100% removal of various pollutants within a short timeframe. This merits further exploration and development to unlock its full potential for environmental remediation applications.

Author Contributions: Conceptualization, O.B.; methodology, O.B. and A.A.A.; software, O.B.; validation, O.B. and H.K.; formal analysis, H.K. and A.E.J.; investigation, O.B. H.K., A.E.J. and M.A.; resources, A.A.A., M.A. and A.E.J.; data curation, O.B.; writing—original draft preparation, O.B.; writing—review and editing, M.B. and A.A.A.; visualization, O.B.; supervision, A.A.A. All authors have read and agreed to the published version of the manuscript.

Funding: This research received no external funding.

Data Availability Statement: Data supporting reported results are available.

Acknowledgments: The authors extend their appreciation to the Deanship of Research and Graduate Studies at King Khalid University for funding this work through the Large Research Project under grant number RGP2/248/45. The authors are also grateful to the University of Quebec in Trois-Rivieres (UQTR), Università degli studi della basilicata (UniBas) and IMAM University for scientific collaboration.

Conflicts of Interest: The authors declare no conflicts of interest.

References

1. Qiao, X.; Pu, Y.; Li, Y.; Huang, Y.; Cheng, H.; Seo, H.J. Structural characteristics and photocatalytic ability of vanadate-sillenite Bi25VO40 nanoparticles. *Powder Technol.* **2016**, *287*, 277–284. [CrossRef]
2. Elgohary, E.A.; Mohamed, Y.M.A.; El Nazer, H.A.; Baaloudj, O.; Alyami, M.S.S.; El Jery, A.; Assadi, A.A.; Amrane, A. A Review of the Use of Semiconductors as Catalysts in the Photocatalytic Inactivation of Microorganisms. *Catalysts* **2021**, *11*, 1498. [CrossRef]
3. Isik, M.; Sarigul, N.; Gasanly, N.M. Thermoluminescence characteristics of $Bi_{12}SiO_{20}$ single crystals. *J. Lumin.* **2020**, *224*, 117280. [CrossRef]
4. Baaloudj, O.; Kenfoud, H.; Badawi, A.K.; Assadi, A.A.; El Jery, A.; Assadi, A.A.; Amrane, A. Bismuth Sillenite Crystals as Recent Photocatalysts for Water Treatment and Energy Generation: A Critical Review. *Catalysts* **2022**, *12*, 500. [CrossRef]
5. Valant, M.; Suvorov, D. Processing and Dielectric Properties of Sillenite Compounds $Bi_{12}MO_{20-\delta}$ (M: Si, Ge, Ti, Pb, Mn, $B_{1/2}P_{1/2}$). *ChemInform* **2010**, *33*, 2900–2904. [CrossRef]
6. Baaloudj, O.; Vu, N.N.; Assadi, A.A.; Le, V.Q.; Nguyen-Tri, P. Recent advances in designing and developing efficient sillenite-based materials for photocatalytic applications. *Adv. Colloid Interface Sci.* **2024**, *327*, 103136. [CrossRef]
7. Wu, X.; Li, M.; Li, J.; Zhang, G.; Yin, S. A sillenite-type $Bi_{12}MnO_{20}$ photocatalyst: UV, visible and infrared lights responsive photocatalytic properties induced by the hybridization of Mn 3d and O 2p orbitals. *Appl. Catal. B Environ.* **2017**, *219*, 132–141. [CrossRef]
8. Zhou, J.; Zou, Z.; Ray, A.K.; Zhao, X.S. Preparation and characterization of polycrystalline bismuth titanate $Bi_{12}TiO_{20}$ and its photocatalytic properties under visible light irradiation. *Ind. Eng. Chem. Res.* **2007**, *46*, 745–749. [CrossRef]
9. Guo, Y.; Guo, W.; Yang, Y.; Guo, Y.; Jia, Y.; Liu, H. Self-assembled hierarchical $Bi_{12}TiO_{20}$-graphene nanoarchitectures with excellent simulated sunlight photocatalytic activity. *Phys. Chem. Chem. Phys.* **2014**, *16*, 2705–2714. [CrossRef]
10. Tho, N.T.M.; Khanh, D.N.N.; Thang, N.Q.; Lee, Y.I.; Phuong, N.T.K. Novel reduced graphene oxide/$ZnBi_2O_4$ hybrid photocatalyst for visible light degradation of 2,4-dichlorophenoxyacetic acid. *Environ. Sci. Pollut. Res.* **2020**, *27*, 11127–11137. [CrossRef]
11. Baaloudj, O.; Nasrallah, N.; Bouallouche, R.; Kenfoud, H.; Khezami, L.; Assadi, A.A. High efficient Cefixime removal from water by the sillenite $Bi_{12}TiO_{20}$: Photocatalytic mechanism and degradation pathway. *J. Clean. Prod.* **2022**, *330*, 129934. [CrossRef]
12. Almeida, R.M.; Paschoal, C.W.A.; Auletta, J.T.; Kann, Z.R.; Lufaso, M.W. Ionic conductivity in $Bi_2Sn_2O_7$ ceramics. *Ceram. Int.* **2012**, *38*, 1275–1279. [CrossRef]
13. Lin, A.; Ming, C.; Sun, Y.Y. Dilute Element Compounds: A Route to Enriching Inorganic Functional Materials. *J. Phys. Chem. Lett.* **2021**, *12*, 8194–8202. [CrossRef]
14. Hussein, A.M.; Mohammed, W.M.; Mohammed, S.J. The Influence of Tin Oxide Doping on Structural and Optical Properties of (Bi_2O_3) Thin Films. *J. Univ. Babylon Pure Appl. Sci.* **2018**, *26*, 97–106. [CrossRef]
15. Huo, G.; Lu, J.; Ren, L. Phase relations in the Bi_2O_3-CoO_y-SnO_2 system at room temperature. *J. Alloys Compd.* **2004**, *368*, 126–129. [CrossRef]
16. Sakoske, G.E.; Sarver, J.E. Bismuth-Containing Laser Markable Compositions and Methods of Making and Using Same. U.S. Patent 6,680,121, 20 January 2004.
17. Assadi, A.A.; Palau, J.; Bouzaza, A.; Wolbert, D. Modeling of a continuous photocatalytic reactor for isovaleraldehyde oxidation: Effect of different operating parameters and chemical degradation pathway. *Chem. Eng. Res. Des.* **2013**, *91*, 1307–1316. [CrossRef]

18. Yang, J.; Zheng, H.Y.; Zhu, Y.L.; Zhao, G.W.; Zhang, C.H.; Teng, B.T.; Xiang, H.W.; Li, Y. Effects of calcination temperature on performance of Cu–Zn–Al catalyst for synthesizing γ-butyrolactone and 2-methylfuran through the coupling of dehydrogenation and hydrogenation. *Catal. Commun.* **2004**, *5*, 505–510. [CrossRef]
19. Shabalina, A.V.; Golubovskaya, A.G.; Fakhrutdinova, E.D.; Kulinich, S.A.; Vodyankina, O.V.; Svetlichyi, V.A. Phase and Structural Thermal Evolution of Bi–Si–O Catalysts Obtained via Laser Ablation. *Nanomaterials* **2022**, *12*, 4101. [CrossRef]
20. Wang, X.; Yang, C.; Zhou, D.; Wang, Z.; Jin, M. Chemical co-precipitation synthesis and properties of pure-phase $BiFeO_3$. *Chem. Phys. Lett.* **2018**, *713*, 185–188. [CrossRef]
21. Iqbal, M.J.; Ashiq, M.N. Physical and electrical properties of Zr–Cu substituted strontium hexaferrite nanoparticles synthesized by co-precipitation method. *Chem. Eng. J.* **2008**, *136*, 383–389. [CrossRef]
22. Chehab, S.; Conflant, P.; Drache, M.; Boivin, J.C.; McDonald, G. Solid-state reaction pathways of Sillenite-phase formation studied by high-temperature X-ray diffractometry and differential thermal analysis. *Mater. Res. Bull.* **2003**, *38*, 875–897. [CrossRef]
23. Liu, A.; Zhu, M.; Dai, B. A novel high-performance SnO_2 catalyst for oxidative desulfurization under mild conditions. *Appl. Catal. A Gen.* **2019**, *583*, 117134. [CrossRef]
24. Xiao, X.; Hu, R.; Liu, C.; Xing, C.; Qian, C.; Zuo, X.; Nan, J.; Wang, L. Facile large-scale synthesis of β-Bi_2O_3 nanospheres as a highly efficient photocatalyst for the degradation of acetaminophen under visible light irradiation. *Appl. Catal. B Environ.* **2013**, *140–141*, 433–443. [CrossRef]
25. Al-Hada, N.M.; Kamari, H.M.; Baqer, A.A.; Shaari, A.H.; Saion, E. Thermal calcination-based production of SnO_2 nanopowder: An analysis of SnO_2 nanoparticle characteristics and antibacterial activities. *Nanomaterials* **2018**, *8*, 250. [CrossRef] [PubMed]
26. Astuti, Y.; Fauziyah, A.; Nurhayati, S.; Wulansari, A.D.; Andianingrum, R.; Hakim, A.R.; Bhaduri, G. Synthesis of α-Bismuth oxide using solution combustion method and its photocatalytic properties. *IOP Conf. Ser. Mater. Sci. Eng.* **2016**, *107*, 012006. [CrossRef]
27. Liu, Y.; Xin, F.; Wang, F.; Luo, S.; Yin, X. Synthesis, characterization, and activities of visible light-driven Bi_2O_3-TiO_2 composite photocatalysts. *J. Alloys Compd.* **2010**, *498*, 179–184. [CrossRef]
28. Gurešić, D.; Đurić, S.; Andrejević, T.; Popsavin, M.; Glišić, B. Synthesis and spectroscopic characterization of polynuclear silver(I) complex with 2,2′-biquinoline. *Bull. Nat. Sci. Res.* **2020**, *10*, 26–30. [CrossRef]
29. Nakamoto, K. *Infrared and Raman Spectra of Inorganic and Coordination Compounds, Part B: Applications in Coordination, Organometallic, and Bioinorganic Chemistry*; John Wiley & Sons: Hoboken, NJ, USA, 2008; pp. 1–408. [CrossRef]
30. Mursyalaat, V.; Variani, V.I.; Arsyad, W.O.S.; Firihu, M.Z. The development of program for calculating the band gap energy of semiconductor material based on UV-Vis spectrum using delphi 7.0. *J. Phys. Conf. Ser.* **2023**, *2498*, 012042. [CrossRef]
31. Baaloudj, O.; Nasrallah, N.; Kenfoud, H.; Algethami, F.; Modwi, A.; Guesmi, A.; Assadi, A.A.; Khezami, L. Application of $Bi_{12}ZnO_{20}$ sillenite as an efficient photocatalyst for wastewater treatment: Removal of both organic and inorganic compounds. *Materials* **2021**, *14*, 5409. [CrossRef]
32. Sadeghzadeh-Attar, A. Efficient photocatalytic degradation of methylene blue dye by SnO_2 nanotubes synthesized at different calcination temperatures. *Sol. Energy Mater. Sol. Cells* **2018**, *183*, 16–24. [CrossRef]
33. Ma, J.; Chu, J.; Qiang, L.; Xue, J. Effect of different calcination temperatures on the structural and photocatalytic performance of Bi-TiO_2/SBA-15. *Int. J. Photoenergy* **2013**, *2013*, 875456. [CrossRef]
34. Ding, Y.; Zhao, C.; Li, Y.; Ma, Z.; Lv, X. Effect of calcination temperature on the structure and catalytic performance of the cu-mcm-41 catalysts for the synthesis of dimethyl carbonate. *Quim. Nova* **2018**, *41*, 1156–1161. [CrossRef]
35. Saikumari, N.; Dev, S.M.; Dev, S.A. Effect of calcination temperature on the properties and applications of bio extract mediated titania nano particles. *Sci. Rep.* **2021**, *11*, 1734. [CrossRef] [PubMed]
36. Wang, M.; Li, C.; Liu, B.; Qin, W.; Xie, Y. Influence of Calcination Temperature on Photocatalyst Performances of Floral Bi_2O_3/TiO_2 Composite. *Catalysts* **2022**, *12*, 1635. [CrossRef]
37. Hassan, S.M.; Ahmed, A.I.; Mannaa, M.A. Preparation and characterization of SnO_2 doped TiO_2 nanoparticles: Effect of phase changes on the photocatalytic and catalytic activity. *J. Sci. Adv. Mater. Devices* **2019**, *4*, 400–412. [CrossRef]
38. Zhang, G.; Hu, L.; Wang, P.; Yuan, Y. The effect of calcination temperature on the performance of Co_3O_4-Bi_2O_3 as a heterogeneous catalyst of peroxymonosulfate. *IOP Conf. Ser. Earth Environ. Sci.* **2017**, *94*, 012029. [CrossRef]

catalysts

MDPI

Article

The Activation of Oxygen Species on the Pt/CeO_2 Catalyst by H_2 for NO Oxidation

Xiangru Li [1], Shuangye Li [1], Yao Cheng [1], Lechen Zheng [2], Liyun Song [1], Xuehong Zi [2,*] and Hongxing Dai [2]

1 Key Laboratory of Beijing on Regional Air Pollution Control, Beijing University of Technology, Beijing 100124, China; lixiangru981026@163.com (X.L.); shuangyu0307@163.com (S.L.); chengyao2620@126.com (Y.C.); songly@bjut.edu.cn (L.S.)
2 Beijing Key Laboratory for Green Catalysis and Separation, College of Materials Science and Engineering, Beijing University of Technology, Beijing 100124, China; tonyzlcbj@163.com (L.Z.); hxdai@bjut.edu.cn (H.D.)
* Correspondence: zixuehong@bjut.edu.cn

Abstract: The Pt/CeO_2 catalyst has attracted significant attention due to its exceptional performance in NO oxidation. This study comprehensively examines the effects of calcination temperature and H_2 pretreatment on the structure and activity of the Pt/CeO_2 catalyst. Experimental findings indicate that the calcination temperature significantly affects the catalyst's redox performance, thereby modulating its efficacy in NO oxidation reactions. H_2 pretreatment facilitates the creation of oxygen vacancies on the catalyst, assisted by the reduction in PtO_x to Pt, enhancing the formation of activated oxygen and thereby improving NO oxidation. This study offers valuable insights into the design and optimization of Pt/CeO_2 catalysts for environmental applications, particularly in the development of exhaust gas after-treatment technologies.

Keywords: NO oxidation; Pt/CeO_2 catalyst; oxygen species activation; H_2 pretreatment

1. Introduction

Nitrogen oxides (NO_x) are among the primary atmospheric contaminants, contributing to various environmental hazards, including photochemical smog, acid rain, the greenhouse effect, and depletion of the stratospheric ozone layer [1,2]. NO oxidation plays a critical role in advanced exhaust gas treatment technologies, particularly in selective catalytic reduction (SCR) [3], non-selective catalytic reduction (NSCR) [4], and NO_x storage reduction (NSR) [5]. These technologies are predominantly employed to mitigate nitrogen oxide emissions, especially from automotive exhaust and power plants. It is widely recognized that the conversion of NO to NO_2 is a critical step in the rapid SCR reaction, significantly enhancing the rate of the SCR reaction [6]

Currently, a common challenge faced by all NO oxidation catalysts is their activity, which is restricted to a relatively narrow high-temperature range. This limitation not only reduces their efficiency in low-temperature environments but also increases energy consumption and operational costs. Consequently, there is an imperative need to innovate and develop catalysts that can perform effectively at lower temperatures. Such advancements would broaden their applicability and significantly enhance overall performance. To address this challenge, researchers are actively exploring various strategies to lower the activation temperature of catalysts. These approaches include optimizing the chemical composition and microstructure of catalysts, such as introducing novel active components, modifying the properties of the support material, incorporating promoters to alter surface and electronic characteristics, and utilizing nanotechnology to fabricate materials with enhanced surface activity and distinctive catalytic performance. Given the complex operating conditions of diesel vehicles, it is essential to investigate composite catalysts that improve the overall efficiency of NO_x reduction. Under fuel-rich conditions, the CO water–gas shift reaction can be employed to generate H_2, which then reacts with NO to produce NH_3, a key component for the downstream SCR unit. This approach aims to achieve a higher NO_x conversion rate than that observed under steady-state experimental conditions, effectively addressing the complexity of diesel exhaust emissions. During lean-burn conditions, diesel engines attain complete combustion, and molecular sieve catalysts adsorb the resulting NO_x. In the subsequent fuel-rich phase, NH_3 produced through the Cu/CeO_2 catalyst can then reduce the adsorbed NO_x. This innovative strategy eliminates the need for urea and platinum-group metals, significantly reducing operational costs. By employing such dual-function catalysts, this approach not only improves NO_x conversion efficiency but also offers a cost-effective solution for managing the fluctuating exhaust conditions of diesel engines. This intricate interplay of reactions ensures the system dynamically adapts to varying conditions, thereby optimizing catalytic performance and minimizing environmental impact [7].

For NO oxidation catalysts, the main categories include multi-metal oxide catalysts, perovskite catalysts, carbon-based catalysts, fast SCR catalysts, and supported catalysts [8]. Additionally, Wu et al. [9] introduced MnO_x-CeO_2 mixed oxide into a Pt/Al_2O_3 catalyst to enhance its NO oxidation activity. The superior redox properties of the MnO_x-CeO_2 mixed oxide significantly improved the catalyst's performance, achieving a maximum NO conversion rate of 80% at 350 °C. Nickolov et al. [10] and Zhuo et al. [11] demonstrated that TiO_2 promotes the superior dispersion of active components on its surface, effectively preventing the agglomeration of catalyst particles. Kim et al. [12] found that Sr-doped perovskite catalysts ($La_{0.9}Sr_{0.1}CoO_3$) exhibit excellent NO oxidation activity, achieving a maximum NO conversion rate of 86% at 300 °C. Guo et al. [13] studied the oxidation of NO using activated carbon (AC) derived from coconut shell, coal tar, and polyacrylonitrile. Under identical conditions, coconut shell-derived AC exhibited the highest activity, maintaining robust NO oxidation even in low concentrations of O_2. Additionally, noble metals exhibit excellent catalytic activity for NH_3-SCR at low temperatures. Kim et al. [14] evaluated the NO oxidation and NH_3-SCR of Pt/Al_2O_3, Cu/Al_2O_3, and Pt-Cu/Al_2O_3 catalysts. The Pt/Al_2O_3 catalyst exhibited the highest efficiency for both NO oxidation efficiency and NO_x removal. However, a common challenge associated with these metal oxide catalysts is their limited capacity to generate and migrate oxygen vacancies, as well as their poor thermal stability. Furthermore, achieving optimal catalytic performance often requires a high loading of precious metals, which limits their activity in NO oxidation reactions.

In the case of supported catalysts, which include noble metals and metal oxides, the active component predominantly resides on the catalyst surface, while the support offers a large surface area for separating the active phase and providing the necessary space for the catalytic reaction. Hernánove et al. [15] found that an Au-TiO_2 photocatalyst with a gold content of 0.5% exhibited the best catalytic performance, attaining an NO conversion rate of 85% within 60 min under UV irradiation. Single-atom catalysts (SACs) are synthesized using atomic-level dispersion techniques to create atomically dispersed catalysts (ADCs). In these catalysts, active metal atoms are uniformly distributed across the support, with each metal atom individually dispersed rather than forming metal nanoparticles [16]. Liu et al. utilized atomic dispersion techniques to load Ru metal onto $MgAl_{1.2}Fe_{0.8}O_4$ and compared its activity with that of conventional Ru nanocatalysts supported on $MgAl_{1.2}Fe_{0.8}O_4$. The results revealed that the atomically dispersed catalyst demonstrated superior N_2O decomposition performance under both low and high N_2O concentration conditions. Characterization data indicated that the atomically dispersed Ru acted as the primary active site responsible for this enhanced performance [17]. Due to its high reducibility and exceptional oxygen storage capacity, CeO_2 has been extensively used as a catalyst support or promoter [18]. Olsson's study confirms that the reaction rate of NO oxidation decreases with decreasing platinum particle sizes [19]. They explain that smaller platinum particles are more prone to the formation of PtO and PtO_2, which exhibit lower activity compared to metallic platinum. When Pt is loaded onto CeO_2, the strong interaction at the Pt-Ce interface often results in significant electron perturbation, thereby greatly enhancing the catalytic activity of the catalyst [20,21]. It is widely recognized that the CO oxidation process on Pt/CeO_2 surfaces involves a reaction between CO adsorbed on Pt and active oxygen species supplied by CeO_2 [22]. This process does not involve competitive adsorption between CO and O_2. Zhang et al. found that the composition of Au species in Au/CeO_2 catalysts remains relatively stable after redox reactions. In this context, Ce^{4+}-OH species are reduced to Ce^{3+} by CO, which are subsequently re-oxidized to Ce^{4+}-OH by H_2O, generating oxygen vacancies (Ce^{3+}-O_v-Ce^{3+}) in the process. They proposed an associative redox mechanism to explain the redox dynamics at the metal-support interface of the catalyst [23]. However, current research has not yet fully elucidated the state of Pt, the interactions between Pt and Ce, the redox cycling mechanisms between Ce^{3+} and Ce^{4+}, and their collective influence on NO oxidation activity.

In this study, a series of 1 wt% Pt/CeO_2 catalysts with varying calcination temperatures were prepared using the impregnation method. The NO oxidation activity of the Pt/CeO_2 catalysts, both before and after H_2 pretreatment, was evaluated. Characterization techniques such as XRD, TPD, TPR, and in situ DRIFTS were employed to investigate the impact of H_2 treatment on the activity of the catalyst annealed at different temperatures. The primary focus was on examining the influence of H_2 reduction treatment on catalyst activity and the synergistic effects between the Pt and CeO_2 surfaces. The role of active oxygen species on the Pt/CeO_2 surface was studied, and the catalytic sites and mechanisms of the NO oxidation reaction were elucidated based on proposed mechanisms in the literature.

2. Results and Discussion

2.1. NO Oxidation Performance

The NO oxidation performance of the Pt/CeO_2 catalyst gradually decreased with the increasing calcination temperature. This trend persisted even after the catalyst underwent H_2 pretreatment. During the H_2 pretreatment process, the total flow rate of 10% H_2/N_2 was 200 mL/min. The catalyst was thermally treated at a heating rate of 5 °C/min at 200 °C for 30 min. Moreover, NO conversion of the catalyst increased following H_2 pretreatment, with a particularly marked improvement observed in the catalysts calcined at 500 and 600 °C. Compared to the NO reaction rate of 0.13 μmol/(g·h) at 325 °C over the 1 wt% Pt/CeO_2 catalyst reported in the literature [24], those over the 1 wt% Pt/CeO_2-500 catalysts before and after H_2 reduction treatment obtained in the present work were 0.76 and 0.87 μmol/(g·h), respectively. Obviously, our catalyst outperformed the catalyst reported in the literature [24].

The catalytic stability test of the 1 wt% Pt/CeO_2-500 catalyst after H_2 pretreatment was carried out at 325 °C for 10 h of NO oxidation, and its result is shown in Figure 1c. Apparently, no significant decreases in NO conversion were observed in 10 h of NO oxidation under the adopted reaction condition.

2.2. Kinetic Study

The kinetics of the NO reaction on a 1 wt% Pt/CeO_2 catalyst were studied at various temperatures. The effects of calcination temperature and H_2 reduction treatment on the kinetics of NO oxidation were investigated. As the calcination temperature increased, the activation energy (E_a) of the catalyst also rose. The E_a of the catalyst decreased from 23.7 to 19.5 kJ/mol after H_2 pretreatment, which is comparable to the reported E_a value of 31.4 kJ/mol for 0.5 wt% Pt/CeO_2 [25]. Table 1 presents the activation energies of the catalysts with different calcination temperatures. As the calcination temperature increased, the E_a of the catalyst gradually rose, which is inconsistent with the activity test results. This indicates that the calcination temperature influences various properties of the catalyst, including the specific surface area, crystalline phase composition, metal dispersion, and metal-support interactions. These properties can affect the catalytic activity, and even if the E_a is higher, the overall performance of the catalyst may still improve. In comparison to Pt/SiO_2 (57.7 kJ/mol) and $Pt/CeZrO_2$ (34.6 kJ/mol) [26], the Pt/CeO_2 catalyst displays a lower activation energy, indicating a reduction in the reaction barrier. This ultimately facilitates the NO oxidation reaction, making it more favorable.

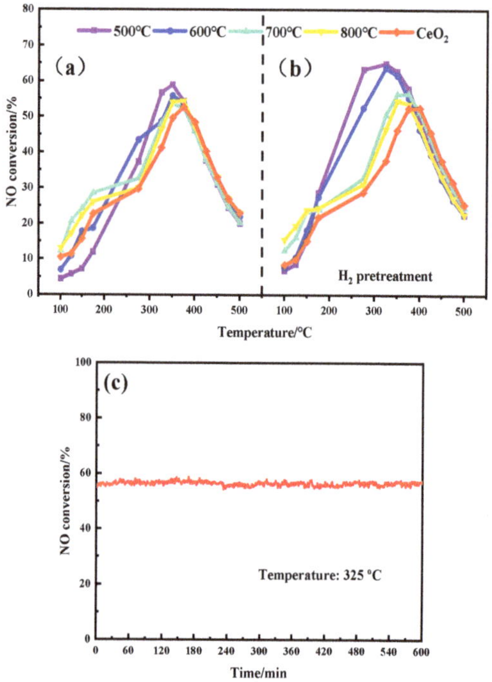

Figure 1. Catalytic activities of the 1 wt% Pt/CeO_2 catalysts (**a**) without hydrogen pretreatment and (**b**) with H_2 pretreatment, and (**c**) the catalytic stability of the 1 wt% Pt/CeO_2-500 catalyst after H_2 pretreatment. Reaction conditions: [NO] = 500 ppm, [O_2] = 10%, N_2 (balance), WHSV = 60,000 mL/(g·h).

Table 1. Summary of kinetic parameters for NO oxidation over Pt/CeO_2.

Catalyst	E_a (kJ/mol)		Feed Gas Composition	Ref.
	Fresh	H_2 Pretreatment		
1 wt% Pt/CeO_2-500	23.7	19.5	500 ppm NO, 10% O_2, N_2 (balance)	This work
1 wt% Pt/CeO_2-600	26.7	20.9		
1 wt% Pt/CeO_2-700	31.1	30.6		
1 wt% Pt/CeO_2-800	32.2	31.8		
0.5 wt% Pt/CeO_2-500	31.4	–	500 ppm NO, 10% O_2, N_2 (balance)	[25]
1.7 wt% Pt/SiO_2-500	57.7	–	250 ppm NO, 3.5%O_2, N_2 (balance)	[26]
0.35 wt% Pt/$CeZrO_2$-500	34.6	–	500 ppm NO, 8% O_2, N_2 (balance)	[26]

During the kinetic studies, it was observed that adjusting the O_2 concentration while keeping the NO concentration constant led to significant changes in the NO conversion of the catalyst. In contrast, varying NO concentration while maintaining a fixed O_2 concentration resulted in only minor changes in NO conversion. The results demonstrate a significant impact of O_2 concentration on the NO conversion efficiency of the catalyst, highlighting the dominant role of the oxygen vacancy mechanism. Notably, the concentration and replenishment of oxygen exhibit a direct correlation with O_2 concentration. When the O_2 concentration is increased, more oxygen molecules are adsorbed and dissociated, generating a greater number of active oxygen atoms. These atoms can quickly fill the oxygen vacancies, thereby enhancing NO conversion [27]. This finding highlights the critical importance of controlling oxygen vacancy dynamics to maximize catalytic performance, suggesting that optimizing O_2 concentration is essential for achieving higher NO conversions.

2.3. Catalyst Characterization

2.3.1. XRD Analysis

The XRD patterns of the catalysts exhibit characteristic peaks at 2θ values of 28.55°, 33.08°, 47.48°, 56.33°, 59.09°, 69.40°, 76.70°, 79.07°, and 88.41°, corresponding to the CeO_2 crystal planes with Miller indices (111), (200), (220), (311), (222), (400), (331), (420), and (422) (JCPDS PDF# 34-0394), respectively.

Figure 2 reveals that the XRD patterns of the Pt-loaded catalysts lack characteristic Pt diffraction peaks at 2θ angles of 39.8°, 46.2°, 67.5°, 81.2°, and 85.7° [28], indicating high dispersion of Pt species on the CeO_2 surface, potentially in an amorphous state or obscured by CeO_2 peaks. This suggests the absence of Pt clusters or particles. Notably, increasing the calcination temperature enhances the sharpness of CeO_2 diffraction peaks.

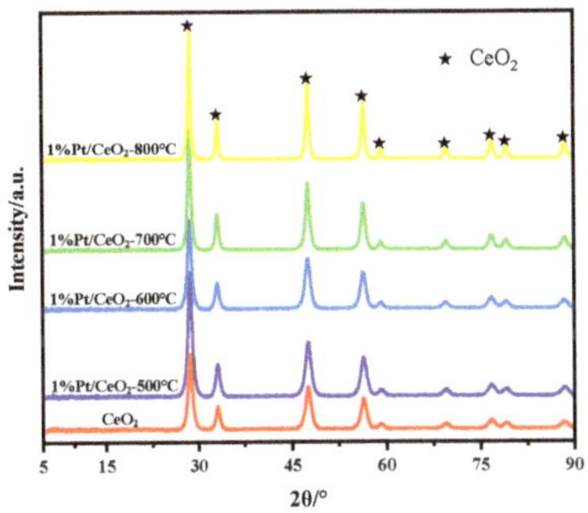

Figure 2. XRD patterns of the Pt/CeO_2 catalysts.

In practical applications, catalysts must withstand high temperatures over extended periods to prevent deactivation or performance degradation due to phase transformations. CeO_2 facilitates effective oxygen storage and release at high temperatures, enhancing the catalyst's redox properties. The stable crystal phase of CeO_2 provides an ideal carrier environment, promoting the uniform dispersion of Pt nanoparticles and strong metal-support interactions. This synergy enhances the catalyst's selectivity, stability, and overall performance.

2.3.2. TEM Analysis

The cubic structure of CeO_2 was identified by the lattice fringes of the (111) and (200) crystal planes with interplanar spacings of d = 0.32 and 0.29 nm, respectively. The lattice fringe of the Pt particles was measured at 0.22 nm, corresponding to the Pt (111) crystal plane. Figure 3e–i displays representative elemental distribution maps obtained via EDS, illustrating the distribution of Pt, Ce, and O elements in the 1 wt% Pt/CeO_2 catalyst. Notably, all three elements are uniformly distributed along the TEM cross-section. The density of Pt species on the CeO_2 support is relatively uniform, and the complementary profiles of O and Ce elements further indicate the uniform distribution of Pt particles on the surface of CeO_2, leading to the formation of more Pt-O-Ce bonds [29]. This enhances both the structural stability and catalytic activity of the catalyst. It can be found from the TEM images (Figure 3a,b) that the lattice fringes of CeO_2 and PtO_2 species are clearly seen in the Pt/CeO_2 catalysts calcined at 500 and 800 °C, indicating the possible formation of a Pt-O-Ce-like bond. Such a deduction was confirmed by the subsequent EDS mappings of Pt,

O, and Ce elements, in which the Pt species are uniformly dispersed on the CeO_2 support (i.e., there might be the formation of a Pt-O-Ce-like bond) [30,31]. The high-resolution TEM images of the 1 wt% Pt/CeO_2-500 catalysts before and after H_2 treatment are shown in Figure 3c,d. Due to the similar contrast of Pt and Ce, it was difficult to distinguish them in the high-resolution TEM images. Therefore, we did not provide the sizes of Pt particles before and after H_2 treatment.

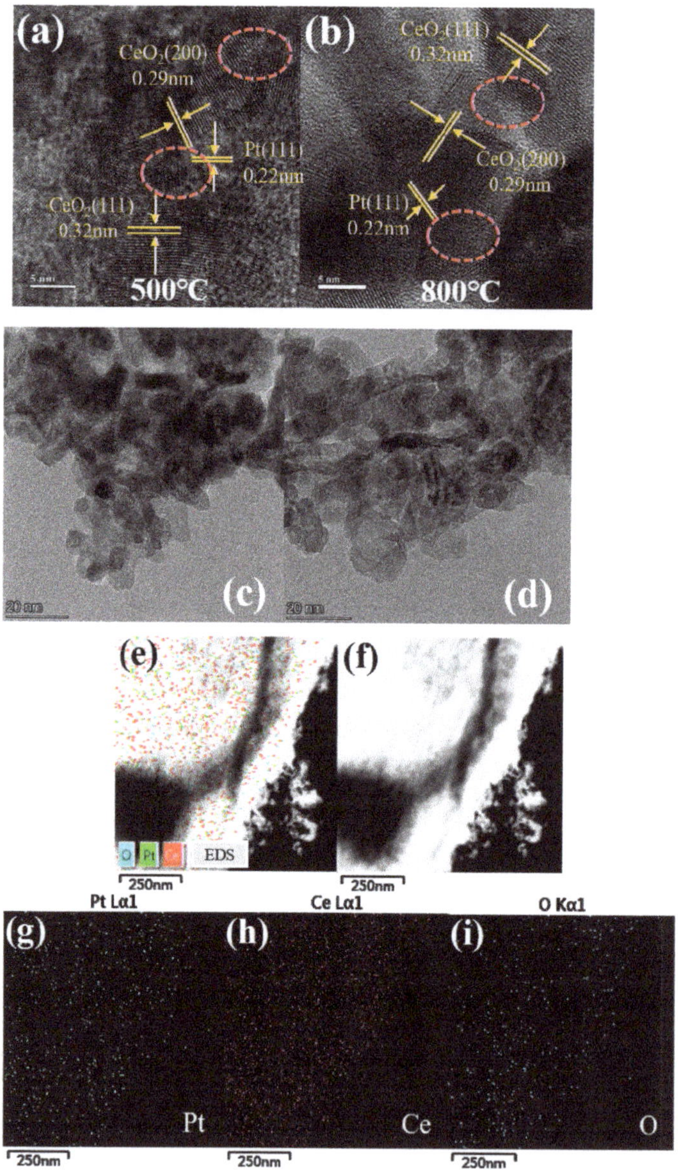

Figure 3. TEM images of the 1 wt% Pt/CeO_2 catalysts calcined at (**a**) 500 °C and (**b**) 800 °C, (**c**,**d**) HRTEM images of 1 wt% Pt/CeO_2-500 before and after H_2 pretreatment, and (**e**–**i**) corresponding EDS mappings of Pt, Ce, and O elements. The red circle in subfigure (**a**,**b**) is the interface between Pt and CeO_2.

2.3.3. H_2-TPR Analysis

Three principal reduction peaks are observed in the Pt/CeO_2 sample shown in Figure 4: a low-temperature reduction peak (α peak) at below 200 °C, attributed to the reduction in PtO_x [32]; a peak at approximately 266 °C (β peak), associated with the reduction in surface-active oxygen species near the Pt/Ce interface [33]; and a peak around 390 °C (γ peak), attributed to the reduction in surface lattice oxygen on CeO_2, away from Pt. Furthermore, this peak gradually shifts to lower temperatures as the catalyst calcination temperature increases. At lower temperatures, the formation of Pt-O-Ce bonds activates the O-Ce bonds through electron transfer, facilitating the reduction in surface oxygen species on the catalyst [27]. Characterization results exhibit a significant decrease in the low-temperature peak area of the catalyst following H_2 reduction treatment. This observation indicates that the H_2 reduction process effectively reduces some of the more readily reducible oxygen species or active sites on the catalyst surface, resulting in a diminished presence of oxygen species on the catalyst.

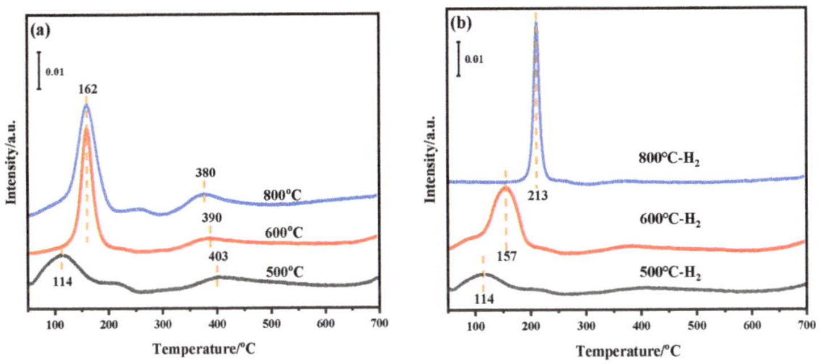

Figure 4. H_2-TPR profiles of the (**a**) fresh and (**b**) H_2 pretreated catalysts.

To further compare the low-temperature reducibility of the Pt/CeO_2 catalysts, the H_2 consumption of the catalysts was calculated, as summarized in Table 2. After air pretreatment, H_2 consumption increased with the rising calcination temperature, although the increase was not significant. However, for catalysts subjected to H_2 reduction treatment, the changes in H_2 consumption were more pronounced and did not follow a clear trend with the increasing calcination temperature. The 1 wt% Pt/CeO_2 catalyst subjected to H_2 reduction at 600 °C exhibited increased H_2 consumption compared with its 500 °C calcined counterpart, indicating a higher density of oxygen vacancies on its surface. However, its catalytic activity was surprisingly lower, suggesting that the mere presence of oxygen vacancies alone is insufficient to enhance NO oxidation. Efficient oxygen migration and supply mechanisms are essential for facilitating this reaction, highlighting that the oxygen vacancy quantity alone is not the sole determining factor.

Table 2. H_2 consumption (μmol) of the catalysts after H_2 or air pretreatment.

Catalyst	1 wt% Pt/CeO_2-500	1 wt% Pt/CeO_2-600	1 wt% Pt/CeO_2-800
Fresh	75.81	81.20	108.62
H_2 pretreatment	51.60	74.94	50.03

2.3.4. NO-TPSR Analysis

The variation in NO_2 product signal intensity on the Pt/CeO_2 catalyst and the temperature are depicted in the graph. As shown in the NO-TPSR profiles of the 1 wt% Pt/CeO_2-500 and 1 wt% Pt/CeO_2-800 samples (Figure 5), the intensity of the NO_2 signal was stronger in the sample treated in H_2 at 500 or 800 °C than that in the sample treated in air at 500 or 800 °C, indicating that H_2 treatment was beneficial for the improvement in catalytic activity of the sample. The worse activity of the sample treated at 800 °C than the sample treated at 500 °C might be due to its sintering. The most significant change in the low-temperature peak intensity suggests that the catalyst displays an increased number of surface Pt active sites and oxygen vacancies following H_2 reduction treatment. This enables the availability of a greater number of oxygen species at lower temperatures, thereby facilitating NO dissociation [34]. Consequently, the catalyst treated with H_2 reduction exhibits enhanced NO oxidation activity, aligning with the findings from catalyst activity testing.

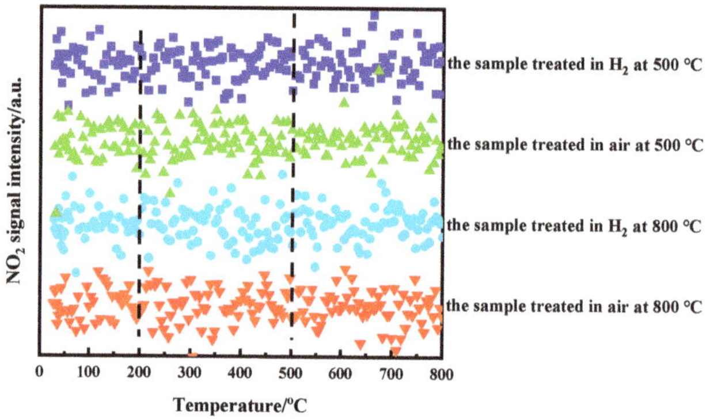

Figure 5. NO-TPSR profiles of the catalysts before and after H_2 pretreatment at 500 and 800 °C.

2.3.5. O_2-TPD Analysis

As shown in Figure 6, the peak at approximately 108 °C, within the low-temperature range (<200 °C), is attributed to the desorption of weakly adsorbed oxygen species, O_2^- (ads), from the catalyst surface. The peak centered at 282 °C is associated with the desorption of chemisorbed oxygen, O^- (sur), which is predominantly bound to the active sites on the CeO_2 surface. Furthermore, the incorporation of Pt enhances the activation and adsorption of oxygen species, although the primary adsorption sites remain on the CeO_2 surface. Notably, the peak at approximately 490 °C is attributed to the desorption of lattice oxygen, O^{2-} (latt), which is strongly bound to the surface and requires elevated temperatures for release [35]. The catalytic activity of metal oxides is strongly influenced by surface-adsorbed oxygen, which possesses greater mobility and reactivity than lattice oxygen, facilitating its desorption and subsequent participation in catalytic reactions.

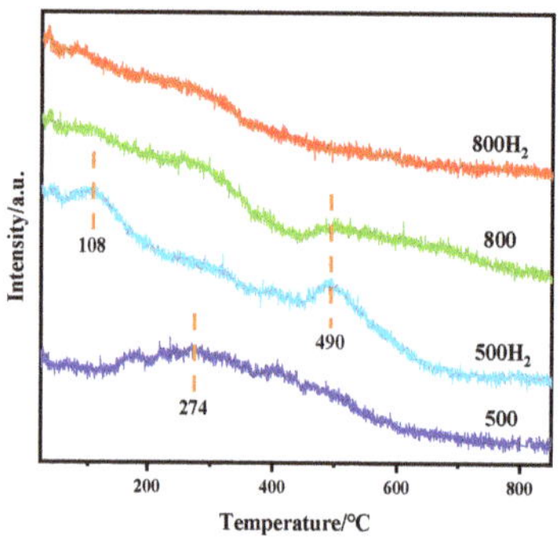

Figure 6. O_2-TPD profiles of the catalysts before and after H_2 pretreatment at calcination temperatures of 500 and 800 °C.

Oxygen vacancies are typically regarded as centers for oxygen adsorption and desorption, acting as a bridge for the adsorption of active gaseous oxygen [36]. Compared to catalysts calcined at 800 °C, those calcined at 500 °C exhibit stronger desorption peaks and larger peak areas for chemisorbed oxygen, indicating that a greater amount of adsorbed oxygen is released from the catalyst. This suggests an increased desorption of adsorbed oxygen from the catalyst, implying an enhanced oxygen supply capability. However, at 800 °C, Pt particle sintering on the catalyst surface likely occurs, leading to increased particle size, reduced surface area, and decreased active sites. Consequently, the O_2 adsorption capacity decreases, yielding a smaller desorption peak area. Furthermore, H_2 reduction treatment modifies the surface structure, leading to reduced active sites and decreased adsorbed oxygen species. The greater the amount of desorbed oxygen, the higher the oxygen migration rate, indicating a larger number of oxygen vacancies and a richer abundance of active sites, which promotes the catalytic oxidation of NO. This is consistent with the catalyst's activity in NO oxidation.

2.3.6. XPS Analysis

The metal chemical valence and adsorbed oxygen species of the catalysts play important roles in catalyzing NO oxidation. XPS characterization experiments for the Pt/CeO_2 catalysts were performed, and the results are shown in Figure 7 and Table 3. Each of the asymmetrical O 1s XPS spectra of the catalysts were divided into three components at binding energies of 529.3, 531.3, and 533.1 eV (Figure 7a), which were attributed to the surface lattice oxygen (O_{latt}), adsorbed oxygen (O_{ads}), and adsorbed water or carbonate species [37], respectively. The O_{ads}/O_{latt} molar ratios of Pt/CeO_2-500 and Pt/CeO_2-600 increased from 0.35 and 0.32 to 0.51 and 0.35 after H_2 pretreatment, respectively. The increase in the O_{ads}/O_{latt} molar ratio enhanced the catalytic activity of the Pt/CeO_2 catalyst for NO oxidation.

Table 3. The surface element compositions of the 1 wt% Pt/CeO_2 catalysts obtained through the XPS technique.

Catalyst	Molar Ratio		
	$Pt^0/(Pt^{2+} + Pt^{4+})$	Ce^{3+}/Ce^{4+}	O_{ads}/O_{latt}
1 wt% Pt/CeO_2-500	0.86	0.18	0.35
1 wt% Pt/CeO_2-600	0.48	0.25	0.32
1 wt% Pt/CeO_2-500-H_2	0.98	0.49	0.51
1 wt% Pt/CeO_2-600-H_2	0.86	0.37	0.35

The Pt 4f XPS spectra of the Pt/CeO_2-500 and Pt/CeO_2-600 catalysts are shown in Figure 7b. Pt was present mainly in a metallic state (Pt^0) and as platinum oxide (Pt^{n+}, i.e., Pt^{2+} and Pt^{4+}). The deconvoluted components at binding energies of 71.5 and 74.8 eV were related to the $4f_{7/2}$ and $4f_{5/2}$ states of the surface Pt^0 species, those at binding energies of 72.6 and 75.9 eV were assigned to the $4f_{7/2}$ and $4f_{5/2}$ of the surface Pt^{2+} species, and the those at binding energies of 74.2 and 77.5 eV were ascribed to $4f_{7/2}$ and $4f_{5/2}$ states of the surface Pt^{4+} species [27]. The signals of Pt $4f_{5/2}$ were 3.3 eV higher than those of Pt $4f_{7/2}$. After H_2 treatment, the Pt^{4+} species nearly disappeared. The $Pt^0/(Pt^{2+} + Pt^{4+})$ molar ratios on the surface of the 1 wt% Pt/CeO_2-500 and 1 wt% Pt/CeO_2-600 catalysts before and after H_2 treatment were from 0.86 and 0.48 to 0.98 and 0.86, respectively. In other words, the rise in the $Pt^0/(Pt^{2+} + Pt^{4+})$ molar ratio increased the activity of the Pt/CeO_2 catalyst after H_2 treatment for NO oxidation [31].

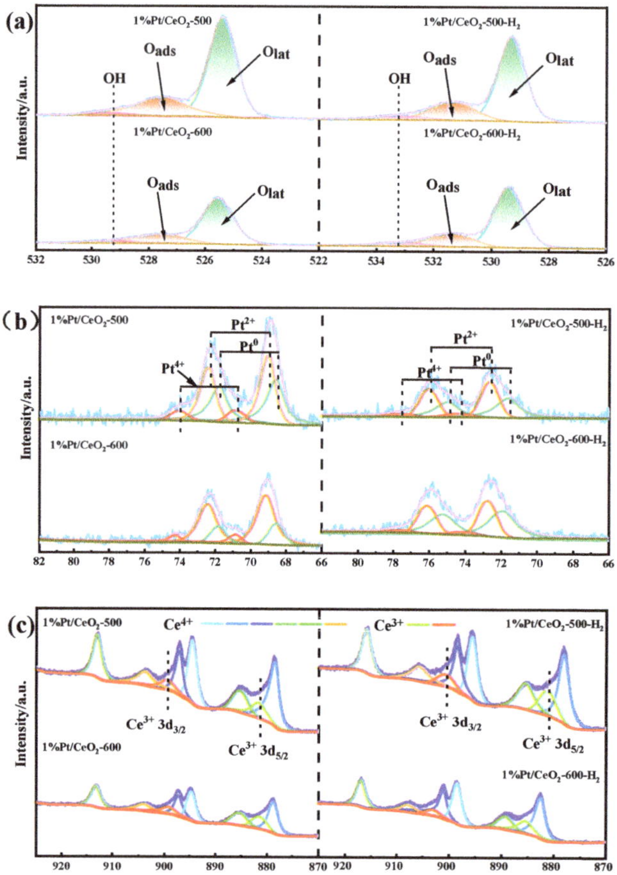

Figure 7. (**a**) O 1s, (**b**) Pt 4f, and (**c**) Ce 3d XPS spectra of the as-prepared catalysts.

The deconvolution of Ce 3d XPS spectra (Figure 7c) indicates that both Ce^{3+} and Ce^{4+} species exist on the surface of Pt/CeO_2. It is generally believed that oxygen vacancies are generated to maintain the electroneutrality of a catalyst, owing to the existence of Ce^{3+}. The higher the content of Ce^{3+}, the more oxygen vacancies. The H_2 pretreatment increased the content of Ce^{3+}, leading to an increase in the oxygen vacancy concentration on the catalyst surface. The Ce^{3+}/Ce^{4+} molar ratios of the 1 wt% Pt/CeO_2-500 and 1 wt% Pt/CeO_2-600 catalysts before and after H_2 treatment were from 0.18 and 0.25 to 0.49 and 0.37, respectively. It is well known that oxygen vacancies on the surface of a catalyst can adsorb O_2 to generate the reactive adsorbed oxygen species [37]. Hence, the more oxygen vacancies, the better the catalytic activity of the Pt/CeO_2 catalyst.

The XPS spectra of the 1 wt% Pt/CeO_2-500 and 1 wt% Pt/CeO_2-600 catalysts before and after H_2 pretreatment are shown Figure 7. The positions of the O 1s, Pt 4f, and Ce 3d peaks of the two catalysts after H_2 pretreatment were significantly shifted, revealing the formation of a Pt-O-Ce-like bond in the Pt/CeO_2 catalysts.

2.3.7. In Situ DRIFTS Analysis

NO Adsorption and Desorption

Figure 8 presents the DRIFTS spectra of NO adsorption on the Pt/CeO_2 catalyst. The prominent band at 1608 cm^{-1} is typically attributed to NO_2 species on the catalyst surface. As the adsorption time increases, the band intensity also increases, indicating the reaction of adsorbed NO with lattice oxygen on the catalyst to form NO_2 species. The bands at 1369 and 1536 cm^{-1} are attributed to monodentate nitro (-NO_2) and bidentate nitrate species, respectively. The band at 1450 cm^{-1} corresponds to linear nitrite [25]. Bands at 1273 and 1238 cm^{-1} are attributed to monodentate nitrate species. The band at 1012 cm^{-1} is associated with N-O stretching vibrations of nitrate or nitrite species, formed through the reaction of NO with surface oxygen. Under continuous N_2 purging, the positions and intensities of these bands remain unchanged, indicating the stability of the relevant surface species and chemical states.

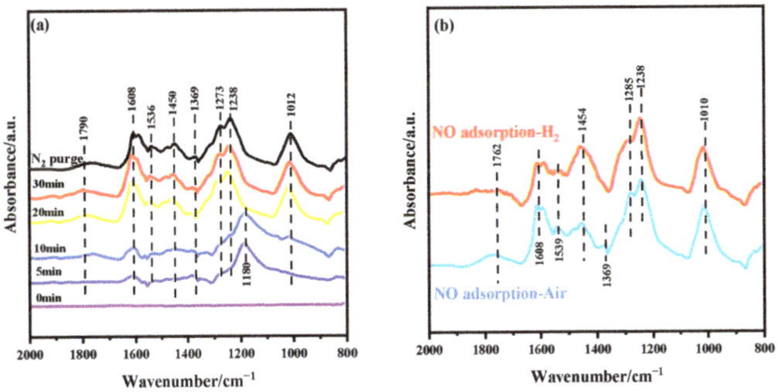

Figure 8. In situ DRIFTS spectra of NO adsorption and desorption after air treatment over the 1 wt% Pt/CeO_2-500 catalyst: (**a**) adsorption and desorption, (**b**) a comparison of the different treatment methods.

As shown in Figures 8b and S1, the NO adsorption spectra following hydrogen and air treatments exhibit similar trends. After air pretreatment, the Pt nanoparticles on the catalyst are oxidized to higher oxidation states, such as Pt^{2+} or Pt^{4+}, resulting in weaker CO adsorption by the oxidized Pt species. This weaker adsorption is typically indicated by higher-frequency vibrational modes. Additionally, air pretreatment creates oxygen vacancies on the surface of CeO_2, which serve as active sites and influence the CO adsorption behavior. Following air treatment, the intensity of the band attributed to NO_2 increases, which may be associated with the rise in surface-adsorbed oxygen. Notably, after H_2 treatment, the bands at 1539 and 1369 cm^{-1} weaken or disappear, while the band attributed to linear nitrite (1450 cm^{-1}) intensifies significantly. The H_2 reduction treatment increases oxygen vacancies on the catalyst surface and enhances surface reducibility, potentially leading to the reduction or conversion of nitrate species to nitrite (NO_2^-).

Pt and Oxygen Species Confirmation Probed by CO over Pt/CeO_2

To clearly identify Pt species and oxygen species on the Pt/CeO_2 catalyst, CO was introduced to a chamber as a probe molecule due to its strong adsorption capability on various active sites of the catalyst surface. This technique facilitates the identification of intermediate species involved in CO-related chemical reactions on the catalyst surface, thereby enhancing the understanding of surface chemical reaction processes. It also enables the inference of the presence and distribution of different active sites and indirectly provides insights into changes in the surface oxidation states of CeO_2.

The CO-DRIFTS spectra of CO adsorption on the catalysts are shown in Figure 9. The band at 2086 cm^{-1} is associated with the linear adsorption of CO on Pt nanoparticles. For the catalyst calcined at 500 °C, the band observed at 2117 cm^{-1} is attributed to the linear adsorption of CO on the oxidized Pt species, the band at 2166 cm^{-1} is linked to CO adsorbed on the Ce^{3+} ions [34,38], and the bands at 1481 and 1566 cm^{-1} are related to the products (e.g., carbonates) of CO oxidation [30].

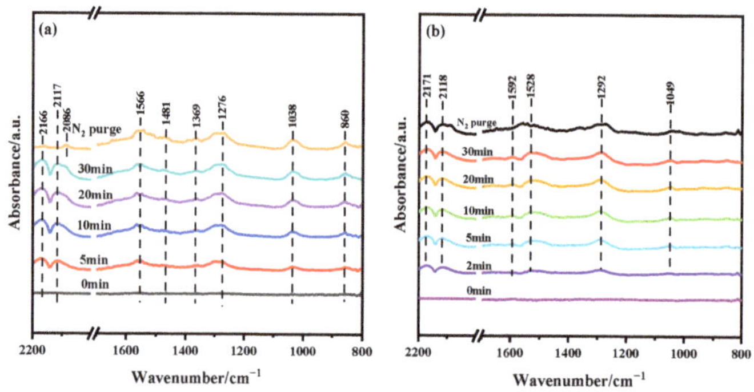

Figure 9. In situ DRIFTS spectra of CO adsorption and desorption after H_2 treatment over (**a**) 1 wt% Pt/CeO_2-500 and (**b**) 1 wt% Pt/CeO_2-800 catalysts.

With continuous purging by N_2, the CO adsorption bands on Pt^{2+} and Ce^{3+} gradually disappear, while the CO adsorption band on Pt^0 weakens but remains at a certain intensity. This suggests that the CO adsorption on Ce^{3+} and Pt^{2+} is relatively weak, making it more easily removed during N_2 purging [39]. Therefore, it can be inferred that the CO adsorption bonds at these sites have lower binding energies. The interaction between CO molecules and the surface is relatively weak, making them more susceptible to disruption during N_2 purging. In contrast, CO adsorption on Pt^0 is stronger and persists even during N_2 purging, indicating that the CO adsorption bond energy on Pt^0 is higher and less easily disrupted. This suggests that different adsorption sites exhibit varying selectivity toward CO adsorption. CO adsorption sites on Pt^0 may possess higher stability and selectivity, while those on Ce^{3+} and Pt^{2+} are more vulnerable to environmental changes, such as N_2 purging.

Compared to the catalyst calcined at 500 °C, the CO-DRIFTS spectrum bands of the catalyst calcined at 800 °C shifted to higher wavenumbers. The bands at 1592 and 1528 cm^{-1} were attributed to the products (e.g., carbonates) of CO oxidation [40]. The bands within the 800–1200 cm^{-1} range were likely associated with adsorption modes linked to oxygen vacancies on the CeO_2 surface or Pt/CeO_2 interface, suggesting that oxygen vacancies generated by H_2 pretreatment significantly impact CO adsorption. Figure S2 shows that after 30 min N_2 purging, the bands at 2117 and 2166 cm^{-1} for the 500 °C-calcined catalyst significantly diminish, whereas those for the 800 °C-calcined catalyst remain largely unchanged [41]. This indicates that the 500 °C-calcined catalyst has a weaker affinity for CO, and the catalyst calcined at 800 °C maintains its surface species more effectively, likely due to stronger binding sites or a more stable surface structure.

H_2 pretreatment can modify the surface chemistry and structure of the catalyst, thereby influencing its adsorption behavior towards different species. For Pt/CeO_2 catalysts, H_2 pretreatment typically reduces a portion of the Pt and CeO_2, leading to the formation of more metallic Pt nanoparticles and oxygen vacancies. These metallic Pt nanoparticles exhibit higher surface activity and stronger CO adsorption capability, which influence the adsorption mode and vibrational frequencies of CO. As the adsorption time increases, CO gradually accumulates onto the catalyst surface, resulting in an increased adsorption amount.

O_2 and Pre-Adsorbed NO

The transient reaction of O_2 with pre-adsorbed NO was investigated, and the corresponding DRIFTS spectra for this process over the 1 wt% Pt/CeO_2-500 catalyst are presented in Figure 10. As shown in Figure 10a, NO pretreatment yields various adsorbed NO species on the catalyst. Specifically, the intense band at ca. 1176 cm^{-1} is attributed to nitrite species, while the weaker band around 1296 cm^{-1} is assigned to monodentate nitrite species [25]. As time progresses, significant changes occur in the intensities of these bands.

This process is believed to involve the reaction of NO with oxygen species on the catalyst surface, oxygen vacancies on CeO_2, or oxygen associated with PtO_x. Notably, the spectral bands corresponding to bridging bidentate nitrate species (1569 cm^{-1}) progressively decrease in intensity over time, indicating a possible conversion to nitrite species. The NO or nitrite species at 1473 and 1386 cm^{-1} gradually transform into nitrate species at 1435 cm^{-1} after the introduction of O_2. The nitrite species at 1176 cm^{-1} are progressively converted into monodentate nitrate or bridging nitrate species at 1227 and 1257 cm^{-1}, as well as bidentate nitrate species at 1015 cm^{-1}. The nitrate species are relatively stable and did not signify notable changes during subsequent cyclic adsorption processes. These observations suggest that NO is adsorbed on Pt/CeO_2 as a mixture of nitrite and nitrate species, a finding that aligns with the results reported by Philipp et al. [42] The decomposition of nitrate species occurs within the temperature range of 300–500 °C, releasing NO into the gas phase. These NO molecules can undergo further oxidation to form NO_2 on the catalyst surface [43]. At appropriate temperatures, the decomposition of nitrate species and the oxidation of NO can synergistically enhance the efficiency of the catalyst. This observation is consistent with the results obtained from activity tests conducted at different temperatures.

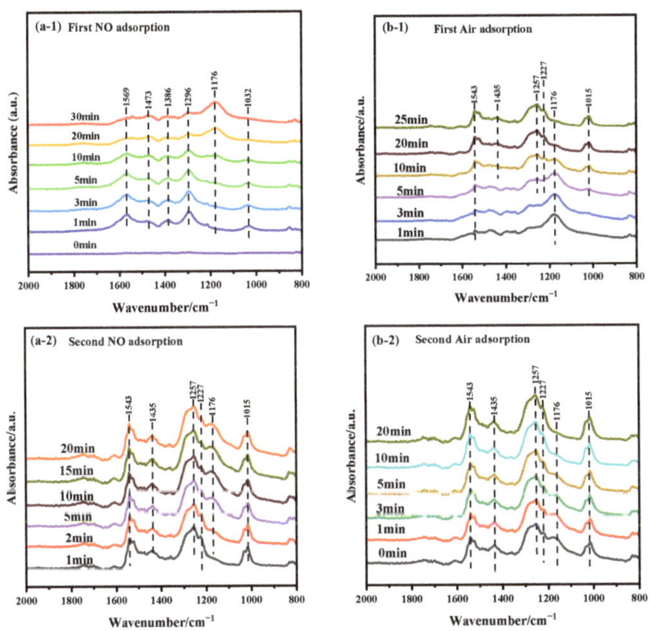

Figure 10. In situ DRIFTS spectra of NO and O_2 adsorption after H_2 treatment over the 1 wt% Pt/CeO_2-500 catalyst: (**a**) NO adsorption and (**b**) O_2 adsorption.

Upon air introduction, the reaction between O_2 and pre-adsorbed NO occurs. At 200 °C, a distinct spectral band emerges at 1543 cm^{-1}, indicating the predominant formation of monodentate nitrate species as a result of NO oxidation. Notably, the monodentate nitrate species exhibit relatively low stability, thereby facilitating NO oxidation. In the second cycle of NO adsorption and O_2 adsorption, the DRIFTS spectra remain largely unchanged, with consistent band positions and intensities. This suggests that the adsorbed species, including nitrate and nitrite, have attained a dynamic equilibrium on the catalyst surface. Subsequent NO and O_2 adsorption does not yield additional reaction product intermediates, indicating equilibrium stabilization.

2.4. Catalytic Mechanism

The kinetic test results indicate that NO oxidation on the catalyst surface proceeds via the oxygen vacancy mechanism. Consequently, the catalytic mechanism for the 1 wt% Pt/CeO_2 catalyst is illustrated in Figure 11. During the initial adsorption phase, NO is more readily adsorbed onto the metallic Pt nanoparticles than on the Pt oxides. This adsorption is notably stronger, resulting in the formation of stable chemical bonds that are essential for the initial activation of NO. After H_2 pretreatment, the Ce^{4+} species in CeO_2 are partially reduced to the Ce^{3+} species, forming a large number of oxygen vacancies on the catalyst surface. Molecular oxygen are first adsorbed at the oxygen vacancies and Pt sites, which then react with the adsorbed NO. The active adsorbed oxygen species might be the most effective factor for the high activity of the Pt/CeO_2 catalyst. This unique property enables efficient oxygen extraction and migration from the Pt-Ce interface to the Pt nanoparticles, resulting in the formation of abundant oxygen vacancies on the catalyst surface. The surface oxygen vacancies play a vital role in enhancing the Pt/CeO_2 system's catalytic performance, especially after H_2 reduction, which enhances reactive oxygen species generation and reactant adsorptions, thereby optimizing NO oxidation.

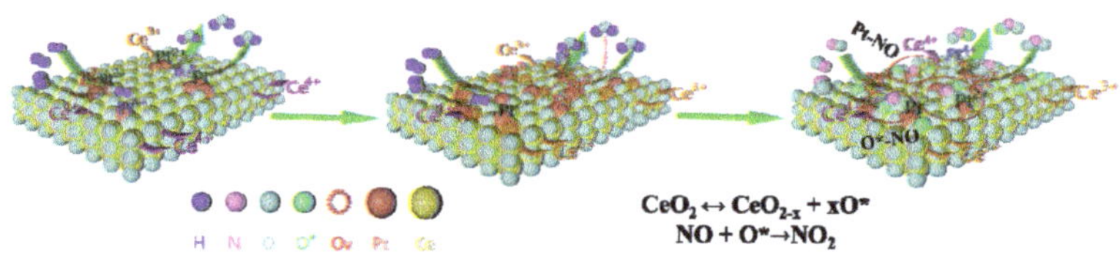

Figure 11. Catalytic mechanism diagram of NO oxidation over Pt/CeO_2.

3. Experimental Section

3.1. Catalyst Synthesis

A total of 1 wt% Pt/CeO_2 noble metal catalysts were prepared at different calcination temperatures using the impregnation method. First, cerium nitrate was calcined at 550 °C in a muffle furnace with a heating rate of 5 °C/min for 4 h to obtain CeO_2. A specific amount of chloroplatinic acid and CeO_2 was then dissolved in water according to the calculated ratio. The mixture was stirred in a water bath until the water completely evaporated and was subsequently dried in an oven at 110 °C. After drying, the solid was ground into a powder using an agate mortar and pestle and then calcined at different temperatures in a muffle furnace with a heating rate of 10 °C/min for 4 h. The mass fraction of Pt was 1 wt%. The resulting catalysts were labeled as 1 wt% Pt/CeO_2-500, 1 wt% Pt/CeO_2-600, 1 wt% Pt/CeO_2-700, and 1 wt% Pt/CeO_2-800, respectively.

3.2. Catalytic Testing

An activity evaluation was conducted in a fixed-bed quartz tubular reactor, with the tested temperature range set between 100 and 500 °C. The NO oxidation performance was assessed using 1.0 g of catalyst in a simulated feed gas containing 500 ppm NO, 10% O_2, and N_2 as the balance gas. The total gas flow rate was 1000 mL/min, and the weight hourly space velocity (WHSV) was controlled at 60,000 mL/(g·h) using mass flow meters. The gas mixture at the reactor outlet included NO, NO_2, and NO_x, which were monitored using an exhaust gas analyzer (Thermo 42i-HL, Waltham, MA, USA). The NO conversion was analyzed using the following equations:

$$\text{NO conversion} = \frac{NO_{2_{out}}}{NO_{x_{out}}} \times 100\% \quad (1)$$

where $[NO_x] = [NO] + [NO_2]$. The subscript "out" represents the concentrations of the corresponding substances at the inlet and outlet, respectively.

Kinetic measurements were carried out at different temperatures while maintaining the NO conversion below 20% to eliminate the effects of heat and diffusion. The feed gas consisted of NO (300–800 ppm), O_2 (5%), and N_2 as the balance gas, with a total flow rate of 400 mL/min. The gas concentrations were continuously monitored using a NO_x analyzer (Thermo, 42i-HL). To better evaluate the catalytic activity of the catalyst, the following equation was used to calculate the kinetic parameters for NO oxidation:

$$k = -\frac{V}{W} \times \ln(1 - x) \tag{2}$$

where k is the reaction rate constant ($cm^3/(g \cdot s)$), V is the total gas flow rate (cm^3/s), W is the actual mass of the catalyst (g), and x is the NO conversion.

The apparent activation energy was calculated using the Arrhenius equation, given by the following equation:

$$K = A \cdot \exp\left(-\frac{E_a}{RT}\right) \tag{3}$$

where k is the rate constant, R is the molar gas constant, T is the thermodynamic temperature, E_a is the apparent activation energy, and A is the pre-exponential factor. This equation can also be expressed as $\ln k = \ln A - E_a/RT$. By plotting the data, a straight line could be obtained. The apparent activation energy was determined from the slope of this line, and the pre-exponential factor was calculated from the intercept.

3.3. Catalyst Characterization Methods

X-ray diffraction (XRD) was conducted using a Bruker/AXS D8 Advance X-ray diffractometer (Bruker, Billerica, MA, USA) with Cu Kα radiation (λ = 1.54 Å) over a 2θ range of 10 to 80°. Detailed morphological and structural information about the Pt/CeO_2 catalyst was obtained through transmission electron microscopy (TEM).

Hydrogen temperature programmed reduction (H_2-TPR) was conducted using an Auto Chem II 2920 chemisorption analyzer (Micromeritics, Norcross, GA, USA). The testing method involved weighing 0.1 g of the catalyst sample with a particle size of 0.3–0.45 mm into a U-shaped reaction tube. The sample was heated under air conditions at a rate of 10 °C/min to 300 °C and maintained at the temperature for 1 h. It was then purged with helium (He) and cooled to room temperature. Subsequently, the sample was exposed to 10% H_2/Ar and heated at a rate of 10 °C/min to 800 °C, with signal detection carried out using a thermal conductivity detector (TCD).

For the nitric oxide temperature-programmed surface reaction (NO-TPSR), 50 mg of the catalyst was preheated in air at 500 °C for 30 min, cooled to room temperature, and then purged with 900 ppm NO/N_2 gas until the baseline stabilized. The temperature was subsequently increased from room temperature to 800 °C at a rate of 10 °C/min. The testing conditions for the air reduction in the samples at 500 and 800 °C are as follows: 50 mg of the catalyst was heated under air conditions at 500 °C for 30 min, then cooled to 200 °C in the same atmosphere. The catalyst was subsequently exposed to 5% H_2/Ar for 30 min, followed by cooling to room temperature. The sample was then purged with 5% NO/N_2 gas until the baseline stabilized, after which the temperature was increased from room temperature to 800 °C at a rate of 20 °C/min. The desorbed gas was detected using a mass spectrometer (QGA, Hiden, Warrington, UK).

For oxygen temperature-programmed desorption (O_2-TPD), a 100 mg sample was pretreated at 300 °C for 1 h. After cooling to 30 °C, the sample was exposed to 10% synthetic air for 1 h until saturation was achieved. The Ar flow (40 mL/min) was then activated for 1 h to remove weakly adsorbed O_2 from the surface. Finally, the desorption process was conducted by ramping the temperature from 30 to 800 °C at a rate of 10 °C/min under an Ar atmosphere, and the desorbed gas was analyzed using a mass spectrometer (MS).

X-ray photoelectron spectroscopy (XPS) was used to investigate the surface elemental compositions of the catalysts using a Thermo Scientific ESCALAB 250xi (Thermo Fisher Scientific, Waltham, MA, USA) with a 0.1 eV step and a 46.95 eV pass energy (PE) rate.

The X-ray source was monochromatized with Al Kα radiation ($h\nu$ = 1486.6 eV), and the contaminant carbon peak (C 1s = 284.8 eV) was used to calibrate the shift in binding energy.

In situ diffuse reflectance infrared Fourier transform spectroscopic (In situ DRIFTS) experiments of the catalysts were conducted using an infrared spectrometer (Nicolet 6700, Thermo Fisher Company, Waltham, MA, USA) with a liquid-nitrogen-cooled MCT detector. The samples were preheated at 300 °C in an N_2 atmosphere for 1 h to remove water and adsorbed gases from the surface. As the sample cooled to 50 °C, the background spectrum was collected. All spectra were recorded by accumulating 32 scans with a resolution of 4 cm^{-1}.

4. Conclusions

This study explores the influence of calcination temperature and H_2 pretreatment on NO oxidation catalysis, revealing the critical role of oxygen vacancies in enhancing reaction rates. However, excessive temperature exposure is detrimental, as it depletes surface-active oxygen and inhibits NO oxidation performance. The H_2 pretreatment-induced augmentation of surface oxygen vacancies in 1 wt% Pt/CeO_2-800 and 1 wt% Pt/CeO_2-500 catalysts substantially enhances NO oxidation. This highlights the pivotal role of oxygen vacancy dynamics in driving catalytic performance, emphasizing the importance of optimizing calcination temperature and oxygen vacancy availability to achieve maximum NO conversion, as the presence and renewal of oxygen vacancies are integral to the catalyst's efficacy.

Supplementary Materials: The following supporting information can be downloaded at https://www.mdpi.com/article/10.3390/catal14110778/s1, Figure S1: In situ DRIFTS spectra of NO and CO adsorption and desorption after H_2 or air treatment over the 1 wt% Pt/CeO_2-500 catalyst: (a) adsorption and (b) desorption; Figure S2: In situ DRIFTS spectra of CO adsorption and desorption after H_2 or air treatment over the 1 wt% Pt/CeO_2-800 catalyst: (a) adsorption and (b) desorption.

Author Contributions: Conceptualization, X.L., S.L., Y.C., L.Z., L.S., X.Z. and H.D.; Methodology, X.L., S.L., Y.C., L.Z., L.S., X.Z. and H.D.; Formal Analysis, X.L., S.L. and L.S.; Investigation, X.L., S.L., Y.C. and L.Z.; Resources, L.S., X.Z. and H.D.; Writing—Original Draft Preparation, X.L. and S.L.; Writing—Review Editing, L.S., X.Z. and H.D.; Visualization, X.L., S.L., Y.C. and L.Z.; Supervision, L.S., X.Z. and H.D.; Funding Acquisition, L.S., X.Z. and H.D. All authors have read and agreed to the published version of the manuscript.

Funding: This work was financially supported by the National Key R&D Program of China (2022YFB3506200) and the National Natural Science Foundation of China (21806005).

Institutional Review Board Statement: Not applicable.

Informed Consent Statement: Not applicable.

Data Availability Statement: Data are contained within the article and Supplementary Materials.

Conflicts of Interest: The authors declare no conflicts of interest.

References

1. Cheng, L.; Sin, S.; Ji, J.; Yang, S.; Tan, C.; Gu, Z.; Song, W.; Huang, C.; Sun, C.; Tang, C.; et al. Selective catalytic reduction of NO with NH_3 over MnO_x-CeO_2 catalysts: The great synergy between CeO_2 and crystalline phase of Mn_3O_4. *Fuel* **2023**, *342*, 127772. [CrossRef]
2. He, X.; Chiu, Y.H.; Chang, T.H.; Lin, T.Y.; Wang, Z. The energy efficiency and the impact of air pollution on health in China. *Healthcare* **2020**, *8*, 29. [CrossRef] [PubMed]
3. Tang, X.; Wu, P.; Wang, Y.; Liu, Y. Recent advances in heavy metal poisoning mechanism and regeneration methods of selective catalytic reduction (SCR) denitration catalyst. *Fuel* **2024**, *355*, 129429. [CrossRef]
4. Javed, M.T.; Irfan, N.; Gibbs, B.M. Control of combustion-generated nitrogen oxides by selective non-catalytic reduction. *J. Environ. Manag.* **2007**, *83*, 251–289. [CrossRef]
5. Shigemoto, A.; Inoda, Y.; Ukai, C.; Higo, T.; Oka, K.; Sekine, Y. Electric field-assisted NSR process for lean NO_x reduction at low temperatures. *Chem. Commun.* **2024**, *60*, 1563–1566. [CrossRef]
6. Bendrich, M.; Scheuer, A.; Hayes, R.E.; Votsmeier, M. Unified mechanistic model for standard SCR, fast SCR, and NO_2 SCR over a copper chabazite catalyst. *Appl. Catal. B* **2018**, *222*, 76–87. [CrossRef]

7. Zhang, N.; Qian, Y.; Toyao, T.; Shimizu, K. Continuous unsteady-state De-NO_x system via tandem water–gas shift, NH_3 synthesis, and NH_3-SCR under periodic lean/rich conditions. *Environ. Sci. Technol.* **2023**, *57*, 19584–19593. [CrossRef]
8. Hong, Z.; Wang, Z.; Li, X. Catalytic oxidation of nitric oxide (NO) over different catalysts: An overview. *Catal. Sci. Technol.* **2017**, *7*, 3440–3452. [CrossRef]
9. Liu, S.; Wu, X.; Weng, D.; Li, M.; Lee, H. Combined promoting effects of platinum and MnO_x–CeO_2 supported on alumina on NO_x-assisted soot oxidation: Thermal stability and sulfur resistance. *Chem. Eng. J.* **2012**, *203*, 25–35. [CrossRef]
10. Nickolov, R.; Stankova, N.; Khristova, M.; Mehandjiev, D. Copper oxide supported on carbon modified alumina as catalyst for reduction of NO with CO. *J. Colloid. Interf. Sci.* **2003**, *265*, 121–128. [CrossRef]
11. An, Z.; Zhuo, Y.; Xu, C.; Chen, C. Influence of the TiO_2 crystalline phase of MnO_x/TiO_2 catalysts for NO oxidation. *Chinese J. Catal.* **2014**, *35*, 120–126. [CrossRef]
12. Kim, C.H.; Qi, G.; Dahlberg, K.; Li, W. Strontium-doped perovskites rival platinum catalysts for treating NO_x in simulated diesel exhaust. *Science* **2010**, *327*, 1624–1627. [CrossRef] [PubMed]
13. Guo, Z.C.; Xie, Y.S.; Hong, I.Y.; Kim, J.Y. Catalytic oxidation of NO to NO_2 on activated carbon. *Energy Conv. Manag.* **2001**, *42*, 2005–2018. [CrossRef]
14. Kang, M.; Kim, D.; Park, E.; Kim, J.; Yie, J.; Kim, S.; Hopeweeks, L.; Eyring, E. Two-stage catalyst system for selective catalytic reduction of NO_x by NH_3 at low temperatures. *Appl. Catal. B* **2006**, *68*, 21–27. [CrossRef]
15. Hernandez-Fernandez, J.; Aguilar-Elguezabal, A.; Castillo, S.; Ceron-Ceron, B.; Arizabalo, R.D.; Moran-Pineda, M. Oxidation of NO in gas phase by Au-TiO_2 photocatalysts prepared by the sol-gel method. *Catal. Today* **2009**, *148*, 115–118. [CrossRef]
16. Li, L.; Zhang, N. Atomic dispersion of bulk/nano metals to atomic-sites catalysts and their application in thermal catalysis. *Nano Res.* **2023**, *16*, 6380–6401. [CrossRef]
17. Liu, K.; Zhao, X.; Ren, G.; Yang, T.; Ren, Y.; Lee, A.F.; Su, Y.; Pan, X.; Zhang, J.; Chen, Z.; et al. Strong metal-support interaction promoted scalable production of thermally stable single-atom catalysts. *Nat. Commun.* **2020**, *11*, 1263. [CrossRef]
18. Ma, Y.; Gao, W.; Zhang, Z.; Zhang, S.; Tian, Z.; Liu, Y.; Ho, J.C.; Qu, Y. Regulating the surface of nanoceria and its applications in heterogeneous catalysis. *Surf. Sci. Rep.* **2018**, *73*, 1–36. [CrossRef]
19. Olsson, L. The influence of Pt oxide formation and Pt dispersion on the reactions $NO_2 \Leftrightarrow NO + 1/2O_2$ over Pt/Al_2O_3 and $Pt/BaO/Al_2O_3$. *J. Catal.* **2002**, *210*, 340–353. [CrossRef]
20. Lykhach, Y.; Kozlov, S.M.; Skála, T.; Tovt, A.; Stetsovych, V.; Tsud, N.; Dvořák, F.; Johánek, V.; Neitzel, A.; Mysliveček, J.; et al. Counting electrons on supported nanoparticles. *Nat. Mater.* **2016**, *15*, 284–288. [CrossRef]
21. Li, S.; Xu, Y.; Chen, Y.; Li, W.; Lin, L.; Li, M.; Deng, Y.; Wang, X.; Ge, B.; Yang, C.; et al. Tuning the selectivity of catalytic carbon dioxide hydrogenation over iridium/cerium oxide catalysts with a strong metal-support interaction. *Angew. Chem. Int. Ed.* **2017**, *56*, 10761–10765. [CrossRef] [PubMed]
22. Nie, L.; Mei, D.; Xiong, H.; Peng, B.; Ren, Z.; Hernandez, X.; Delariva, A.; Wang, M.; Engelhard, M.H.; Kovarik, L.; et al. Activation of surface lattice oxygen in single-atom Pt/CeO_2 for low-temperature CO oxidation. *Science* **2017**, *358*, 1419–1423. [CrossRef] [PubMed]
23. Zhang, N.; Li, L.; Jing, Y.; Qian, Y.; Chen, D.; Maeda, N.; Murayama, T.; Toyao, T.; Shimizu, K. In situ/operando spectroscopic evidence on associative redox mechanism for periodic unsteady-state water–gas shift reaction on Au/CeO_2 catalyst. *J. Catal.* **2024**, *433*, 115500. [CrossRef]
24. Khivantsev, K.; Jaegers, N.R.; Aleksandrov, H.A.; Song, I.; Pereira-Hernandez, X.I.; Engelhard, M.H.; Tian, J.; Chen, L.; Motta Meira, D.; Kovarik, L.; et al. Single Ru(II) ions on ceria as a highly active catalyst for abatement of NO. *J. Am. Chem. Soc.* **2023**, *145*, 5029–5040. [CrossRef] [PubMed]
25. Ji, Y.; Toops, T.J.; Graham, U.M.; Jacobs, G.; Crocker, M. A kinetic and DRIFTS study of supported Pt catalysts for NO oxidation. *Catal. Lett.* **2006**, *110*, 29–37. [CrossRef]
26. Marques, R.; Darcy, P.; Costa, P.D.; Mellottée, H.; Trichard, J.; Djéga-Mariadassou, G. Kinetics and mechanism of steady-state catalytic NO + O_2 reactions on Pt/SiO_2 and $Pt/CeZrO_2$. *J. Mol. Catal. A* **2004**, *221*, 127–136. [CrossRef]
27. Xie, J.; Wang, S.; Wang, F. Photo-induced thermal effect on Pt/CeO_2 surface for promoting formaldehyde oxidation performance by photothermal catalysis. *Appl. Surf. Sci.* **2024**, *644*, 158709. [CrossRef]
28. Zhang, J.; Qin, X.; Chu, X.; Chen, M.; Chen, X.; Chen, J.; He, H.; Zhang, C. Tuning metal-support interaction of Pt-CeO_2 catalysts for enhanced oxidation reactivity. *Environ. Sci. Technol.* **2021**, *55*, 16687–16698. [CrossRef]
29. Li, M.; Li, W.; Yang, Y.; Yu, D.; Lin, J.; Wan, R.; Zhu, H. Remarkably efficient Pt/CeO_2–Al_2O_3 catalyst for catalytic hydrodeiodination of monoiodoacetic acid: Synergistic effect of Al_2O_3 and CeO_2. *Chemosphere* **2023**, *327*, 138515. [CrossRef]
30. Lv, Y.; Li, A.; Ye, J.; Wang, H.; Hu, P.; Wang, K.; Guo, Y.; Tang, X.; Dai, S. Exploring the facet-dependent structural evolution of Pt/CeO_2 catalysts induced by typical pretreatments for CO oxidation. *ACS Appl. Mater. Interfaces* **2024**, *16*, 43556–43564. [CrossRef]
31. Wang, J.; Shi, X.; Chen, L.; Li, H.; Mao, M.; Zhang, G.; Yi, H.; Fu, M.; Ye, D.; Wu, J. Enhanced performance of low pt loading amount on Pt-CeO_2 catalysts prepared by adsorption method for catalytic ozonation of toluene. *Appl. Catal. A* **2021**, *625*, 118342. [CrossRef]
32. Peng, R.; Li, S.; Sun, X.; Ren, Q.; Chen, L.; Fu, M.; Wu, J.; Ye, D. Size effect of Pt nanoparticles on the catalytic oxidation of toluene over Pt/CeO_2 catalysts. *Appl. Catal. B* **2018**, *220*, 462–470. [CrossRef]

33. Yang, Q.; Li, L.; Wang, X.; Ma, Y. Tunable metal-support interaction of Pt/CeO_2 catalyst via surfactant-assisted strategy: Insight into the total oxidation of CO and toluene. *J. Hazard. Mater.* **2022**, *424*, 127601. [CrossRef] [PubMed]
34. Wang, K.; Zhang, R.; Wu, S.; Chen, M.; Tang, J.; Zhao, L.; Liu, Y.; Fan, Y. The tuned selective catalytic reduction behaviors of NO over the Pt/CeO_2 catalysts by preferentially exposing different facets of CeO_2 supports. *Catal. Commun.* **2022**, *172*, 106530. [CrossRef]
35. Chen, B.; Wang, B.; Sun, Y.; Wang, X.; Fu, M.; Wu, J.; Chen, L.; Tan, Y.; Ye, D. Plasma-assisted surface interactions of Pt/CeO_2 catalyst for enhanced toluene catalytic oxidation. *Catalysts* **2019**, *9*, 2. [CrossRef]
36. Zhou, G.; Lan, H.; Gao, T.; Xie, H. Influence of Ce/Cu ratio on the performance of ordered mesoporous CeCu composite oxide catalysts. *Chem. Eng. J.* **2014**, *246*, 53–63. [CrossRef]
37. Ma, L.; Ding, C.; Wang, J.; Xu, H.; Zhang, K. Lanthanide modified pt/CeO_2-based catalysts for methane partial oxidation: Relationship between catalytic activity and structure. *Int. J. Hydrogen Energy* **2023**, *48*, 19074–19086. [CrossRef]
38. Zhang, N.; Miyazaki, S.; Qian, Y.; Jing, Y.; Toyao, T.; Shimizu, K. Mechanism of the periodic unsteady-state water–gas shift reaction on highly dispersed Cu-loaded CeO_2 catalysts. *ACS Catal.* **2023**, *13*, 8503–8515. [CrossRef]
39. Mihaylov, M.Y.; Ivanova, E.Z.; Vayssilov, G.N.; Hadjiivanov, K.I. Revisiting ceria-NO_x interaction: FTIR studies. *Catal. Today* **2020**, *357*, 613–620. [CrossRef]
40. Lu, Y.; Lin, F.; Zhang, Z.; Thompson, C.; Zhu, Y.; Doudin, N.; Kovarik, L.; García Vargas, C.E.; Jiang, D.; Fulton, J.L.; et al. Enhancing activity and stability of Pd-on-TiO_2 single-atom catalyst for low-temperature CO oxidation through *in situ* local environment tailoring. *J. Am. Chem. Soc.* **2024**, *146*, 28141–28152. [CrossRef]
41. Meunier, F.C. Relevance of IR spectroscopy of adsorbed CO for the characterization of heterogeneous catalysts containing isolated atoms. *J. Phys. Chem. C* **2021**, *125*, 21810–21823. [CrossRef]
42. Philipp, S.; Drochner, A.; Kunert, J.; Vogel, H.; Theis, J.; Lox, E.S. Investigation of NO adsorption and NO/O_2 co-adsorption on NO_x -storage-components by DRIFT-spectroscopy. *Top. Catal.* **2004**, *30–31*, 235–238. [CrossRef]
43. Urán, L.; Gallego, J.; Ruiz, W.; Bailón-García, E.; Bueno-López, A.; Santamaría, A. Monitoring intermediate species formation by DRIFT during the simultaneous removal of soot and NO_x over $LaAgMnO_3$ catalyst. *Appl. Catal. A* **2019**, *588*, 117280. [CrossRef]

catalysts

MDPI

Article

UiO-67 Metal–Organic Framework as Advanced Adsorbent for Antiviral Drugs from Water Environment

Sitah Almotiry [1,2], Basma G. Alhogbi [1,*], Mohamed Abdel Salam [1,*] and Mariusz Jaremko [3]

1 Chemistry Department, Faculty of Science, King Abdulaziz University, P.O. Box 80200, Jeddah 21589, Saudi Arabia; sitah_saad@yahoo.com
2 Department of Chemistry, College of Science, Qassim University, P.O. Box 1162, Buraidah 51452, Saudi Arabia
3 Environmental Science and Engineering (BESE) Division, King Abdullah University of Science and Technology (KAUST), P.O. Box 4700, Thuwal 23955, Saudi Arabia; mariusz.jaremko@kaust.edu.sa
* Correspondence: balhogbi@kau.edu.sa or alhogbib@gmail.com (B.G.A.); masalam16@hotmail.com (M.A.S.)

Abstract: Metal–organic frameworks (MOFs) have attained significant usage as adsorbents for antiviral medicines in contemporary times. This study focused on synthesizing a UiO-67 metal–organic framework using the hydrothermal method. The synthesized framework was then characterized using Fourier transform infrared spectroscopy (FTIR), scanning electron microscopy (SEM), thermogravimetric analyses (TGA), and zeta potential measurements. The UiO-67 was then employed for the purpose of assessing the efficiency of various adsorption factors in the removal of antiviral medicines from aqueous solutions, including drugs such as ritonavir (RTV) and lopinavir (LPV), which were extensively used for the treatment of coronavirus (COVID-19) during the pandemic. The variables examined were the quantity of adsorbent used, different pH of the solution, temperature, and contact duration. The experimental findings indicate that the highest level of RTV elimination was 91.2% and of experimental adsorption capacity ($q_{e,\mathrm{exp}}$) was 9.7 mg/g and for LPV this was 85.9%, and ($q_{e,\mathrm{exp}}$) 8.9 mg/g, using 50 mg of UiO-67 at a pH 8, temperature of 298 K, and for 120 min. The impact of contact duration and temperature on the kinetics of adsorption was examined by employing pseudo-first-order and pseudo-second-order kinetic models. The experimental results showed a good match with the pseudo-second-order kinetic model with value of R^2 0.99 and the $q_{e,\mathrm{calc}}$ was 9.7 RTV and 8.9 mg/g LPV, which is a good match with $q_{e,\mathrm{exp}}$. Also, based on diffusion kinetic studies, the adsorption was confirmed to be catalytic in nature on the surface of the UiO-67 MOFs. A thermodynamic analysis of adsorption was conducted, whereby calculations for the Gibbs free energy change (ΔG), enthalpy change (ΔH), and entropy change (ΔS) were performed. The positive ΔH values confirm the endothermic nature of the adsorption of RTV and LPV by UiO-67. The ΔG values exhibited negativity across all temperatures, suggesting the spontaneous nature of the adsorption process of RTV and LPV by UiO-67 from an aqueous solution. UiO-67 was shown to be highly effective in extracting RTV and LPV from real environmental samples.

Keywords: COVID-19 drugs; ritonavir; lopinavir; metal–organic frameworks UiO-67; catalytic adsorption

1. Introduction

Over the past few decades, there has been a growing recognition of pharmaceutical compounds and their metabolites as a major source of water pollution, exerting a considerable influence on the worldwide environment [1,2]. The substantial utilization of some pharmaceutical compounds in recent years has resulted in an escalation of their levels in urban wastewater [3,4]. Among the widely used antivirals are ritonavir (RTV) are lopinavir (LPV) [5,6]. These pharmaceuticals have gained attention as emerging organic contaminants (EOCs) because of their recent utilization in the treatment of COVID-19 [7–9]. Also, LPV and RTV were detected in the wastewater effluent (41.0 ng/L and 4.8 ng/L), downstream

of municipal wastewater treatment plants (31.5 ng/L and 5.8 ng/L), and surface water in Wuhan, China (4.7 ng/L and 4.2 ng/L) [10], and South Africa sewage treatment plants (maximum mean concentration 2.5 μg/L and 3.2 μg/L, respectively [11]. The existence of EOCs in water has been found to have adverse effects on the reproductive, neurological, and immunological systems of both aquatic organisms and human beings [12], especially RTV, and LPV [13,14]. The buildup of these medications in aquatic creatures presents a significant environmental concern, as indicated by previous studies [15,16]. Hence, the elimination of pharmaceutical compounds from water sources presents an imperative task, necessitating the exploration and implementation of alternative methodologies. There are many technologies used for the elimination of antiviral drugs such as RTV and LPV from aquatic environment, such as adsorption by sludge [17], photocatalysis [18], photolosis (UV) [19], membrane bioreactor and ozonation [20], electrochemical advanced oxidation process [21], and activated sludge [22]. The efficiency of adsorption technology in the removal of pharmaceutical pollutants from discarded water has been recognized, along with its economic advantages stemming from its efficient target absorption, high removal rate, and ease of operation [23,24]. Many adsorptive materials have been extensively studied for the purpose of pharmaceutical removal. These materials encompass silica-based materials [25,26], resins [27], metal oxides [28], activated carbon [29], and graphene oxide [30]. In recent times, there has been a growing interest in metal–organic frameworks (MOFs) as very promising adsorbents for liquid-phase adsorption [31–34]. MOFs are a highly sophisticated category of porous crystalline materials that exhibit intriguing structures and features. These include the presence of several aromatic organic ligands and Lewis's acid sites, a large surface area, the ability for surface functionalization, and a diverse range of shapes and topologies [35–37].

In 1995, the first MOF with a three-dimensional open skeleton structure was developed by Yaghi et al. [38]. Since then, various excellent MOFs have been constructed in large quantities and there are various reports about water remediation by MOFs. Karimi and Namazi et al. (2024) utilized hydrogel beads of a magnetic graphene oxide@MIL-88 metal–organic framework (GO@Fe_3O_4@MIL-88@Alg) for the absorption of treatments used for COVID-19 such as doxycycline (DOX), hydroxychloroquine (HCQ), naproxen (NAP), and dipyrone (DIP) from aquatic environment [39]. Pharmaceutical drugs and personal care products (PPCPs) were treated by the manufacture of a Pd@ MIL-100 (Fe) nanocomposite [40]. The widespread use of MOFs is mostly related to the stability of the water in environmental remediation [41,42]. The University of Oslo (UiO) series, the Material of Institute Lavoisier (MIL) series, and the Zeolitic Imidazolate Framework (ZIF) series materials are common water-stable MOFs [38]. The MOFs of UiO-67 are a member of the UiO Zr-MOF family, which is characterized by its microporous nature [43–46]. This type of MOF has wide applications due to its thermal and chemical composition and mechanical stability [47–50].

The main objective of the current study is aquatic environmental sustainability through the remediation of water polluted with organic pollutants such as antiviral drugs. The novelty of the current study involved the synthesis of UiO-67 metal–organic frameworks using the hydrothermal technique and their application, for the first time, to remove RTV and LPV as antivirals drugs from aqueous solutions.

The morphological and structural characteristics of the synthesized UiO-67 will be investigated. The adsorption process will be studied with different parameters such as pH of the solution, different temperatures, varied UiO-67 mass, and contact duration. The adsorption process will be studied kinetically and thermodynamically. The treatment and removal of RTV and LPV from real water samples will also be explored by UiO-67 via an adsorption procedure.

2. Results and Discussion

2.1. Characterization of UiO-67

Solvothermal preparation was used to create the UiO-67. The powder XRD pattern in Figure 1A confirms that the synthesized materials display a typical cubic close-packed structure that is consistent with the work that has been previously reported [51,52]. A thermal gravimetric analysis (TGA) was conducted on a sample of UiO-67. As depicted in Figure 1B, the thermal gravimetric analysis (TGA) of the specimen exhibits three distinct phases: initial evaporation of the solvent at around 100 °C, subsequent removal of the compensatory agent at around 450 °C, and eventual disintegration of the framework within the temperature range of 450 °C to 600 °C. Considering the structural characteristics of the sample UiO-67, this study examines the percentage of mass loss (% ZrO_2) in the temperature range of 450 °C to 600 °C, which was around 65%. The same observations were reported previously for UiO-67 [53], UiO-66 MOFs [54], and zeolitic tetrazolate imidazole frameworks (ZTIFs) [55]. The UiO-67 material had an average zeta potential value of +20.8 mV, indicating a satisfactory level of stability Figure 1C in accordance with previous reported work [56]. A higher magnitude of zeta potential signifies enhanced physical colloidal stability as a result of electrostatic repulsion among the constituent particles. In the N_2 adsorption–desorption isotherms shown in Figure 1D, UiO-67 demonstrates a type-I isotherm, indicating that it is a characteristic microporous material [57], with a BET specific surface area of 1415 m^2/g, whereas the average pore width and the pore volume of the UiO-67 channels were 2.06 nm and 0.801 cm^3/g, respectively, which agreed well with previous reported value for different MOFs such as UiO-67 [56], MIL-53(Al) MOFs modified with rice husk [58], MIL-53(Al) modified by biomass hydrochar [59], and sulfonic-functionalized MIL-100-Fe MOF [60]. In the FTIR spectra presented in Figure 2A, the absorption bands at 3382 cm^{-1} are ascribed to hydrogen bonding. The peaks detected at 1600 and 1581 cm^{-1} are assigned to the coordinated carboxylate groups (COO-). The strong peak detected at 1418 cm^{-1} is assigned to the skeletal vibrations of the benzene ring (C=C). The morphology of UiO-67 depicted by SEM in Figure 2B,C, where the octahedral structure of Zr-MOF crystals is visible with an average crystal dimension of 450 nm, is in accordance with a previous reported study [61].

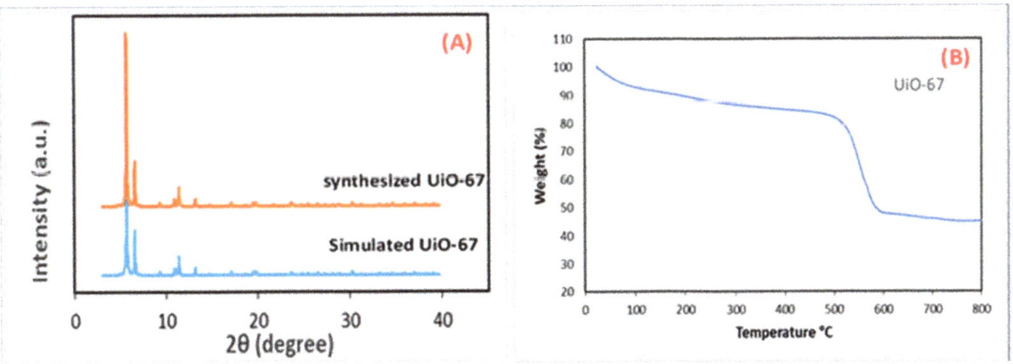

Figure 1. *Cont.*

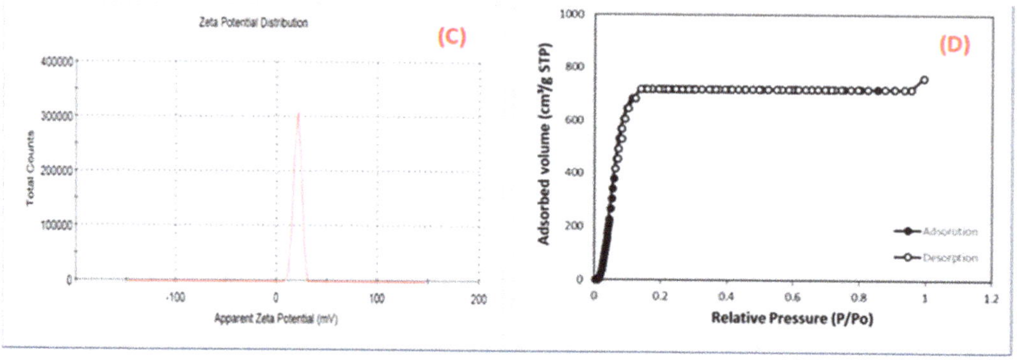

Figure 1. XRD patterns of the simulated and synthesized UiO-67 (**A**), the mass changes in UiO-67 with temperature measured by TGA (**B**), the zeta potential measurement of UiO-67 (**C**), and N_2 adsorption–desorption isotherms of UiO-67 at 77K (**D**).

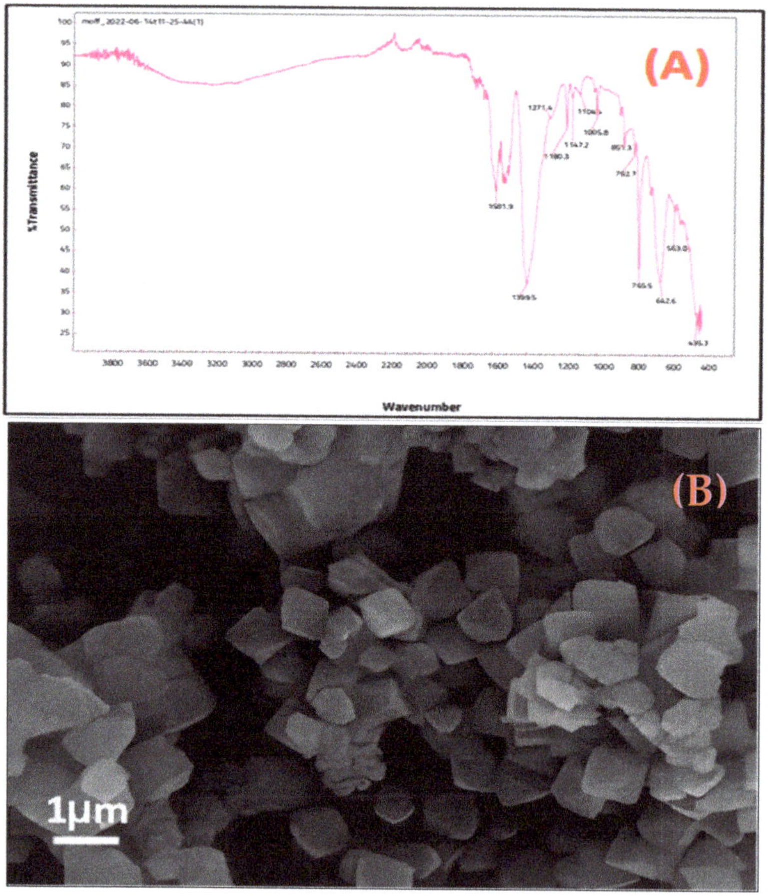

Figure 2. *Cont.*

Figure 2. TIR spectra of UiO-67 (**A**) and SEM images of UiO-67 (**B**,**C**).

2.2. Adsorption Parameter

The parameters encompassed in this study are contact time, UiO-67 mass, and solution conditions including pH and temperature. Therefore, the effect of different factors on the interaction between UiO-67 and RTV and LPV was explored and studied. Figure 3A shows the effect of a different UiO-67 mass on the removal of RTV and LPV. The mass added started with 5 mg of UiO-67; the percentage removal was 7.9% of RTV and 4.6% of LPV after 60 min. The removal percentage of RTV and LPV gradually increased with the increasing amount of UiO-67 added: 50 mg resulted in 92.3% of the RTV and 86.5% of LPV. These percentages of RTV and LPV removed increased by adding 70 mg UiO-67 to the solution until they reached 94.2% and 97.6% removal of RTV and LPV, respectively. A further increase in the amount of UiO-67 to 100.0 mg increased the % removal of RTV to almost 99.2%, and of LPV to 94.9%, after 60 min. It is clear from the findings that there is an increase in the % removal of RTV and LPV with the increase in UiO-67 mass.

The influence of solution pH on the removal process is a significant physico-chemical component that affects the adsorption mechanism at the interface between the solid and solution. Hence, an investigation was conducted to examine the impact of solution pH on the elimination of RTV and LPV by the utilization of UiO-67. The pH levels ranged from 2.0 to 12.0, and the corresponding results in Figure 3B show that the maximum elimination was seen at pH 8 as 95.6% for RTV and 79.9% for LPV, which is the optimal pH value. Figure 3C shows the impact of solution temperature on the adsorption process for RTV and LPV. Additionally, as the temperature of the solution increased, there was a progressive increase in the removal % of RTV and LPV.

Figure 4 shows the dependence of the % removed of RTV and LPV by UiO-67 on the contact time at different temperatures, 296 K, 308 K, and 322 K. In general, Figure 4 shows that the percentage of adsorption of the RTV and LPV by UiO-67 after 5.0 min at 296 K was 91.2% for RTV and 85.9% for LPV, and at 308 K the removal percent was increased to 91.7% for RTV and 86.9% for LPV. The results showed that by increasing the temperature to 322 K, the removal percentage also increased to 92.7% for RTV and 89.3% for LPV. Meanwhile, increasing the contact time by more than 10.0 min slowly increased the removal percentage until it reached 97.4% and 89.46% at 296 K, 96.9% and 94.4% at 308 K, and 99.5% and 95.3% at 322 K for RTV and LPV, respectively. Figure 4 shows the effect of contact time on the removal of RTV and LPV by the UiO-67 as a function of amount removed. The amount removed after 5.0 min by UiO-67 was 9.2, 9.0, and 9.3 mg/g of RTV at 296 K, 302 K, and 322 K, respectively. The amount removed for LPV after 5 min was 8.6, 8.7, and 8.9 mg/g LPV by UiO-67 at 296 K, 302 K, and 323 K. Increasing the contact time was associated with a slight increase in the amount removed until it became constant after 60 min with amount removed of RTV being 9.6, 9.7, and 9.9 mg/g at 296 K, 302 K, and 323 K, respectively, and the amount removed of LPV being 8.9, 9.4, and 9.5 mg/g at 296 K, 308 K, and 322 K, respectively. Overall, the data shown in the results indicate that the process of adsorption exhibited endothermic behavior.

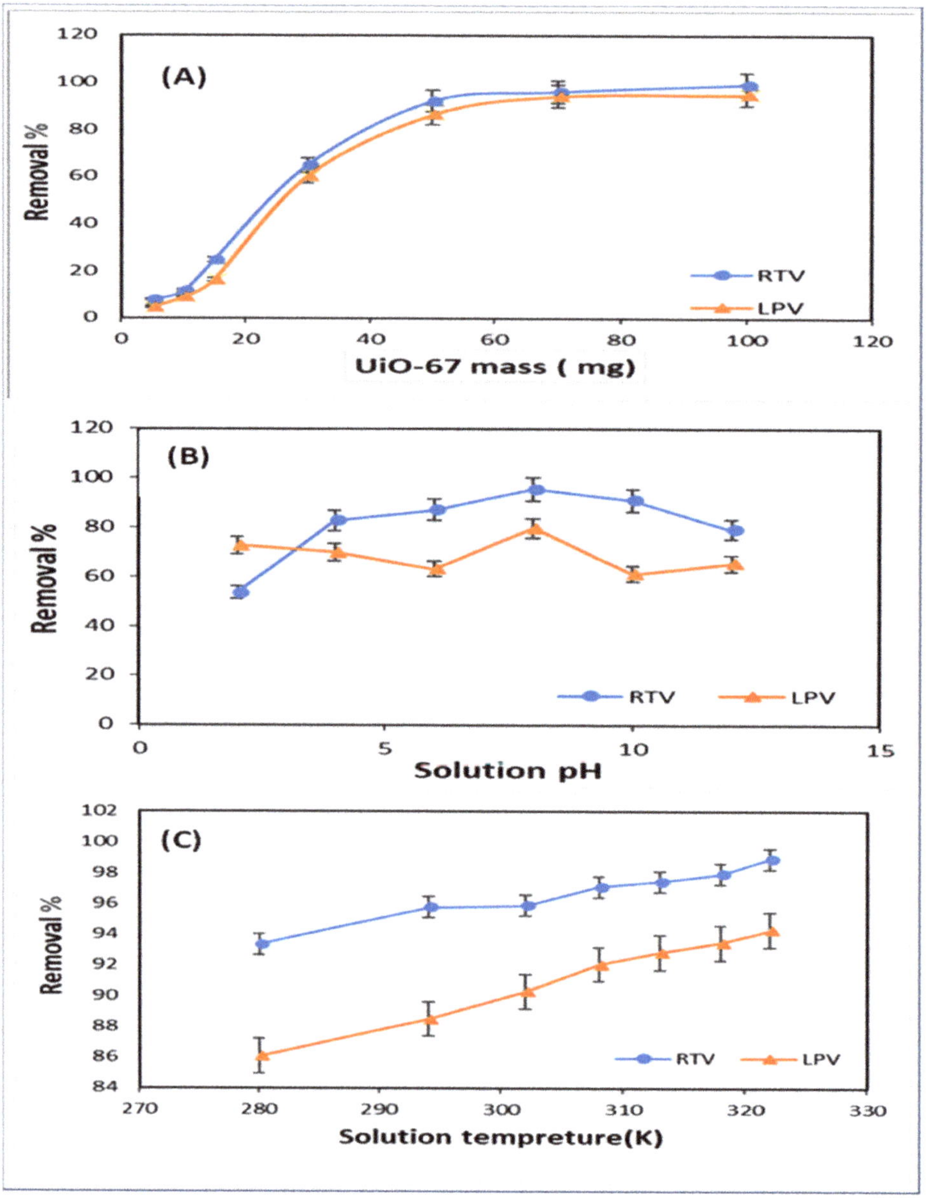

Figure 3. Effect of different adsorbent mass (**A**), pH of solution (**B**), and solution temperature (**C**) on the elimination of RTV and LPV by UiO-67 from aqueous solution.

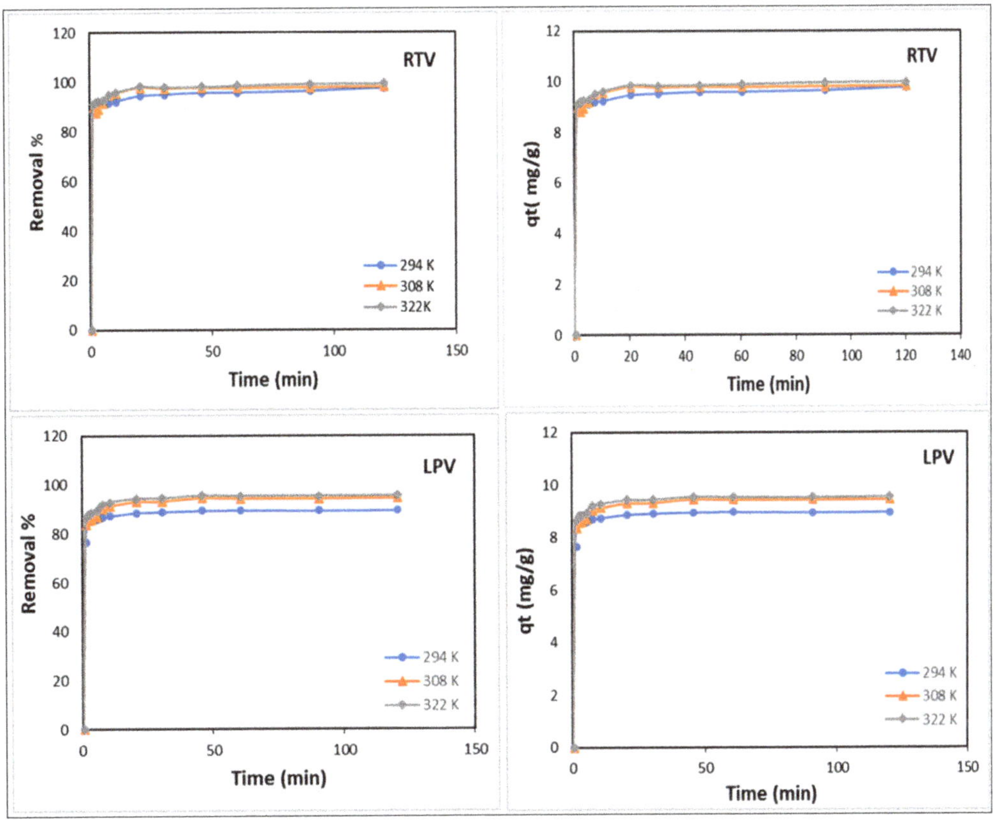

Figure 4. The effect of contact time on the % removal and adsorption capacity of RTV and LPV from aqueous solution at 60 min by 50 mg UiO-67.

2.3. Adsorption Kinetics Models

To obtain a comprehensive understanding of the adsorption and remediation process, it is imperative to conduct a thorough investigation of the kinetic adsorption process. This entails closely monitoring the experimental variables that may influence the rate of adsorption, thereby facilitating the achievement of equilibrium within a suitable period. These investigations provide insights into the potential adsorption process and the transitional phases involved in the production of the ultimate adsorbate–adsorbent complex. Once the reaction rates and their corresponding dependent parameters have been definitively determined, this knowledge might apply to the development of adsorbent materials for commercial purposes. Additionally, it will contribute to a deeper understanding of the intricate dynamics involved in the process of adsorption.

The adsorption of RTV and LPV into UiO-67 was applied to pseudo-first-order model Equation (8) by plotting ln $(q_e - q_t)$ vs. t; RTV and LPV at different temperatures did not converge well and did not give straight lines with an R^2 value of 0.66 as is clear from Figure 5, with unacceptable regression coefficients, as is presented in Tables 1 and 2. Also, the estimated values of the amount adsorbed at equilibrium ($q_{e,\text{calc}}$) were not close to the experimental values ($q_{e,\text{exp}}$). This indicated that the adsorption of the RTV and LPV by UiO-67 could not described by the pseudo-first-order Lagergren equation.

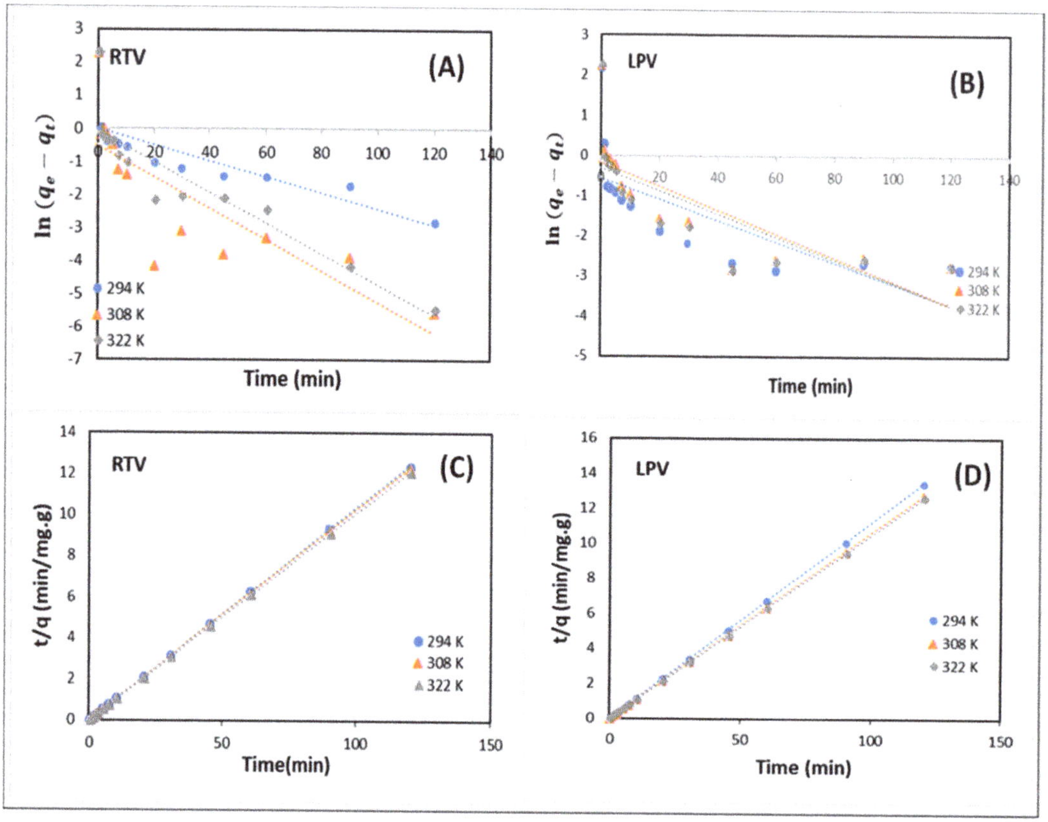

Figure 5. Pseudo–first–order (**A**,**B**) and pseudo–second–order model (**C**,**D**) for the adsorption of RTV and LPV by UiO-67 at different temperatures.

The linearized form of pseudo-second-order rate Equation (9) for the adsorption of RTV and LPV onto UiO-67 by plotting t/q_t vs. t is presented in Figure 5A; good converging was obtained to the experimental data and the straight lines obtained gave excellent regression coefficients ($R^2 > 0.99$), as is presented in Tables 1 and 2.

These findings confirm the suitability of the pseudo-second-order rate equation for the description of RTV and LPV adsorption by UiO-67 from aqueous solution. Also, it is clear from the results that the amount adsorbed per unit mass of UiO-67 at equilibrium ($q_{e,\text{calc}}$) calculated from the slope of the pseudo-second-order plot was in good agreement with the experimental values ($q_{e,\text{exp}}$), as is presented in Tables 1 and 2. These findings agreed well with previous studies and showed the suitability of the pseudo-second-order kinetic model to describe the removal process of different pharmaceuticals, personal care products, dyes, and endocrine-disrupting chemicals (EDCs) in aqueous solution, such as graphene nanoplatelets for the removal of aspirin, acetaminophen, and caffeine [62], Multiwalled Carbon Nanotubes for the decolorization of methylene blue cationic dye [63], the removal of Ketoprofen and Ibuprofen by an Al/Ni-layered double-hydroxide onto polyaniline-wrapped sisal fibers [64], the removal of hydroxychloroquine sulfate by cocoa shell activated carbon [65], the adsorption of metronidazole antibiotic from water with a stable Zr(IV)-MOFs [54], adsorption of the antiviral drugs (oseltamivir and ritonavir) by zeolitic tetrazolate imidazole frameworks (ZTIFs) [55], adsorption of glyphosate by a MIL-53(Al)@rice husk hybrid [58], MIL-53(Al) modified by biomass hydrochar for the removal of ketorolac and naproxen [59], and the removal of diclofenac by the sulfonic-functionalized MIL-100-Fe MOF from aqueous solution [60].

Table 1. Different kinetic models for the adsorption of ritonavir on UiO-67 at different temperatures.

Pseudo-First-Order Kinetic Model					
Temperature	$q_{e,exp}$ (mg/g)	$q_{e,calc}$ (mg/g)	k_1 $\left(\text{min}^{-1}\right)$		R^2
294 K	9.707	1.014	0.024		0.620
308 K	9.718	0.611	0.047		0.662
322 K	9.952	0.921	0.046		0.826
Pseudo-Second-Order Kinetic Model					
Temperature	$q_{e,exp}$ (mg/g)	$q_{e,calc}$ (mg/g)	k_2 (g/mg·min)	h	R^2
294 K	9.707	9.71	0.274	25.859	0.999
308 K	9.718	9.80	0.539	51.897	0.999
322 K	9.952	9.95	0.372	36.882	0.999
Liquid Film Diffusion Model					
Temperature	k_{fd} (min^{-1})	R^2			
294 K	0.011	0.863			
308 K	0.014	0.540			
322 K	0.023	0.847			
Intra-Particle Diffusion Model					
Temperature	k_{id} (mg/g·min$^{1/2}$)	C (mg/g)	R^2		
294 K	0.357	7.038	0.224		
308 K	0.384	6.918	0.225		
322 K	0.367	7.250	0.223		

Table 2. Different kinetic models for the adsorption of lopinavir on UiO-67at different temperatures.

Pseudo-First-Order Kinetic Model					
Temperature	$q_{e,exp}$ (mg/g)	$q_{e,calc}$ (mg/g)	K_1 $\left(\text{min}^{-1}\right)$		R^2
294 K	9.0	0.581	0.026		0.495
308 K	9.5	0.866	0.029		0.594
322 K	9.6	0.739	0.029		0.563
Pseudo-Second-Order Kinetic Model					
Temperature	$q_{e,exp}$ (mg/g)	$q_{e,calc}$ (mg/g)	K_2 (g/mg·min)	h	R^2
294 K	9.0	8.949	0.719	57.581	0.999
308 K	9.5	9.454	0.398	35.531	0.999
322 K	9.6	9.554	0.493	44.972	0.999
Liquid Film Diffusion Model					
Temperature	k_{fd} (min^{-1})	R^2			
294 K	0.003	0.405			
308 K	0.008	0.613			
322 K	0.008	0.613			
Intra-Particle Diffusion Model					
Temperature	k_{id} (mg/g·min$^{1/2}$)	C (mg/g)	R^2		
294 K	0.329	6.569	0.219		
308 K	0.367	6.743	0.249		
322 K	0.356	6.744	0.227		

2.4. Adsorption Rate-Controlling Mechanism

The adsorption process usually occurs through a series of consecutive steps. Initially, the adsorbate molecules are transported from the bulk aqueous phase to the surface of the stagnant-boundary liquid film surrounding the solid adsorbent. Subsequently, diffusion takes place through the boundary layer, allowing the adsorbate molecules to reach the external surface of the solid adsorbent. At this point, adsorption takes place at an active site on the surface of the solid adsorbent. Finally, intra-particle diffusion and adsorption occur through the pores and aggregates of the solid adsorbent. This section describes an investigation that was conducted to determine the rate-controlling mechanism for the adsorption of RTV and LPV by UiO-67 from an aqueous solution. Two distinct models, namely the liquid film diffusion model and the intra-particle diffusion model, were employed for this purpose. The liquid film diffusion model is another kinetic model that assumes the flow of adsorbed molecules through a liquid film surrounding the solid adsorbent, as in the following Equation (1):

$$ln(1 - \mathrm{F}) = -k_{fd} \times t \tag{1}$$

where F is the fractional attainment of equilibrium (q_t/q_e) and k_{fd} (min^{-1}) is the film diffusion rate coefficient. A linear plot of $-ln(1 - \mathrm{F})$ vs. t with zero intercept suggests that the kinetics of the adsorption process is controlled by liquid film. The rate of diffusion within particles is expressed by the following Equation (2):

$$q_e = k_{id}t^{1/2} + C_i \tag{2}$$

where k_{id} (mg/g·min) is the intra-particle diffusion constant. As shown in Figure 6A,B, the experimental data for absorption of RTV and LPV by UiO-67 were analyzed using an intra-particle diffusion model.

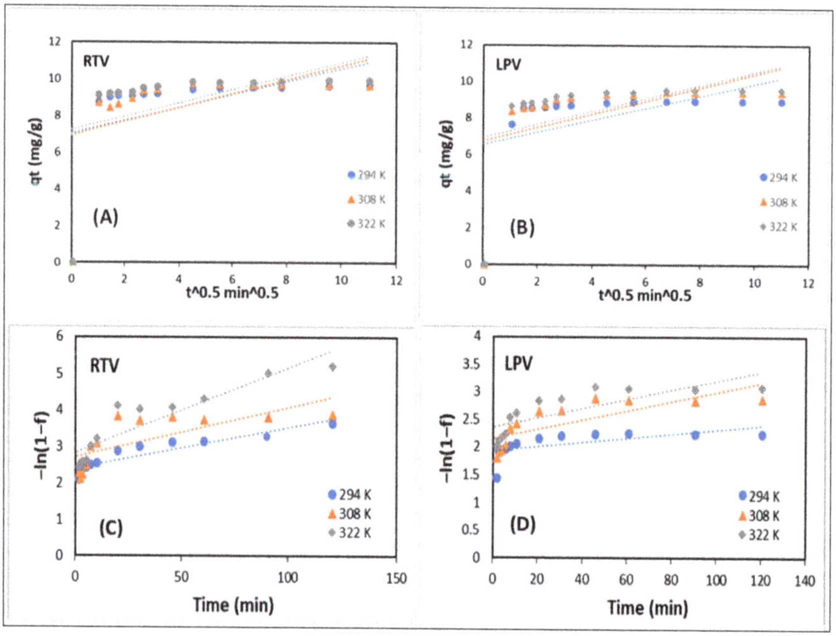

Figure 6. Intra-particle diffusion (**A**,**B**) and liquid film diffusion model plots (**C**,**D**) for the adsorption of RTV and LPV by UiO-67 at different temperatures. (Experimental conditions: 10 mL sample volume and 50 mg UiO-67).

The use of the liquid film diffusion model to analyze the adsorption data of RTV and LPV by UiO-67 exhibited moderate convergence and yielded moderate linear regression coefficients, as depicted in Figure 6C,D and Tables 1 and 2. However, the use of the intra-particle diffusion model to analyze the experimental data did not provide straight lines passing through the origin, resulting in unsatisfactory linear regression coefficients, as shown in Tables 1 and 2. The convergence and the linear regression of the liquid film diffusion model are much better than the intra-particle diffusion model, which may indicate that the adsorption processes of RTV and LPV by UiO-67 occur mainly on the surface of the UiO-67 active sites, and not inside the pores of the UiO-67. These results suggested that the adsorption was catalytic in nature on the surface of the UiO-67 MOFs due to the interaction of the RTV and LPV with the active sites present on the UiO-67 surface. It is well known in heterogeneous catalysis reactions that reactants diffuse from the bulk fluid phase to adsorb on the catalyst surface, which in this case is the solid adsorbent UiO-67. So, the adsorbate molecules (RTV and LPV), must migrate across the surface to an active site of the UiO-67, where RTV and LPV molecules are adsorbed and immobilized physically to form physical bonds with the UiO-67 (product) by following a more energetically facile path through catalytic intermediates, as the adsorption takes place in different intermediate steps, and the adsorption rate is affected by different factors including solution temperature and mass transfer. After the desorption process, the active site on the MOF adsorbent is available for another cycle of adsorption [49].

2.5. Thermodynamic Studies

The adsorption of RTV and LPV by UiO-67 was studied thermodynamically to explore the spontaneity of the process. Different thermodynamic parameters were calculated, Gibbs free energy change (ΔG), enthalpy change (ΔH), and entropy change (ΔS) from the variation in the thermodynamic distribution coefficient D with a change in temperature according to Equation (3), and the ΔH and ΔS were calculated according to the Equation (4):

$$D = \frac{q_e}{C_e} \tag{3}$$

$$lnD = \frac{\Delta S^{\circ}}{R} - \frac{\Delta H^{\circ}}{2.303\ RT} \tag{4}$$

where R is the gas constant (8.314 J/(mol K)), T is the absolute temperature (K), and D is the distribution coefficient of the adsorbate.

When plotting lnD vs. $1/T$ for the adsorption of RTV and LPV by UiO-67, straight lines were obtained, as presented in Figure 7. The ΔH and ΔS values were calculated from the slope and the intercept of the straight line. The ΔH values were positive, +23.5 kJ/mole and +24.4 kJ/mole for the adsorption of RTV and LPV, respectively, by UiO-67 from aqueous solution. The positive ΔH values confirm the endothermic nature of the adsorption of RTV and LPV by UiO-67, which explains the increase in adsorption with the rising solution temperature. Also, the values of ΔH suggests a strong affinity between RTV and LPV for UiO-67, and the physical nature of the adsorption. The positive values of ΔS, +36.2 KJ/mole for RTV and +40.3 KJ/mole for LPV, suggested an increase in the degree of randomness at the UiO-67 solution interface during the adsorption and immobilization of RTV and LPV. The ΔG values were calculated for 296 K using the following Equation (5):

$$\Delta G^{\circ} = \Delta H^{\circ} - T\Delta S^{\circ} \tag{5}$$

The adsorption of RTV and LPV by UiO-67 was spontaneous, as indicated by the negative values of the calculated free energy change, ΔG −6.9 kJ/mole and −7.2 kJ/mole for the adsorption of RTV and LPV, respectively, by UiO-67 from aqueous solution. Generally, the negative values of ΔG positive values of ΔH and ΔS suggested that the adsorption of RTV and LPV by UiO-67 process is an entropy-driven process.

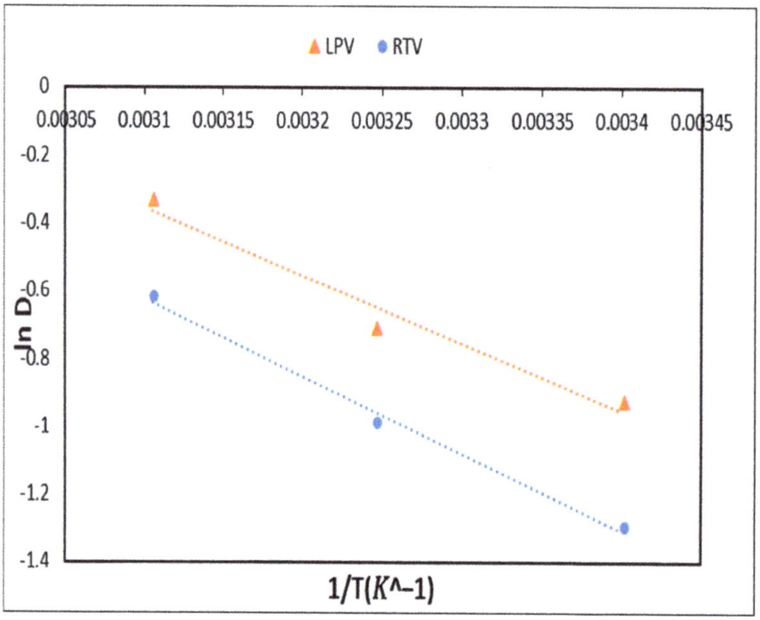

Figure 7. Thermodynamic parameters for the adsorption of RTV and LPV by UiO-67. (Experimental conditions: 10 mL sample volume and 50 mg UiO-67).

3. Environmental Applications

The present study aimed to assess the effectiveness of UiO-67 as a solid adsorbent for the removal of RTV and LPV from real water samples. For this investigation, a selection of five samples was chosen, including sea water, wastewater, well water, tap water, and bottled water. The amounts of RTV and LPV were measured first, and it was found that they were below the detection limit of the HPLC method. This suggests that the selected water samples were free from these pollutants. Consequently, the real water samples were spiked with RTV and LPV, with a resultant concentration of 50.0 mg/L for both RTV and LPV, and then the real samples were equilibrated overnight.

Following this, an amount of 50.0 mg of UiO-67 was added to the spiked samples and left for 60 min. The amount adsorbed was calculated by comparing the concentrations of RTV and LPV before and after the adsorption, and the results are presented in Figure 8. The data shown in Figure 8 demonstrate that around 90% of RTV and LPV was removed from the spiked real water samples, confirming the applicability of UiO-67 for environmental remediation, especially in the case of pollution with antiviral drugs such as RTV and LPV.

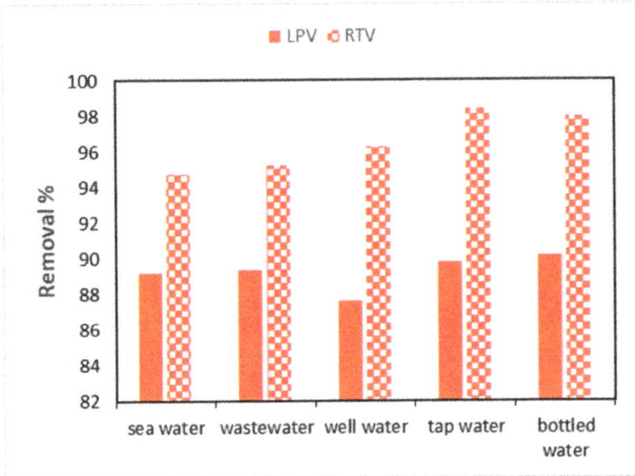

Figure 8. Removal of RTV and LPV by UiO-67 from spiked environmental water samples.

4. Materials and Methods

4.1. Materials

The chemical reagents employed in the production of UiO-67 include Zirconium tetrachloride ($ZrCl_4$) 98%, which was purchased from Energy Chemical (Shanghai, China), 4,4′-biphenyldicarboxylic acid (BPDC) 98%, obtained from Tianjin Hewns Biochem LLC. (Tianjin, China), acetic acid, hydrochloric acid (HCl) at a concentration of 37%, dimethylformamide (DMF) 99%, and acetone 96%, which were obtained from Shenzhen Biocomma Biotech (Shenzhen, China). The pharmaceutical drugs, ritonavir (RTV) and lopinavir (LPV), the chemical composition is shown in Figure 9, were purchased from Sigma–Aldrich (Darmstadt, Germany) at an analytical purity of 98%. All solutions were prepared using deionized water.

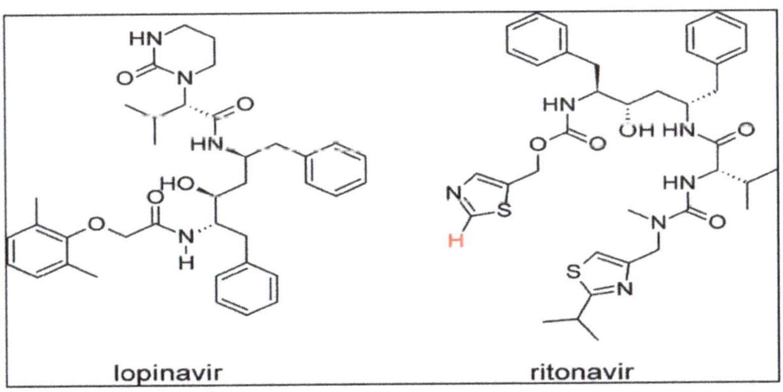

Figure 9. The chemical structures of antivirals lopinavir (LPV) and ritonavir (RTV).

4.2. Preparation of UiO-67

The UiO-67 metal–organic framework (MOF) was synthesized as shown in Scheme 1 using a mixture of $ZrCl_4$ (233 mg, 1 mmol), 4,4′-biphenyldicarboxylic acid (BPDC, 242 mg, 1 mmol), acetic acid (0.6 g, 10 mmol), and HCl (37%, 12 M; 0.16 mL, 2 mmol) mixed in 30 mL of N, N-dimethylformamide (DMF) using the process of sonication. The mixture was placed in a 100 mL Teflon-lined stainless-steel autoclave and heated at 120 °C for 48 h. Subsequently, the UiO-67 white product was obtained by the process of centrifugation and subjected to three rounds of washing with DMF. The UiO-67 material, in its original form, was immersed in DMF at ambient temperature for a duration of 6 h in order to eliminate any residual BPDC that was not bound. Subsequently, the particles that were acquired underwent several washes with acetone to facilitate the displacement of the entrapped DMF, followed by a drying process at a temperature of 100 °C for a duration of 24 h, as shown in Scheme 1.

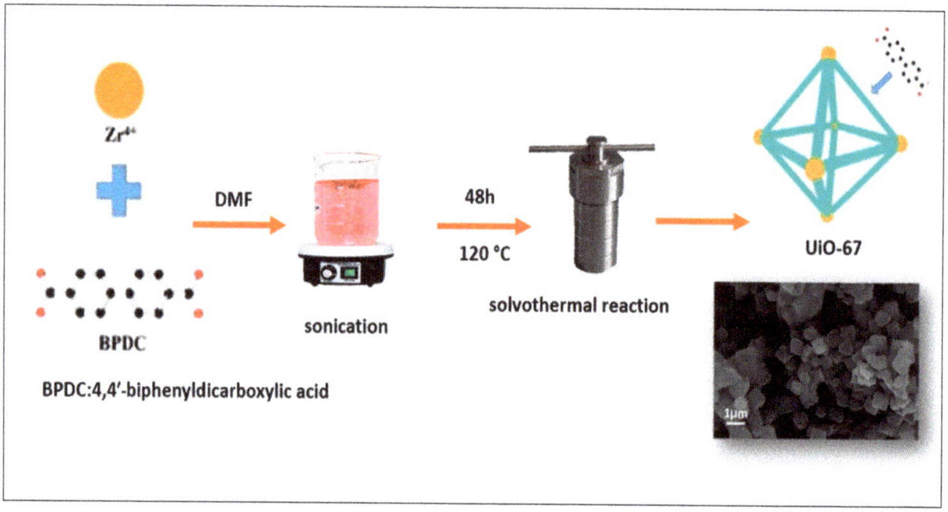

Scheme 1. Synthesis UiO-67 metal-organic framework (MOF).

4.3. Characterization of UiO-67

Powder XRD (X-ray diffraction) patterns were obtained using a D8 ADVANCE ECO (Bruker, Billerica, MA, USA) with Cu Kα radiation (40 kV, 40 mA) at a rate of 6° min^{-1} over a range of 4–40° (2θ). The thermogravimetric analyses (TGA) were carried out using a TG/SDTQ600 (USA) instrument in atmosphere within a temperature range of 0.0–900 K at a heating rate of 10 °C min^{-1} under a nitrogen flow of 100 mL min^{-1} with open alumina pans. The morphology of the samples was examined by using a scanning electron microscope (SEM, JSM-6701F, Tokyo, Japan). The FTIR spectra were recorded using a FTIR spectrophotometer (Spectrum 100, Perkin Elmer, Shelton, CT, USA). The zeta potential was obtained through measurement of the electrophoretic mobility of the particles using a Zeta sizer (Malvern Instruments GmbH, Malvern, WR14 1XZ, UK). Nitrogen adsorption–desorption isotherms were measured by using a N_2 porosimeter (Tristar 3020, Micromeritics, Norcross, GA, USA) at 77 K; the samples were degassed at 150 °C for 12 h under vacuum before this analysis.

4.4. Analytical Method

The concentrations of RTV and LPV were determined using high-performance liquid chromatography (HPLC) with a Hewlett Packard 1100 series liquid chromatograph (Avondale, CA, USA). The high-performance liquid chromatography (HPLC) system was comprised of two pumps and a UV detector. The separations were conducted using an analytical reversed-phase C18 column with dimensions of 4.6 mm × 150 mm and a particle size of 5 μm. The mobile phase flow rate was set at 1.5 mL/min, and gradient conditions were employed. The mobile phase consisted of a combination of deionized water with 1% formic acid and acetonitrile, with a volume ratio of 40:60. A volume of 10 μL was utilized for the injected sample size, while the detection wavelength for UV analysis was set at 254 nm. The chromatogram results are shown in Figure 10.

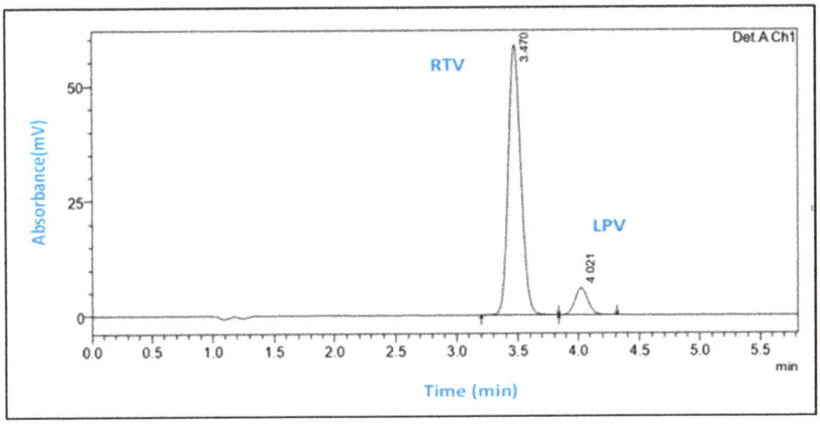

Figure 10. Chromatogram data of ritonavir and lopinavir at 254 nm.

4.5. Adsorption Studies

Adsorption tests were conducted in the study to investigate the impact of several adsorption factors, including adsorbent mass, solution pH, solution temperature, and adsorption time. A set of 10.0 mL solutions, each containing 50.0 mg/L of lopinavir and ritonavir, were created in 25.0 mL glass bottles and maintained at a certain temperature. A test for the effect of UiO-67 mass was performed by using different masses (0, 5, 10, 15, 30, 50, 70, and 100 mg) in 10 mL of diluted RTV and LPV drug mixture and 60 min contact time at 296 K. A test for the effect of solution pH was performed by using different pH values (2, 4, 6, 8, and 11) that were adjusted using HCl solution and NaOH solution and using 50 mg of UiO-67 in 10 mL of diluted drug mixture and 60 min contact time at 296 K. A test for the effect of temperature was performed by measuring the amount adsorbed at different temperatures (280, 294, 302, 308, 313, 318, and 322 K) using 10 mg of UiO-67 in 10 mL of diluted drug mixture and 60 min contact time. A test for the effect of adsorption time was performed by measuring the amount adsorbed at different times (1, 2, 3, 5, 7, 10, 15, 20, 30, 45, 60, 90, and 120 min) using 10 mg of UiO-67 in 10 mL of diluted drug mixture and 60 min contact time at 296 K. Subsequently, the solution was promptly filtered using a filter paper (specifically, Whatman® quantitative filter paper, ashless, Grade 42, 2.5 μm) to collect the supernatant. The concentrations of lopinavir and ritonavir in the aqueous solution were subsequently analyzed using high-performance liquid chromatography (HPLC). Prior to injection, the solution was filtered through a 0.45 μm syringe filter made of cellulose acetate. This filtration step was implemented to ensure that the injected solution did not contain any UiO-67, which could potentially obstruct the HPLC column and lead to inaccurate findings. The percentage of removal in the solution was determined by utilizing Equation (6) [35]:

$$Removal\ \% = \frac{C_i - C_e}{C_i} \times 100 \tag{6}$$

where C_i is the initial concentration and C_e is the equilibrium concentration of RTV and LPV in solution (mg/L).

The adsorption capacity at any time was calculated according to the following Equation (7):

$$q_e = \frac{C_i - C_e}{m} \times V \tag{7}$$

where q_e is the amount of RTV and LPV adsorbed by the UiO-67 (mg/g) at equilibrium, V is the initial solution volume (L), and m is the mass of UiO-67 (g). It is noteworthy to mention that each experiment was performed in triplicate and the reported values are the average of three measurements.

The adsorption process of RTV and LPV by solid adsorbents such as UiO-67 was applied to the most used kinetic models: Lagergren pseudo-first-order model Equation (8) and the pseudo-second-order Equation (9) [50,51]:

$$ln(q_e - q_t) = \log q_e - \frac{k_1}{2.303}t \tag{8}$$

$$\frac{t}{q_t} = \frac{1}{k_2 q_{e2}} + \frac{1}{q_e}t \tag{9}$$

where k_1 (min^{-1}) is the pseudo-first-order rate coefficient, k_2 (g/(mg·min)) is the pseudo-second-order rate coefficient, and q_e and q_t are the values of the amount absorbed per unit mass mg/g at equilibrium and at any time t, respectively.

4.6. Real Water Samples

Five environmental water samples were used to evaluate the efficiency of UiO-67 in the removal of lopinavir and ritonavir. The water samples include seawater, sewage water, well water, tap water, and bottled water. The samples underwent filtration using a 0.45 L Millipore filter paper and after that were kept in Teflon bottles at a temperature of 5 °C in the dark.

5. Conclusions

This study investigated the adsorption of antiviral drugs such as RTV and LPV compounds from aqueous solutions onto the metal–organic framework UiO-67. The adsorption experimental findings show that the highest level of adsorption capacity for RTV and LPV was ($q_{e,exp}$) 9.9 mg/g, which was attained with an efficiency of 91.2% for RTV and 85.9% for LPV using 50 mg of UiO-67 at a temperature of 298 K, pH 8, and duration time 120 min. The adsorption of RTV and LPV on the UiO-67 MOFs followed the second-order-model with R^2 0.99, and the data depict that the adsorption capacity at equilibrium was 9.9 mg/g ($q_{e,calc}$), which is in good agreement with the experimental values ($q_{e,exp}$). Moreover, the rate-determining step of adsorption was mainly due to the liquid film diffusion, which suggested the catalytic nature of the adsorption mechanism at the surface of the UiO-67 MOFs due to the interaction of the RTV and LPV molecules with the active sites present on the UiO-67 surface. The adsorption was spontaneous, as indicated by the negative values of the calculated free energy change, ΔG −6.9 kJ/mole. Generally, the negative values of ΔG and the positive values of ΔH +24.2 kJ/mole endothermic nature and the ΔS +39.5 KJ/mole suggested that the adsorption of RTV and LPV by UiO-67 process is one of random disorder. The utilization of UiO-67 as an adsorbent has shown significant efficacy in the removal of RTV and LPV from real environmental samples. Therefore, it is evident that UiO-67 has high efficiency and adsorption capabilities as a composite material for the removal of RTV and LPV from aqueous solutions. In future studies, UiO-67's exhibited potential as a promising adsorbent, capable of effectively facilitating the removal of diverse components through the utilization of adsorption techniques, will be further studied.

Author Contributions: B.G.A.: project administration, supervision, writing—review and editing; S.A.: methodology, formal analysis, writing—original draft; M.A.S.: validation, data assignment, supervision, writing—review and editing; M.J.: data curation, formal analysis. All authors have read and agreed to the published version of the manuscript.

Funding: This research received no external funding.

Data Availability Statement: The data presented in this study are available on request.

Acknowledgments: The authors acknowledge King Abdulaziz University for facilitated the practical work and technical support and would like to thank King Fahd Medical Research Center (KFMRC), King Abdulaziz University for technical support.

Conflicts of Interest: The authors declare no conflicts of interest.

References

1. Lozano, I.; Pérez-Guzmán, C.J.; Mora, A.; Mahlknecht, J.; Aguilar, C.L.; Cervantes-Avilés, P. Pharmaceuticals and personal care products in water streams: Occurrence, detection, and removal by electrochemical advanced oxidation processes. *Sci. Total Environ.* **2022**, *827*, 154348. [CrossRef] [PubMed]
2. Kujawska, A.; Kiełkowska, U.; Atisha, A.; Yanful, E.; Kujawski, W. Comparative analysis of separation methods used for the elimination of pharmaceuticals and personal care products (PPCPs) from water—A critical review. *Sep. Purif. Technol.* **2022**, *290*, 120797. [CrossRef]
3. Van Boeckel, T.P.; Gandra, S.; Ashok, A.; Caudron, Q.; Grenfell, B.T.; Levin, S.A.; Laxminarayan, R. Global antibiotic consumption 2000 to 2010: An analysis of national pharmaceutical sales data. *Lancet Infect. Dis.* **2014**, *14*, 742–750. [CrossRef] [PubMed]
4. Yin, L.; Wang, B.; Yuan, H.; Deng, S.; Huang, J.; Wang, Y.; Yu, G. Pay special attention to the transformation products of PPCPs in environment. *Emerg. Contam.* **2017**, *3*, 69–75. [CrossRef]
5. De Clercq, D. Anti-HIV drugs: 25 compounds approved within 25 years after the discovery of HIV. *Int. J. Antimicrob. Agents* **2009**, *33*, 307–320. [CrossRef]
6. Chandwani, A.; Shuter, J. Lopinavir/ritonavir in the treatment of HIV-1 infection: A review. *Ther. Clin. Risk Manag.* **2008**, *4*, 1023–1033.
7. Patel, T.K.; Patel, P.B.; Barvaliya, M.; Saurabh, M.K.; Bhalla, H.L.; Khosla, P.P. Efficacy and safety of lopinavir-ritonavir in COVID-19: A systematic review of randomized controlled trials. *J. Infect. Public. Health* **2021**, *14*, 740–748. [CrossRef]
8. Çinarka, H.; Günlüoğlu, G.; Çörtük, M.; Yurt, S.; Kiyik, M.; KOŞAR, F.; Tanriverdi, E.; Arslan, M.A.; Baydili, K.N.; Koç, A.S.; et al. The comparison of favipiravir and lopinavir/ritonavir combination in COVID-19 treatment. *Turk. J. Med. Sci.* **2021**, *51*, 1624–1630. [CrossRef]
9. Gasmi, A.; Noor, S.; Menzel, A.; Khanyk, N.; Semenova, Y.; Lysiuk, R.; Beley, N.; Bolibrukh, L.; Gasmi Benahmed, A.; Storchylo, O.; et al. Potential drugs in COVID-19 management. *Curr. Med. Chem.* **2024**, *20*, 3245–3264. [CrossRef]
10. Zhang, Z.; Zhou, Y.; Han, L.; Guo, X.; Wu, Z.; Fang, J.; Hou, B.; Cai, Y.; Jiang, J.; Yang, Z. Impacts of COVID-19 pandemic on the aquatic environment associated with disinfection byproducts and pharmaceuticals. *Sci. Total Environ.* **2022**, *811*, 151409. [CrossRef]
11. Nannou, C.; Ofrydopoulou, A.; Evgenidou, E.; Heath, D.; Heath, E.; Lambropoulou, D. Antiviral drugs in aquatic environment and wastewater treatment plants: A review on occurrence, fate, removal and ecotoxicity. *Sci. Total Environ.* **2020**, *699*, 134322. [CrossRef] [PubMed]
12. Brillas, E. A critical review on ibuprofen removal from synthetic waters, natural waters, and real wastewaters by advanced oxidation processes. *Chemosphere* **2022**, *286*, 131849. [CrossRef] [PubMed]
13. Kumari, M.; Kumar, A. Environmental and human health risk assessment of mixture of COVID-19 treating pharmaceutical drugs in environmental waters. *Sci. Total Environ.* **2022**, *812*, 152485. [CrossRef] [PubMed]
14. Kuroda, K.; Li, C.; Dhangar, K.; Kumar, M. Predicted occurrence, ecotoxicological risk and environmentally acquired resistance of antiviral drugs associated with COVID-19 in environmental waters. *Sci. Total Environ.* **2021**, *776*, 145740. [CrossRef] [PubMed]
15. Lee, C.-C.; Hsieh, C.-Y.; Chen, C.S.; Tien, C.-J. Emergent contaminants in sediments and fishes from the Tamsui River (Taiwan): Their spatial-temporal distribution and risk to aquatic ecosystems and human health. *Environ. Pollut.* **2020**, *258*, 113733. [CrossRef]
16. Marchant, J. When antibiotics turn toxic. *Nature* **2018**, *555*, 431–433. [CrossRef]
17. Krasucka, P.; Rombel, A.; Yang, X.J.; Rakowska, M.; Xing, B.; Oleszczuk, P. Adsorption and desorption of antiviral drugs (ritonavir and lopinavir) on sewage sludges as a potential environmental risk. *J. Hazard. Mater.* **2022**, *425*, 127901. [CrossRef]
18. Thi, L.-A.P.; Panchangam, S.C.; Do, H.-T.; Nguyen, V.-H. Prospects and challenges of photocatalysis for degradation and mineralization of antiviral drugs. *Nanostruct. Photocatal.* **2021**, *17*, 489–517.
19. Ngumba, E.; Gachanja, A.; Tuhkanen, T. Removal of selected antibiotics and antiretroviral drugs during post-treatment of municipal wastewater with UV, UV/chlorine and UV/hydrogen peroxide. *Water Environ. J.* **2020**, *34*, 692–703. [CrossRef]
20. Mascolo, G.; Laera, G.; Pollice, A.; Cassano, D.; Pinto, A.; Salerno, C.; Lopez, A. Effective organics degradation from pharmaceutical wastewater by an integrated process including membrane bioreactor and ozonation. *Chemosphere* **2010**, *78*, 1100–1109. [CrossRef]
21. Zhou, Y.; Li, W.; Kumar, V.; Necibi, M.C.; Mu, Y.-J.; Shi, C.-Z.; Chaurasia, D.; Chauhan, S.; Chaturvedi, P.; Sillanpää, M.; et al. Synthetic organic antibiotics residues as emerging contaminants waste-to-resources processing for a circular economy in China: Challenges and perspective. *Environ. Res.* **2022**, *211*, 113075. [CrossRef] [PubMed]
22. Yan, X.; Zheng, S.; Huo, Z.; Shi, B.; Huang, J.; Yang, J.; Ma, J.; Han, Y.; Wang, Y.; Cheng, K.; et al. Effects of exogenous N-acyl-homoserine lactones on nutrient removal, sludge properties and microbial community structures during activated sludge process. *Chemosphere* **2020**, *255*, 126945. [CrossRef]
23. Rosman, N.; Salleh, W.; Mohamed, M.A.; Jaafar, J.; Ismail, A.F.; Harun, Z. Hybrid membrane filtration-advanced oxidation processes for removal of pharmaceutical residue. *J. Colloid. Interface Sci.* **2018**, *532*, 236–260. [CrossRef] [PubMed]
24. Park, J.; Yamashita, N.; Park, C.; Shimono, T.; Takeuchi, D.M.; Tanaka, H. Removal characteristics of pharmaceuticals and personal care products: Comparison between membrane bioreactor and various biological treatment processes. *Chemosphere* **2017**, *179*, 347–358. [CrossRef]
25. Piane, M.D.; Vaccari, S.; Corno, M.; Ugliengo, P. Silica-based materials as drug adsorbents: First principle investigation on the role of water microsolvation on ibuprofen adsorption. *J. Phys. Chem. A* **2014**, *118*, 5801–5807. [CrossRef] [PubMed]

26. Jin, J.; Feng, T.; Gao, R.; Ma, Y.; Wang, W.; Zhou, Q.; Li, A. Ultrahigh selective adsorption of zwitterionic PPCPs both in the absence and presence of humic acid: Performance and mechanism. *J. Hazard. Mater.* **2018**, *348*, 117–124. [CrossRef]
27. Jin, J.; Feng, T.; Ma, Y.; Wang, W.; Wang, Y.; Zhou, Q.; Li, A. Novel magnetic carboxyl modified hypercrosslinked resins for effective removal of typical PPCPs. *Chemosphere* **2017**, *185*, 563–573. [CrossRef]
28. Zhang, Y.; Zhu, H.; Szewzyk, U.; Lübbecke, S.; Geissen, S.U. Removal of emerging organic contaminants with a pilot-scale biofilter packed with natural manganese oxides. *Chem. Eng. J.* **2017**, *317*, 454–460. [CrossRef]
29. Tian, W.; Zhang, H.; Duan, X.; Sun, H.; Tade, M.O.; Ang, H.M.; Wang, S. Nitrogen-and sulfur-codoped hierarchically porous carbon for adsorptive and oxidative removal of pharmaceutical contaminants. *ACS Appl. Mater. Interfaces* **2016**, *8*, 7184–7193. [CrossRef]
30. Zambianchi, M.; Durso, M.; Liscio, A.; Treossi, E.; Bettini, C.; Capobianco, M.L.; Aluigi, A.; Kovtun, A.; Ruani, G.; Corticelli, F.; et al. Graphene oxide doped polysulfone membrane adsorbers for the removal of organic contaminants from water. *Chem. Eng. J.* **2017**, *326*, 130–140. [CrossRef]
31. Zeng, Z.; Lyu, J.; Bai, P.; Guo, X. Adsorptive separation of fructose and glucose by metal–organic frameworks: Equilibrium, kinetic, thermodynamic, and adsorption mechanism studies. *Ind. Eng. Chem. Res.* **2018**, *57*, 9200–9209. [CrossRef]
32. Peng, Y.; Zhang, Y.; Huang, H.; Zhong, C. Flexibility induced high-performance MOF-based adsorbent for nitroimidazole antibiotics capture. *Chem. Eng. J.* **2018**, *333*, 678–685. [CrossRef]
33. Zhao, X.; Wei, Y.; Zhao, H.; Gao, Z.; Zhang, Y.; Zhi, L.; Wang, Y.; Huang, H. Functionalized metal-organic frameworks for effective removal of rocephin in aqueous solutions. *J. Colloid. Interface Sci.* **2018**, *514*, 234–239. [CrossRef] [PubMed]
34. Zhao, X.; Wang, K.; Gao, Z.; Gao, H.; Xie, Z.; Du, X.; Huang, H. Reversing the dye adsorption and separation performance of metal–organic frameworks via introduction of –SO3H groups. *Ind. Eng. Chem. Res.* **2017**, *56*, 4496–4501. [CrossRef]
35. Zhu, X.; Li, B.; Yang, J.; Li, Y.; Zhao, W.; Shi, J.; Gu, J. Effective adsorption and enhanced removal of organophosphorus pesticides from aqueous solution by Zr-based MOFs of UiO-67. *ACS Appl. Mater. Interfaces* **2015**, *7*, 223–231. [CrossRef]
36. Zhou, Z.; Ke, Q.; Wu, M.; Zhang, L.; Jiang, K. Pore space partition approach of ZIF-8 for pH responsive codelivery of ursolic acid and 5-fluorouracil. *ACS Mater. Lett.* **2023**, *5*, 466–472. [CrossRef]
37. Yue, D.; Li, G.; Li, F.; Chen, D.; Li, W.; Qin, B.; Zhang, B.; Zhang, Y.; Wang, Y.; Wang, Z. Zinc-based metal-organic frameworks as fluorescent probes for the detection of Vitamin B6. *Dye. Pigment.* **2023**, *222*, 111857. [CrossRef]
38. Zhang, S.; Wang, J.; Zhang, Y.; Ma, J.; Huang, L.; Yu, S.; Chen, L.; Song, G.; Qiu, M.; Wang, X. Applications of water-stable metal-organic frameworks in the removal of water pollutants: A review. *Environ. Pollut.* **2021**, *291*, 118076. [CrossRef]
39. Karimi, S.; Namazi, H. Efficient adsorptive removal of used drugs during the COVID-19 pandemic from contaminated water by magnetic graphene oxide/MIL-88 metal-organic framework/alginate hydrogel beads. *Chemosphere* **2024**, *352*, 141397. [CrossRef]
40. Liang, R.; Luo, S.; Jing, F.; Shen, L.; Qin, N.; Wu, L. A simple strategy for fabrication of Pd@ MIL-100 (Fe) nanocomposite as a visible-light-driven photocatalyst for the treatment of pharmaceuticals and personal care products (PPCPs). *Appl. Catal. B* **2015**, *176*, 240–248. [CrossRef]
41. Shah, S.S.A.; Khan, N.A.; Rehman, A.U.; Murtaza, G.; Ashiq, S.; Nazir, M.A.; Najam, T.; Sohail, M. Removal of antibiotics from wastewater via metal-organic frameworks (MOFs) as adsorbents. *Toxin Rev.* **2024**, *43*, 277–299. [CrossRef]
42. Yu, S.; Pang, H.; Huang, S.; Tang, H.; Wang, S.; Qiu, M.; Chen, Z.; Yang, H.; Song, G.; Fu, D.; et al. Recent advances in metal-organic framework membranes for water treatment: A review. *Sci. Total Environ.* **2021**, *800*, 149662. [CrossRef] [PubMed]
43. Johnson, B.A.; Bhunia, A.; Fei, H.; Cohen, S.M.; Ott, S. Development of a UiO-type thin film electrocatalysis platform with redox-active linkers. *J. Am. Chem. Soc.* **2018**, *140*, 2985–2994. [CrossRef] [PubMed]
44. Hu, J.; Liu, Y.; Liu, J.; Gu, C.; Wu, D. High CO_2 adsorption capacities in UiO type MOFs comprising heterocyclic ligand. *Microporous Mesoporous Mater.* **2018**, *256*, 25–31. [CrossRef]
45. Ma, J.; Chen, Z.; Diao, Y.; Ye, M.; Liu, X.; Cui, S.; Zhong, M.; Nezamzadeh-Ejhieh, J.; Liu, J.; Ouyang, J. Current and promising applications of UiO-based MOFs in breast cancer therapy. *React. Funct. Polym.* **2024**, *200*, 105918. [CrossRef]
46. Hang, C.; Akpinar, I.; Qin, Y.; Huang, L.; Wang, L.; Li, W.; Wu, J. A Review on Adsorption of Organic Pollutants from Water by UiO-67 and Its Derivatives. *J. Nanoelectron. Optoelectron.* **2021**, *16*, 1861–1873. [CrossRef]
47. Ahmad, K.; Nazir, M.A.; Qureshi, A.K.; Hussain, E.; Najam, T.; Javed, M.S.; Shah, S.S.A.; Tufail, M.K.; Hussain, S.; Khan, N.A. Engineering of Zirconium based metal-organic frameworks (Zr-MOFs) as efficient adsorbents. *Mater. Sci. Eng. B* **2020**, *262*, 114766. [CrossRef]
48. Flores-Cervantes, D.X.; Medina-Montiel, C.; Ramirez-Corona, N.; Navarro-Amador, R. Zirconium Based MOFs and Their Potential Use in Water Remediation: Current Achievements and Possibilities. *Air Soil Water Res.* **2022**, *15*, 11786221221080184. [CrossRef]
49. Boudart, M. Adsorption assisted desorption in catalytic cycles. *J. Mol. Catal. A Chem.* **1999**, *141*, 1–7. [CrossRef]
50. Lin, J.; Wang, L. Comparison between linear and non-linear forms of pseudo-first-order and pseudo-second-order adsorption kinetic models for the removal of methylene blue by activated carbon. *Front. Environ. Sci. Eng. China* **2009**, *3*, 320–324. [CrossRef]
51. Ho, Y.-S.; McKay, G. Pseudo-second order model for sorption processes. *Process Biochem.* **1999**, *34*, 451–465. [CrossRef]
52. Schaate, A.; Roy, P.; Godt, A.; Lippke, J.; Waltz, F.; Wiebcke, M.; Behrens, P. Modulated synthesis of Zr-based metal–organic frameworks: From nano to single crystals. *Chem. A Eur. J.* **2011**, *17*, 6643–6651. [CrossRef]
53. Yan, F.; Wang, Q.; Wang, F.; Huang, Z. Study on energy storage properties of Metal-organic frameworks nanofluids (UIO-67/Water and UIO-67/Methanol) by an experimental and theoretical method. *J. Mater. Sci.* **2021**, *56*, 10008–10017. [CrossRef]

54. Talha, K.; Wang, B.; Liu, J.-H.; Ullah, R.; Feng, F.; Yu, J.; Chen, S.; Li, J.-R. Effective adsorption of metronidazole antibiotic from water with a stable Zr (IV)-MOFs: Insights from DFT, kinetics and thermodynamics studies. *J. Environ. Chem. Eng.* **2020**, *8*, 103642.
55. Saghir, S.; Wang, Y.; Xiao, Z. In situ synthesis of multivariant zeolitic tetrazolate imidazole frameworks (ZTIFs) with uncoordinated N-heteroatom sites for efficient adsorption of antiviral drugs. *J. Clean. Prod.* **2023**, *414*, 137654. [CrossRef] [PubMed]
56. Dong, X.; Lin, Y.; Ma, Y.; Zhao, L. Ce-doped UiO-67 nanocrystals with improved adsorption property for removal of organic dyes. *RSC Adv.* **2019**, *9*, 27674–27683. [CrossRef] [PubMed]
57. Thommes, M.; Kaneko, K.; Neimark, A.V.; Olivier, J.P.; Rodriguez-Reinoso, F.; Rouquerol, J.; Sing, K.S.W. Physisorption of gases, with special reference to the evaluation of surface area and pore size distribution (IUPAC Technical Report). *Pure Appl. Chem.* **2015**, *87*, 1051–1069. [CrossRef]
58. Zambrano-Intriago, L.A.; Daza-López, E.V.; Fernández-Andrade, A.; Luque, R.; Amorim, C.G.; Araújo, A.N.; Rodríguez-Díaz, J.M.; Montenegro, M.C. Application of a novel hybrid MIL-53 (Al)@ rice husk for the adsorption of glyphosate in water: Characteristics and mechanism of the process. *Chemosphere* **2023**, *327*, 138457. [CrossRef]
59. Rivadeneira-Mendoza, B.F.; Quiroz-Fernández, L.S.; da Silva, F.F.; Luque, R.; Balu, A.M.; Rodríguez-Díaz, J.M. Biomass@ MOF nanohybrid materials for competitive drug adsorption: Analysis by conventional macroscopic models and statistical physical models. *Environ. Sci. Nano* **2024**, *11*, 1543–1558. [CrossRef]
60. Sánchez, N.C.; Palomino, G.T.; Cabello, C.P. Sulfonic-functionalized MIL-100-Fe MOF for the removal of diclofenac from water. *Microporous Mesoporous Mater.* **2023**, *348*, 112398. [CrossRef]
61. Zhang, X.; Zhang, W.; Li, G.; Liu, Q.; Xu, Y.; Liu, X. A ratiometric fluorescent probe for determination of the anthrax biomarker 2, 6-pyridinedicarboxylic acid based on a terbium (III)−functionalized UIO-67 metal-organic framework. *Microchim. Acta* **2020**, *187*, 1–7. [CrossRef] [PubMed]
62. Al-Khateeb, L.A.; Almotiry, S.; Salam, M.A. Adsorption of pharmaceutical pollutants onto graphene nanoplatelets. *Chem. Eng. J.* **2014**, *248*, 191–199. [CrossRef]
63. Ahmed, S.H.; Rasheed, E.A.; Rasheed, L.A.; Abdulrahim, F.R. Decolorization of cationic dye from aqueous solution by Multiwalled Carbon Nanotubes. *J. Ecol. Eng.* **2024**, *25*, 72–84. [CrossRef]
64. Negarestani, M.; Reisi, S.; Sohrabi, M.; Shayesteh, H.; Farimaniraad, H.; Mollahosseini, A.; Hosseinzadeh, M.; Tavassoli, S. In-situ growth of Al/Ni layered double hydroxide onto polyaniline-wrapped sisal fibers for highly efficient removal of pharmaceutical Ketoprofen and Ibuprofen contaminants: Batch and fixed-bed column studies. *J. Water Process Eng.* **2024**, *57*, 104657. [CrossRef]
65. Finnouche, F.; Hamadi, A.; Kais, H.; Yeddou-Mezenner, N.; Nekkaa, Z. Efficient removal of hydroxychloroquine sulfate using cocoa shell activated carbon: A green approach investigating equilibrium, kinetics, and thermodynamics. *J. Dispers. Sci. Technol.* **2024**, 1–11. [CrossRef]

catalysts

MDPI

Article

Boosting Benzene's Ozone Catalytic Oxidation at Mild Temperatures over Highly Dispersed Ag-Doped Mn_3O_4

Hao Guo [1,2,3,4], Liwei Cen [1,2], Kui Deng [1,5], Wenlong Mo [1,2], Hojo Hajime [3], Di Hu [1,2], Pan Zhang [1], Wenfeng Shangguan [4,*], Haibao Huang [1,2,6] and Hisahiro Einaga [3,*]

1 School of Chemical Engineering and Technology, Xinjiang University, Urumqi 830017, China; guohao@xju.edu.cn (H.G.); seabao8@gmail.com (H.H.)
2 Xinjiang Key Laboratory of Coal Clean Conversion & Chemical Engineering Process, Urumchi 830017, China
3 Department of Energy and Material Sciences, Faculty of Engineering Sciences, Kyushu University, Fukuoka 816-8580, Japan
4 Research Center for Combustion and Environmental Technology, School of Mechanical Engineering, Shanghai Jiao Tong University, Shanghai 200240, China
5 Xinjiang Academy of Environmental Protection Sciences, Urumqi 830011, China
6 College of Ecology and Environment, Xinjiang University, Urumqi 830017, China
* Correspondence: shangguan@sjtu.edu.cn (W.S.), einaga.hisahiro.399@m.kyushu-u.ac.jp (H.E.)

Abstract: Transition metal oxides show high activity while still facing the challenges of low mineralization and poor durability in the ozone catalytic oxidation (OCO) of volatile organic compounds (VOCs). Improving the oxygen mobility and low-temperature reducibility of transition metal oxides was found to be an effective way to address the above challenges. Here, highly dispersed Ag was added to Mn_3O_4 via the co-precipitation oxalate route, and the obtained Ag/Mn_3O_4 exhibited higher mineralization and stability in benzene catalytic ozonation at room temperature. Compared to Mn_3O_4, the concentration of CO_2 formed from benzene oxidation over Ag/Mn_3O_4 was significantly increased, from 585.4 ppm to 810.9 ppm, while CO generation was greatly suppressed to only one tenth of its original value (194 ppm vs. 19 ppm). In addition, Ag/Mn_3O_4 exhibited higher catalytic stability than Mn_3O_4. The introduction of Ag obviously improved the oxygen mobility and low-temperature reducibility of Mn_3O_4. Moreover, the highly dispersed Ag also promoted the activity of surface oxygen species and the chemisorption of benzene on Mn_3O_4. The above physicochemical properties contributed to the excellent catalytic performance and durability of Ag/Mn_3O_4. This research could shed light on the improvement in VOC mineralization via ozone catalytic oxidation.

Keywords: ozone catalytic oxidation; VOCs; Ag/Mn_3O_4; benzene mineralization

1. Introduction

Volatile organic compounds (VOCs), typically emitted from paint shops, oil refineries, etc., are considered precursors of photochemical smog and haze and could cause serious harm to both humans and the eco-environment [1,2]. VOCs not only have a high potential to form ozone and particulate matter ($PM_{2.5}$) but also exhibit toxicity and carcinogenicity [3,4]. Therefore, the governments in many countries have implemented strict measures and extremely low emission limits to control VOC pollution. Currently, there are various treatment methods for VOCs, such as adsorption, catalytic oxidation, combustion, and so on [5,6]. Among the numerous VOC removal technologies, ozone catalytic oxidation (OCO) exhibited several advantages. For example, during the OCO process, O_3 can be decomposed into oxygen atoms and further transformed into a variety of powerful reactive oxygen species (ROS), which can precisely and effectively react with the VOCs at near-room temperature, thus ensuring energy efficiency and safety [7,8]. Consequently, OCO has been extensively investigated for the removal of various types of VOCs, such as formaldehyde, methyl mercaptan, dichloroethane, aromatic VOCs, etc. [9–12].

Catalysts play an important role in the OCO of VOCs. In general, noble metal catalysts such as Pt, Pd, and Rh always show excellent performance, but the high price and the disadvantage of easy poisoning limit their practical application [13]. Compared to noble metal catalysts, transition metal oxides (Mn, Co, Ni, etc.) show several advantages, such as a preferential reducibility, abundant oxygen species, a high anti-poisoning ability, and so on. Moreover, the cost of these transition metal oxides is also lower than that of noble metals [14]. Therefore, transition metal oxides have been widely applied in the OCO of VOCs.

Among the various transition metal oxides, MnO_x usually exerts multiple manganese oxidation states, resulting in fruit oxygen vacancies as well as strong low-temperature reducibility, which are key factors for the redox steps of VOC ozonation [15,16]. Compared to other MnO_x such as Mn_2O_3 and MnO_2, Mn_3O_4 was reported to facilitate the reactions between VOC molecules and ozone to form surface-adsorbed oxygen atomic species, which could be more easily and deeply oxidized [17]. However, Mn_3O_4 usually suffered from the incomplete oxidation of VOC molecules, low ozone utilization, and severe instability, which prevented its further application in OCO technology [18]. As a result, further research should be conducted to improve both the catalytic activity and stability of Mn_3O_4.

It is widely accepted that the addition of another suitable metal could always promote the catalytic activity and durability of single-metal oxides, which would be attributed to the "electronic effect" and "geometric effect" [19,20]. Thus, many transition metals have been introduced to modify the Mn_3O_4 in ozone decomposition, such as Cu, Fe, Ce, and so on [21,22]. For example, Fu et al. found that the partial substitution of Mn by Fe in Mn_3O_4 could result in abundant oxygen species and active sites of Mn^{3+}, thus greatly promoting ozone decomposition [23]. Of all the transition metals, Ag has distinctive characteristics. It has been reported that Ag-containing catalysts usually have quantities of active sites with high activity in ozone decomposition or the OCO of VOCs [24]. In addition, the application of Ag could also promote the chemisorption of VOCs, which is a crucial step in the OCO process [10]. Therefore, Ag was introduced into manganese oxides, and the synthesized Ag-Mn catalysts were applied to ozone decomposition. Xie et al. argued that the deposition of Ag on Mn_3O_4 nanosheets noticeably improved the reducibility and generated more highly active oxygen vacancies; thus, the obtained Mn_3O_4@Ag exhibited excellent and stable catalytic performance in O_3 decomposition [25]. Moreover, Liu et al. reported that the OCO of VOCs involves ozone decomposition and the oxidation of VOC molecules [7]. Therefore, the higher catalytic performance of Ag/Mn_3O_4 in ozone decomposition also indicated its potential in the OCO of VOCs.

In this paper, highly dispersed Ag was added to Mn_3O_4 via the co-precipitation oxalate route, and the obtained Ag/Mn_3O_4 catalyst was then applied to the OCO of benzene. The introduction of Ag greatly improved benzene mineralization, with much less by-products, and the catalytic stability of Mn_3O_4 at room temperature, which was attributed to the promoted oxygen mobility, enhanced low-temperature reducibility, activated surface oxygen species, and enhanced chemisorption of benzene. This work provides us with a new way to improve VOC mineralization with low by-products and catalytic durability in the OCO of VOCs.

2. Results and Discussions

2.1. Crystal Phase and Textures of Mn_3O_4-Based Catalysts

The XRD patterns and N_2 adsorption/desorption isotherms and pore size distributions are shown in Figure 1. From Figure 1a, the Mn_3O_4, Ag/Mn_3O_4 somewhat showed the same XRD patterns as standard cubic Mn_3O_4 (PDF # 18-0803) according to the analysis results of Jade 6.5, indicating that Ag doping did not change the crystal phase of manganese oxide. In addition, the diffraction peaks assigned to Ag or Ag_2O could hardly be observed, indicating the high dispersion of Ag in Mn_3O_4 [26].

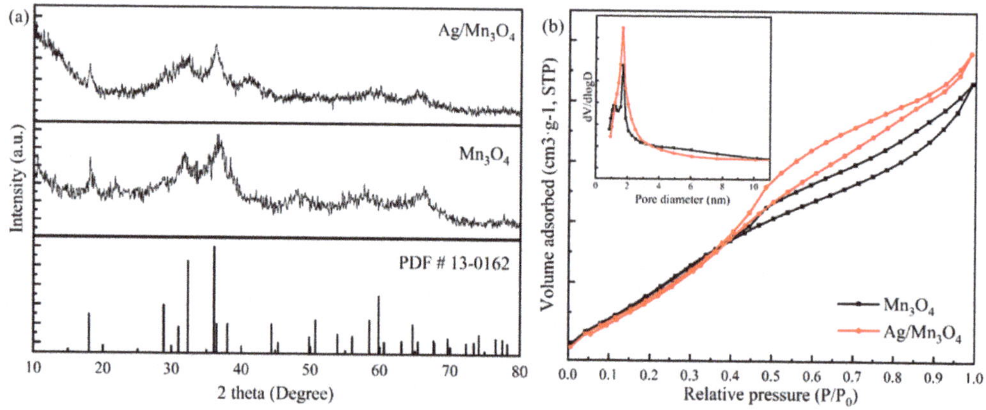

Figure 1. XRD spectra (**a**) and N_2-adsorption/desorption isotherms (**b**), inset pore size distribution curves) of the obtained catalysts.

The N_2 adsorption/desorption isotherms of the two catalysts in Figure 1b are both type IV with H3 hysteresis loops within a wide relative pressure range of 0.4–1.0, implying the development of mesoporous structures after calcination [27]. Notably, Mn_3O_4 showed a specific surface area of 385.2 $m^2 \cdot g^{-1}$, while Ag doping resulted only in a slight decrease in the specific surface area of Ag/Mn_3O_4 (see Table 1). The results were beneficial to the adsorption of ozone and benzene, and the higher specific surface area would also provide more active sites to promote the OCO of benzene.

Table 1. Summary of BJH parameters of Mn_3O_4 and Ag/Mn_3O_4.

Sample	Specific Surface Area/$m^2 \cdot g^{-1}$	Pore Volume/$cm^3 \cdot g^{-1}$	Average Pore Size/nm
Mn_3O_4	385.2	0.40	4.2
Ag/Mn_3O_4	372.4	0.44	4.8

Figure 2 shows the high-resolution TEM (HRTEM) image and high-angle annular dark-field STEM images of Ag/Mn_3O_4, respectively. As shown in Figure 2b, the interplanar spacings were 0.299 nm and 0.253 nm, which were very close to the (2 2 0) and (3 1 1) crystal planes of Mn_3O_4 (PDF # 13-0162), respectively. Furthermore, the dispersion of Ag was uniform, without any agglomeration, as shown in Figure 2b,d, which was in somewhat of an agreement with the XRD results in Table 1. Summary of BJH parameters of Mn_3O_4 and Ag/Mn_3O_4.

2.2. Surface Chemical Properties of the Obtained Catalysts

The surface chemical compositions, apparent element concentration and valence, and oxygen species of the two samples were determined by the XPS technique. The Mn 2p, Ag 3d, and O 1s spectra of Mn_3O_4 and Ag/Mn_3O_4 are shown in Figure 3, and the quantitative analysis results are summarized in Table 2.

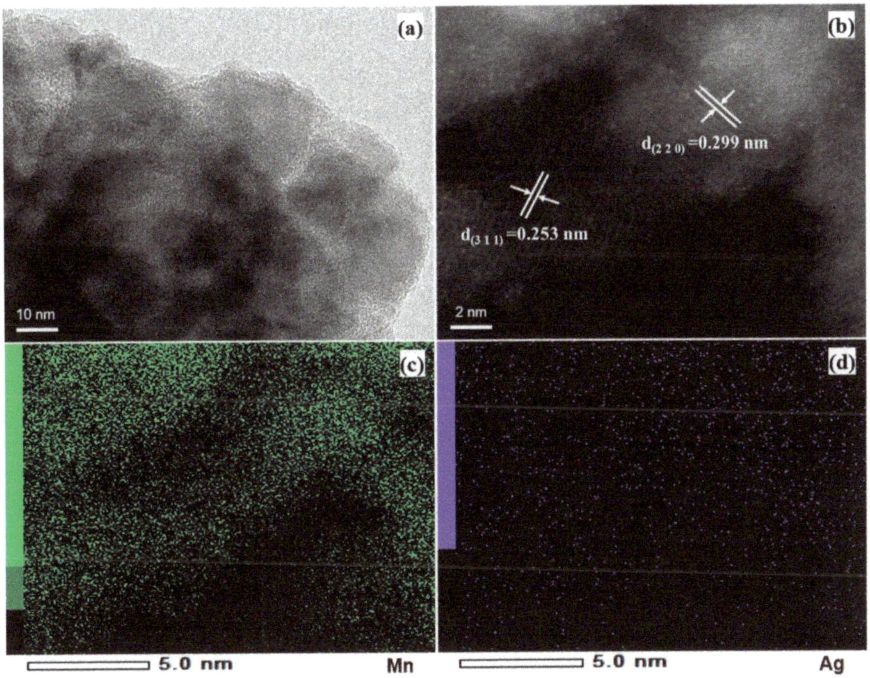

Figure 2. HRTEM image (**a**), HAADF-STEM image (**b**), and elemental mapping ((**c**): Mn, (**d**): Ag) of Ag/Mn_3O_4.

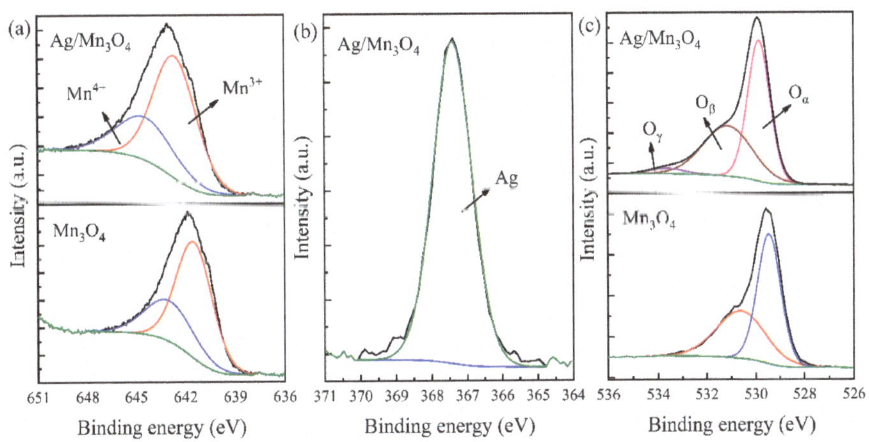

Figure 3. XPS curves of Mn_3O_4 and Ag/Mn_3O_4 ((**a**): Mn 2p 3/2; (**b**): Ag 2d 5/2; and (**c**): O 1 s).

Table 2. Summary of the XPS analysis results.

Catalyst	Mn^{3+}		Mn^{4+}		Ag^{+}		O_α		$O_\beta + O_\gamma$	
	BE/eV	%	BE/eV	%	BE/eV	% §1	BE/eV	%	BE/eV	%
Mn_3O_4	641.8	79.81	643.3	20.19	-	-	529.8	66.1	531.2	33.9
Ag/Mn_3O_4	641.5	75.70	643.0	24.30	367.4	5.01	529.5	57.0	530.6	43.0

§1: The ratio of surface Ag^{+} to the total surface metal ions.

As seen in Figure 3a, the asymmetric Mn 2p $_{3/2}$ spectra of the two catalysts can be decomposed into two symmetric peaks ranging from 641.5 to 641.8 eV and from 643.0 to 643.3 eV, which should be attributed to the surface Mn^{3+} and Mn^{4+}, respectively [28,29]. As shown in Table 2, Ag doping resulted in a higher proportion of surface Mn^{4+} compared to Mn_3O_4, and the result was favorable for the formation of oxygen vacancies due to the charge compensation principle [30]. Furthermore, the higher-valence manganese ions were reported to be in the form of Mn^{4+}-$O_{surf.}$ (surface oxygen species) Lewis acid–base pairs, which could efficiently activate O_3 [14]. Notably, the Ag 3d $_{5/2}$ spectra in Figure 3b show only one peak, centered at 367.4 eV, which could correspond to Ag^+ [31]. As shown in Table 2, the surface ratio of Ag^+ to Mn^{n+} was very close to the designed value, indicating that the co-precipitation method via the oxalate route could make the Ag uniformly dispersed in both the bulk and apparent phases. Ag or other noble metal nanoparticles with higher surface energy often agglomerate and form large-sized particles as a result of Ostwald ripening [32]. Considering the analysis results of XRD, HAADF-STEM, and XPS, it can be concluded that Ag doping through the oxalate route could avoid the formation of Ag agglomeration.

The O 1s XPS spectra of Mn_3O_4 and Ag/Mn_3O_4 shown in Figure 3c could be fitted with two main symmetrical peaks, and the peaks centered at 529.5–529.8 eV should be surface lattice oxygen species (denoted as O_α), while the peaks at 530.6–531.2 eV are assigned as surface adsorbed oxygen species or oxygen vacancies (denoted as O_α) [33,34]. In addition, the O 1s spectra of Ag/Mn_3O_4 also showed another peak centered at about 533.7 eV, which was widely accepted as surface-adsorbed OH/oxy-carbonate species [35]. Noticeably, the total amount of surface-adsorbed oxygen species/vacancies (including both O_β and O_γ) of Ag/Mn_3O_4 was larger than that of Mn_3O_4 displayed in Table 2, indicating that Ag doping not only made the surface oxygen species more active but also increased the ratio of more active adsorbed oxygen species. The above analysis results of the O 1s XPS spectra are somewhat consistent with those of the Mn 2p XPS spectra. Many researchers have proven that surface adsorbed oxygen species/vacancies and surface OH/oxy-carbonate species are beneficial for both O_3 activation and VOC catalytic oxidation [36,37]. Thus, we can expect a satisfactory catalytic performance of Ag/Mn_3O_4.

2.3. Evolution of Oxygen Species and Reducibility of the Gained Catalysts

The evolution of the surface oxygen species of the two catalysts was investigated by the O_2-TPD technique, and the results are shown in Figure 4a. In general, the desorption of adsorbed oxygen species occurred at temperatures ranging from 200 to 400 °C, while the surface lattice oxygen species usually desorbed at temperatures ranging from 400 to 600 °C [38,39]. For the Mn_3O_4 sample, the peak centered at 276 and 498 °C should represent the desorption of adsorbed oxygen species and lattice oxygen species, respectively. In comparison, the desorption peak of adsorbed oxygen species from Ag/Mn_3O_4 at 325 °C, with an obvious shoulder at 241 °C, indicated that the introduction of Ag into Mn_3O_4 resulted in surface oxygen species with a higher reactivity. It is widely accepted that active surface oxygen species would not only generate unsaturated chemical bonds but also lead to unbalanced charges, which would ultimately make the decomposition of ozone as well as the redox reactions occur more easily [40,41]. Since the surface oxygen species of Ag/Mn_3O_4 showed a higher activity than that of Mn_3O_4, the former sample would exert a better catalytic performance in the OCO of benzene.

The reducibility, which could be detected by the H_2-TPR technique, played an important role in the catalytic oxidation of VOCs, and the H_2-TPR curves of Mn_3O_4 and Ag/Mn_3O_4 are shown in Figure 4b. The reduction of Mn^{n+} by H_2 usually followed the order of $Mn^{4+} \rightarrow Mn^{3+} \rightarrow Mn^{2+}$ [42,43]. For sample Mn_3O_4, the first reduction peak at 275 °C with a shoulder at 180 °C should represent the reduction of the Mn^{3+} located in the lattice sites to Mn^{2+}, and the latter peak, centered at 431 °C, could be assigned to the final reduction of Mn^{3+} to Mn^{2+} [44]. With Ag doping, the reduction peaks of Ag/Mn_3O_4 were shifted to lower temperatures compared to those of Mn_3O_4. In other words, the Ag/Mn_3O_4

catalyst in this research exerted a much better low-temperature reducibility, bringing a positive effect to the subsequent OCO of benzene.

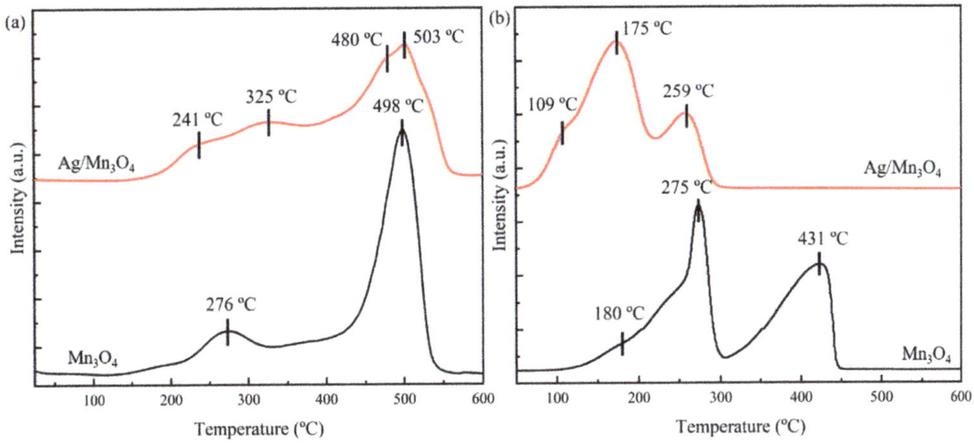

Figure 4. O_2-TPD (**a**) and H_2-TPR (**b**) profiles of as-prepared catalysts.

2.4. Temperature-Programmed Surface Reaction on the Obtained Catalysts

To illustrate the activity of surface oxygen species and the adsorption/desorption behavior of benzene on the as-prepared catalysts, the C_6H_6-TPSR technique was introduced, and the profiles are shown in Figure 5.

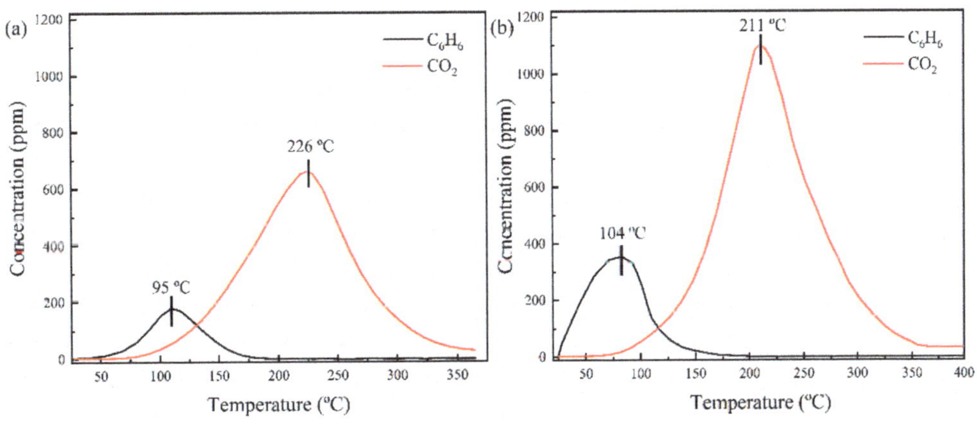

Figure 5. C_6H_6-TPSR profiles of Mn_3O_4 (**a**) and Ag/Mn_3O_4 (**b**).

As shown in Figure 5, the desorption of benzene from Ag/Mn_3O_4 had a larger peak than that of Mn_3O_4, which was inconsistent with their pore volumes in Table 1. The CO_2 was probably formed from the reaction between chemisorbed benzene and adsorbed surface oxygen species, considering both the C_6H_6-TPSR process and the O_2-TPD results [45,46]. Apparently, sample Ag/Mn_3O_4 showed a higher CO_2 desorption but a lower peak temperature than Mn_3O_4 under the N_2 atmosphere, indicating that the surface oxygen species of the former catalyst was more active than that of the latter. In addition, the introduction of Ag also provided more active sites, which was beneficial for the chemisorption of benzene [47]. Researchers have demonstrated that active surface oxygen species could easily react with VOC molecules and that the formed oxygen vacancies would be favorable for ozone decomposition, which, in turn, would be beneficial for benzene ozonation [48].

2.5. Catalytic Performance of Benzene OCO

The performance of benzene OCO over the two obtained catalysts is shown in Figure 6, and some important related data are summarized in Table 3.

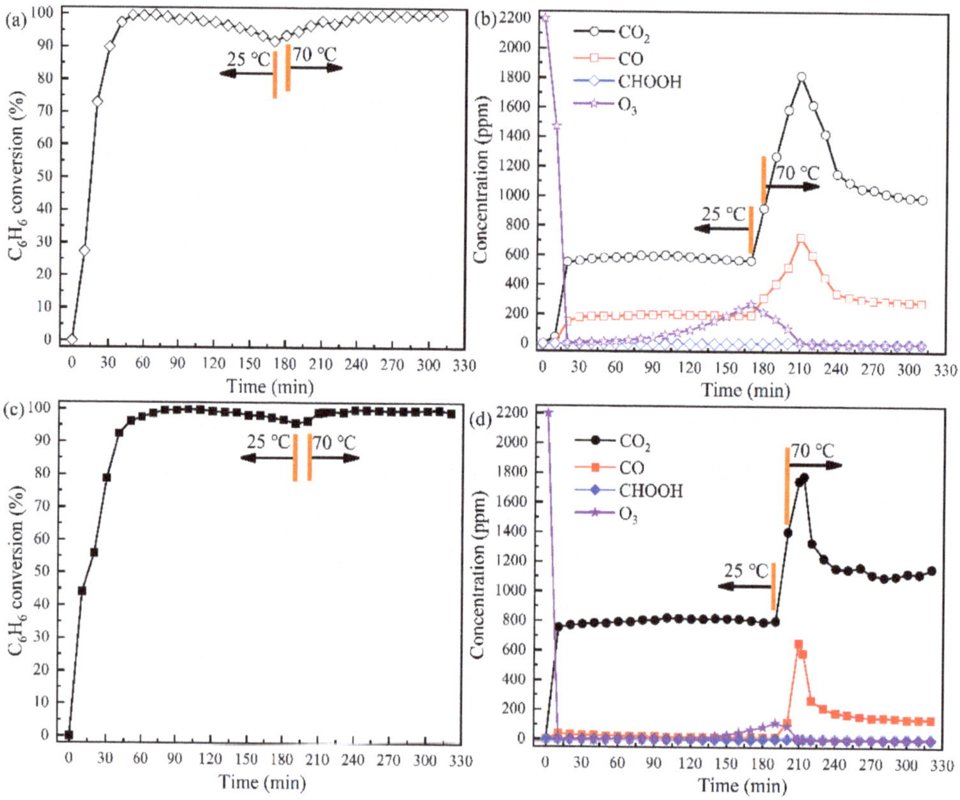

Figure 6. Benzene conversion and concentrations of CO_2, CO, and O_3 as a function of time at 25 °C and 70 °C, respectively ((**a**,**b**): Mn_3O_4; (**c**,**d**): Ag/Mn_3O_4).

Table 3. Summary of benzene OCO with two samples at 25 °C and 70 °C, respectively.

Sample	25 °C				70 °C			
	C_6H_6 rem./% §2	CO_2 /ppm	CO /ppm	Res. O_3/ppm	C_6H_6 rem./% §3	CO_2 /ppm	CO /ppm	Res. O_3/ppm
Mn_3O_4	91.8	585.4	194.7	270.2	99.8	1046.4	301.4	0
Ag/Mn_3O_4	97.4	810.9	19.2	67.2	99.8	1165.4	157.5	0

§2 The value was measured after the OCO of benzene lasted for 170 min at 25 °C. §3 The benzene removal efficiency was detected after the OCO of benzene lasted for 70 min at 70 °C.

From Figure 6a,c, it can be seen that the benzene removal efficiencies over Mn_3O_4 and Ag/Mn_3O_4 rapidly increased from 0 to nearly 100% but then gradually decreased to 91.8% and 97.4% at 25 °C after 170 min (Table 3), respectively. This is because benzene can hardly be completely mineralized by ozone at low temperatures, and some inorganic intermediates would be formed during this process on the surface active sites, leading to the gradual deactivation of the catalysts [49]. According to Einaga et al., some intermediates such as weakly bound formic acid and strongly bound formates (carboxylic acid) may be formed and accumulated on Mn-based catalysts during the OCO of benzene, and some of the above intermediates could be further oxidized to CO_2 or CO at relatively higher temperatures [11]. With the consumption of these intermediates, the activity of the catalyst could be recovered, and this would be consistent with the catalytic performance of the two samples in Figure 6. Noticeably, Ag doping improved the catalytic activity and stability of Mn_3O_4, which might have been due to the active oxygen species on Ag/Mn_3O_4 which could easily react with the intermediates at low temperatures, and the results were in fair agreement with the O_2-TPD and XPS data. Consequently, more CO_2 and much less CO as a by-product were generated when Ag/Mn_3O_4 was applied (see Figure 6d), which might have been mainly due to more active oxygen species, according to the O_2-TPD and C_6H_6-TPSR results. In Table 3, we can clearly see that the CO_2 generation at 25 °C was about 585.4 ppm and 810.9 ppm with Mn_3O_4 and Ag/Mn_3O_4, respectively, while CO production was suppressed to only one tenth of the original value when the latter catalyst was used (19.2 ppm with Ag/Mn_3O_4 vs. 194.7 ppm with Mn_3O_4). That is to say, introducing Ag is beneficial for the complete mineralization of benzene, while also reducing the formation of by-products.

When the temperature was increased to 70 °C, the benzene conversion rates recovered rapidly to nearly 100%, and no deactivation was observed in the last 70 min (Figure 6a,c). Moreover, the elevated temperature also favored the production of CO_2 and CO, which showed a sharp increase in the first 30 min and then decreased to stable, normal values. The sharp peaks for both CO_2 and CO were attributed to the oxidation of some organic intermediates simulated in the catalytic oxidation of benzene at 25 °C [44]. The benzene mineralization percentage for both samples at 70 °C in the last 60 min was very close to 100%, according to the data in Table 3, but Ag/Mn_3O_4 also showed higher CO_2 production and lower CO generation than Mn_3O_4. Meanwhile, nearly no O_3 could be detected in the outlet, further demonstrating that the two samples exerted a satisfying catalytic performance at 70 °C.

Briefly, the Ag doping of Mn_3O_4 improved mineralization and durability in the OCO of benzene at mild temperatures. CO_2 production was promoted while CO generation was noticeably suppressed when Ag/Mn_3O_4 was applied. Considering the above analysis results of the catalysts' characterization, it can be concluded that the highly dispersed Ag in Mn_3O_4 resulted in more active oxygen species, which improved the oxygen mobility and low-temperature reducibility of Mn_3O_4. Moreover, the active oxygen species were also beneficial for the decomposition and activation of ozone, generating reactive oxygen species which could easily and rapidly react with the benzene molecules or intermediates on Ag/Mn_3O_4. As a result, the inorganic intermediates accumulated on the surface of Mn_3O_4 were quickly and deeply oxidized with Ag doping. The above factors can likely be attributed to the preferred mineralization and catalytic durability of Ag/Mn_3O_4 in the OCO of benzene.

3. Experimental

3.1. Catalyst Preparation

Three reagents were used in this research without further purification. The manganese nitrate solution (50 wt%, AR) and ammonium oxalate monohydrate (AR) were purchased from Sinopharm Chemical Reagent Co., Ltd., Shanghai, China. Silver nitrate (AR) was purchased from Lingfeng Chemical Reagent Co., Ltd., Shanghai, China.

The MnO_x and AgMn bimetallic catalysts were synthesized by the (co)precipitation method. First, 3.58 g of manganese nitrate solution and 1.42 g of ammonium oxalate monohydrate were diluted in 100 mL of deionized water. Then, the manganese nitrate solution was added drop by drop to the oxalic acid solution while stirring, and the mixed solution was stirred for another 30 min to ensure the complete precipitation of manganese ions. Then, the mixed solution was filtered and washed three times with deionized water and once with absolute alcohol. The filtered sediment was then dried at 80 °C overnight.

Finally, the dried sediment was calcined at 250 °C for 5 h under static air, with a heating rate of 2 °C·min^{-1}. After calcination, the resulting catalyst was Mn_3O_4. The AgMn bimetallic catalyst with a Ag/Mn molar ratio of 1:19 was obtained by the co-precipitation method. The molar ratio of metal ions to oxalate ions was the same as that used in the preparation of Mn_3O_4, and it was designated as Ag/Mn_3O_4.

3.2. Characterizations

The XRD patterns of the obtained catalysts were recorded using a Rigaku RINT 2200 system (Rigaku Co. Ltd., Sevenoaks, UK) with a Cu Kα radiation, and the 2θ ranged from 10 to 80°. In addition, N_2 adsorption/desorption isotherms were performed using a Micromeritics TriStar II, and the sample was degassed at 200 °C for 1 h prior to measurement. High-resolution transmission electron microscopy (JEM-ARM200F, JEOL) was used to determine the microstructures and interplanar spacings of the as-prepared samples. Oxygen temperature-programmed desorption was carried out using a BELCAT-30 flow-type catalyst analyzer (BEL JAPAN, Inc., Haradanaka, Toyonaka-shi, Japan); both obtained catalysts were pretreated under Ar flow at 250 °C for 1 h before testing. In the O_2-TPD test, the atmosphere was changed to 5% O_2/N_2 when the pretreated catalyst was cooled to r.t. The Ar flow was turned on again when the adsorption of oxygen on the pretreated catalyst reached dynamic equilibrium. Finally, the data were recorded at a temperature ranging from r.t. to 600 °C. The pretreatment of the catalyst during the H_2-TPR test was the same as that of O_2-TPD. The H_2-TPR signal was marked down when the Ar flow was replaced by 5% H_2/N_2, and the operating temperature also ranged from r.t. to 600 °C. X-ray photoelectron spectroscopy (XPS, AXIS UltraDLD, Kratos group, Spring Valley, NV, USA) was used to determine the metal oxidation states, chemical composition, and surface oxygen species of the as-prepared catalysts. In addition, a C_6H_6 temperature-programmed surface reaction (C_6H_6-TPSR) was performed using FTIR (Perkin-Elmer, Spectrum 100, Waltham, MA, USA), with a gas cell at a resolution of 4 cm^{-1}.

3.3. Catalytic Activity Measurements

The catalytic performance test of the synthesized catalysts (40–60 mesh) was carried out in a U-type quartz tube fixed-bed reactor with an inner diameter of 8 mm. In each test, 100 mg of the selected sample was used and heated at 250 °C in simulated air for 1 h. The temperature was maintained at 25 °C for some time and then increased to 70 °C until the end of the test. The weight hourly space velocity (WHSV) in this research was 60,000 mL·$g_{cat}{}^{-1}$·h^{-1}, and the reaction air flow was 100 mL·min^{-1}, with an initial benzene concentration of 200 ppm. The ozone was generated by a UV lamp, and the initial concentration was 2200 ppm. The compositions (C_6H_6, CO, CO_2, etc.) of the downstream flow were analyzed online using an FTIR spectrometer (Perkin-Elmer, Waltham, MA, USA, resolution of 4 cm^{-1}) equipped with a gas cell (2 m path length), while the residual ozone was measured using an ozone analyzer (EG-600, Ebara Jitsugyo, Japan). Prior to the tests, the potential compositions were calibrated with standard C_6H_6, CO, CO_2, etc. In addition, the benzene removal efficiency, CO_2 selectivity, and CO_x (CO_2 + CO) selectivity were calculated using the following equations [44]:

$$\eta_{C_6H_6} = \frac{200 - C_{C_6H_6}}{200} \times 100\% \tag{1}$$

$$S_{CO_2} = \frac{C_{CO_2}}{6 \times (200 - C_{C_6H_6})} \times 100\% \tag{2}$$

$$S_{CO_x} = \frac{C_{CO_2} + C_{CO}}{6 \times (200 - C_{C_6H_6})} \times 100\% \tag{3}$$

where C_{C6H6}, C_{CO2}, and C_{CO} were the initial concentrations of benzene, CO_2, and CO, respectively, after the reaction became stabilized.

4. Conclusions

Ag/Mn_3O_4 was synthesized by the co-precipitation oxalate route and applied to benzene OCO. Ag doping greatly improved crucial physicochemical properties of Mn_3O_4, such as better low-temperature reducibility, oxygen mobility, and surface active oxygen species, and promoted the chemisorption of benzene. As a result, Ag/Mn_3O_4 exhibited an excellent catalytic performance in the OCO of benzene at mild temperatures. For example, CO_2 generation was 585.4 ppm with Mn_3O_4, while it increased to 810.9 ppm when Ag/Mn_3O_4 was used. Meanwhile, CO generation was strongly suppressed to one tenth of its original value (194 ppm vs. 19 ppm). The addition of Ag species to Mn_3O_4 could lead to higher benzene mineralization during the OCO process. Moreover, Ag/Mn_3O_4 also exerted an improved catalytic durability, indicating that the surface intermediates oxidized faster than those on Mn_3O_4.

Author Contributions: Conceptualization, W.S., H.H. (Haibao Huang) and H.E.; methodology, H.G.; investigation, L.C., K.D. and W.M.; writing—original draft preparation, H.G. and P.Z.; writing—review and editing, D.H., H.H. (Haibao Huang) and H. Hojo; supervision, W.S., H.H. (Haibao Huang) and H.E.; funding acquisition, H.G. All authors have read and agreed to the published version of the manuscript.

Funding: This research was funded by Natural Science Foundation of Xinjiang Uygur Autonomous Region (2021D01C036).

Data Availability Statement: No data were created.

Acknowledgments: This work was financially supported by the Natural Science Foundation of Xinjiang Uygur's Autonomous Region (2021D01C036).

Conflicts of Interest: We declare that we have no commercial or associative interests that would constitute conflicts of interest in connection with the submitted work.

References

1. Wang, P.; Ma, X.; Hao, X.; Tang, B.; Abudula, A.; Guan, G. Oxygen vacancy defect engineering to promote catalytic activity toward the oxidation of VOCs: A critical review. *Catal. Rev.* **2024**, *66*, 586–639. [CrossRef]
2. Lin, F.; Xiang, L.; Zhang, Z.; Li, N.; Yan, B.; He, C.; Hao, Z.; Chen, G. Comprehensive review on catalytic degradation of Cl-VOCs under the practical application conditions. *Crit. Rev. Environ. Sci. Technol.* **2022**, *52*, 311–355. [CrossRef]
3. He, C.; Cheng, J.; Zhang, X.; Douthwaite, M.; Pattisson, S.; Hao, Z. Recent advances in the catalytic oxidation of volatile organic compounds: A review based on pollutant sorts and sources. *Chem. Rev.* **2019**, *119*, 4471–4568. [CrossRef]
4. Wei, W.; Chen, S.; Wang, Y.; Cheng, L.; Wang, X.; Cheng, S. The impacts of VOCs on $PM_{2.5}$ increasing via their chemical losses estimates: A case study in a typical industrial city of China. *Atmos. Environ.* **2022**, *273*, 118978. [CrossRef]
5. Zhu, L.; Shen, D.; Luo, K.H. A critical review on VOCs adsorption by different porous materials: Species, mechanisms and modification methods. *J. Hazard. Mater.* **2020**, *389*, 122102. [CrossRef]
6. Zhao, Z.; Ma, S.; Gao, B.; Bi, F.; Qiao, R.; Yang, Y.; Wu, M.; Zhang, X. A systematic review of intermediates and their characterization methods in VOCs degradation by different catalytic technologies. *Sep. Purif. Technol.* **2023**, *314*, 123510. [CrossRef]
7. Liu, B.; Ji, J.; Zhang, B.; Huang, W.; Gan, Y.; Leung, D.Y.C.; Huang, H. Catalytic ozonation of VOCs at low temperature: A comprehensive review. *J. Hazard. Mater.* **2022**, *422*, 126847. [CrossRef]
8. Dai, W.; Zhang, B.; Ji, J.; Liu, B.; Xie, R.; Gan, Y.; Xie, X.; Zhang, J.; Huang, P.; Huang, H. Exceptional ozone decomposition over δ-MnO_2/AC under an entire humidity environment. *Environ. Sci. Technol.* **2023**, *57*, 17727–17736. [CrossRef] [PubMed]
9. Lin, F.; Zhang, Z.; Xiang, L.; Zhang, L.; Cheng, Z.; Wang, Z.; Yan, B.; Chen, G. Efficient degradation of multiple Cl-VOCs by catalytic ozonation over MnO_x catalysts with different supports. *Chem. Eng. J.* **2022**, *435*, 134807. [CrossRef]
10. Xia, D.; Xu, W.; Wang, Y.; Yang, J.; Huang, Y.; Hu, L.; He, C.; Shu, D.; Leung, D.Y.C.; Pang, Z. Enhanced performance and conversion pathway for catalytic ozonation of methyl mercaptan on single-atom Ag deposited three-dimensional ordered mesoporous MnO_2. *Environ. Sci. Technol.* **2018**, *52*, 13399–13409. [CrossRef]
11. Einaga, H.; Teraoka, Y.; Ogata, A. Catalytic oxidation of benzene by ozone over manganese oxides supported on USY zeolite. *J. Catal.* **2013**, *305*, 227–237. [CrossRef]
12. Fan, G.; Guo, Y.; Chai, S.; Zhang, L.; Guan, J.; Ma, G.; Han, N.; Chen, Y. Synthesis of δ-MnO_2 via ozonation routine for low temperature formaldehyde removal. *J. Environ. Sci.* **2025**, *147*, 642–651. [CrossRef]
13. Hu, P.; Amghouz, Z.; Huang, Z.; Xu, F.; Chen, Y.; Tang, X. Surface-confined atomic silver centers catalyzing formaldehyde oxidation. *Environ. Sci. Technol.* **2015**, *49*, 2384–2390. [CrossRef]

14. Chen, W.; He, H.; Liang, J.; Wei, X.; Li, X.; Wang, J.; Li, L. A comprehensive review on metal based active sites and their interaction with O_3 during heterogeneous catalytic ozonation process: Types, regulation and authentication. *J. Hazard. Mater.* **2023**, *443*, 130302. [CrossRef] [PubMed]
15. Gopi, T.; Swetha, G.; Chandra, S.S.; Ramakrishna, C.; Saini, B.; Krishna, R.; Rao, P.V.L. Catalytic decomposition of ozone on nanostructured potassium and proton containing δ-MnO_2 catalysts. *Catal. Commun.* **2017**, *92*, 51–55. [CrossRef]
16. Jia, J.; Zhang, P.; Chen, L. Catalytic decomposition of gaseous ozone over manganese dioxides with different crystal structures. *Appl. Catal. B* **2016**, *189*, 210–218. [CrossRef]
17. Wang, H.C.; Chang, S.H.; Hung, P.C.; Hwang, J.F.; Chang, M.B. Synergistic effect of transition metal oxides and ozone on PCDD/F destruction. *J. Hazard. Mater.* **2009**, *164*, 1452–1459. [CrossRef]
18. Wang, Y.; Arandiyan, H.; Scott, J.; Bagheri, A.; Dai, H.; Amal, R. Recent advances in ordered meso/macroporous metal oxides for heterogeneous catalysis: A review. *J. Mater. Chem. A* **2017**, *5*, 8825–8846. [CrossRef]
19. Liu, S.; Wu, X.; Weng, D.; Ran, R. Ceria-based catalysts for soot oxidation: A review. *J. Rare Earth* **2015**, *33*, 567–590. [CrossRef]
20. Liu, L.; Wu, N.; Ouyang, M.; Xing, Y.; Tian, J.; Chen, P.; Wu, J.; Hu, Y.; Niu, X.; Fu, M.; et al. Enhancement effect induced by the second metal to promote ozone catalytic oxidation of VOCs. *Environ. Sci. Technol.* **2024**, *58*, 6725–6735. [CrossRef]
21. Kim, J.; Lee, J.E.; Lee, H.W.; Jeon, J.K.; Song, J.; Jung, S.C.; Tsang, Y.F.; Park, Y.K. Catalytic ozonation of toluene using Mn-M bimetallic H-ZSM-5 (M: Fe, Cu, Ru, Ag) catalysts at room temperature. *J. Hazard. Mater.* **2020**, *397*, 122577. [CrossRef] [PubMed]
22. Zhang, Y.; Chen, M.; Zhang, Z.; Jiang, Z.; Shangguan, W.; Einaga, H. Simultaneously catalytic decomposition of formaldehyde and ozone over manganese cerium oxides at room temperature: Promotional effect of relative humidity on the $MnCeO_x$ solid solution. *Catal. Today* **2019**, *327*, 323–333. [CrossRef]
23. Fu, Y.; Zhong, L.; Li, Z.; Jin, H.; Liu, X.; Tong, W.; Li, X.; Zhao, S. Spinel $(Mn, Fe)_3O_4$ nanocatalyst for the catalytic ozone decomposition under humid conditions. *ACS Appl. Nano Mater.* **2024**, *7*, 7111–7721. [CrossRef]
24. Nikolov, P.; Genov, K.; Konova, P.; Milenova, K.; Batakliev, T.; Georgiev, V.; Kumar, N.; Sarker, D.K.; Pishev, D.; Rakovsky, S. Ozone decomposition on Ag/SiO_2 and Ag/clinoptilolite catalysts at ambient temperature. *J. Hazard. Mater.* **2010**, *184*, 16–19. [CrossRef]
25. Xie, Y.; Wang, J.; Zhou, Z.; Wu, Y.; Cheng, G.; Li, Y.; Sun, C.; Sun, M.; Yu, L. Engineering of Mn_3O_4@Ag microspheres assembled from nanosheets for superior O_3 decomposition. *Dalton Trans.* **2022**, *51*, 16612–16619. [CrossRef] [PubMed]
26. Jiang, N.; Qiu, C.; Guo, L.; Shang, K.; Lu, N.; Li, J.; Zhang, Y.; Wu, Y. Plasma-catalytic destruction of xylene over Ag-Mn mixed oxides in a pulsed sliding discharge reactor. *J. Hazard. Mater.* **2019**, *369*, 611–620. [CrossRef]
27. Guo, H.; Zhang, Z.; Jiang, Z.; Chen, M.; Einaga, H.; Shangguan, W. Catalytic activity of porous manganese oxides for benzene oxidation improved via citric acid solution combustion synthesis. *J. Environ. Sci.* **2020**, *98*, 196–204. [CrossRef]
28. Li, Z.; Gao, R.; Hou, Z.; Yu, X.; Dai, H.; Deng, J.; Liu, Y. Tandem supported Pt and ZSM-5 catalyst with separated catalytic functions for promoting multicomponent VOCs oxidation. *Appl. Catal. B* **2023**, *339*, 123131. [CrossRef]
29. Jiang, T.; Wang, X.; Zhang, J.; Mai, Y.; Chen, J. Highly efficient MnO_x catalysts derived from Mn-MOFs for chlorobenzene oxidation: The influence of MOFs precursors, oxidant and doping of Ce metal. *Mol. Catal.* **2023**, *551*, 113653. [CrossRef]
30. Tang, W.; Li, J.; Wu, X.; Chen, Y. Limited nanospace for growth of Ni-Mn composite oxide nanocrystals with enhanced catalytic activity for deep oxidation of benzene. *Catal. Today* **2015**, *258*, 148–155. [CrossRef]
31. Li, H.; Qiao, D.; Chu, M.; Guo, L.; Sun, Z.; Fan, Y.; Ni, S.Q.; Tung, C.H.; Wang, Y. Accumulation of Ag(0) single atoms at water/mineral interfaces during sunlight-induced reduction of ionic Ag by phenolic DOM. *Environ. Sci. Technol.* **2023**, *57*, 20822–20829. [CrossRef]
32. Hansen, T.W.; DeLaRiva, A.T.; Challa, S.R.; Datye, A.K. Sintering of catalytic nanoparticles: Particle migration or Ostwald ripening? *Acc. Chem. Res.* **2013**, *46*, 1720–1730. [CrossRef]
33. Liu, B.; Yang, Y.; Tan, Q.; Zhou, K.; Xu, X.; Ding, Y.; Han, Y.; Fan, X.; Tao, R. Cr doped Mn_3O_4 thermal catalytic isopropanol degradation at low-temperature and catalytic mechanism research. *Colloid Surf. A* **2022**, *640*, 128390. [CrossRef]
34. Yang, X.; Yu, X.; Lin, M.; Ma, X.; Ge, M. Enhancement effect of acid treatment on Mn_2O_3 catalyst for toluene oxidation. *Catal. Today* **2019**, *327*, 254–261. [CrossRef]
35. Shao, S.; Li, Z.; Zhang, J.; Gao, K.; Liu, Y.; Jiao, W. Preparation of Ce-MnO_x/γ-Al_2O_3 by high gravity-assisted impregnation method for efficient catalytic ozonation. *Chem. Eng. Sci.* **2022**, *248*, 117246. [CrossRef]
36. Lu, J.; Hao, H.; Zhang, L.; Xu, Z.; Zhong, L.; Zhao, Y.; He, D.; Liu, J.; Chen, D.; Pu, H.; et al. The investigation of the role of basic lanthanum (La) species on the improvement of catalytic activity and stability of HZSM-5 material for eliminating methanethiol-(CH_3SH). *Appl. Catal. B* **2018**, *237*, 185–197. [CrossRef]
37. Lu, Y.; Deng, H.; Pan, T.; Liao, X.; Zhang, C.; He, H. Effective toluene ozonation over δ-MnO_2: Oxygen vacancy-induced reactive oxygen species. *Environ. Sci. Technol.* **2023**, *57*, 2918–2927. [CrossRef]
38. Zhang, Y.; Shi, J.; Fang, W.; Chen, M.; Zhang, Z.; Jiang, Z.; Shangguan, W.; Einaga, H. Simultaneous catalytic elimination of formaldehyde and ozone over one-dimensional rod-like manganese dioxide at ambient temperature. *J. Chem. Technol. Biotechnol.* **2019**, *94*, 2305–2317. [CrossRef]
39. Zhang, X.; Zhao, H.; Song, Z.; Zhao, J.; Ma, Z.; Zhao, M.; Xing, Y.; Zhang, P.; Tsubaki, N. Influence of hydrothermal synthesis temperature on the redox and oxygen mobility properties of manganese oxides in the catalytic oxidation of toluene. *Transit. Metal Chem.* **2019**, *44*, 663–670. [CrossRef]

40. Zhang, B.; Ji, J.; Liu, B.; Zhang, D.; Liu, S.; Huang, H. Highly efficient ozone decomposition against harsh environments over long-term stable amorphous MnO_x catalysts. *Appl. Catal. B* **2022**, *315*, 121552. [CrossRef]
41. Li, M.; Liu, X.; Niu, X.; Zhu, Y. Regulating the mobility of lattice oxygen on hollow cobalt-manganese sub-nanospheres for enhanced catalytic oxidation of toluene and o-xylene. *J. Colloid Interfaces Sci.* **2024**, *671*, 192–204. [CrossRef] [PubMed]
42. Ji, F.; Men, Y.; Wang, J.; Sun, Y.; Wang, Z.; Zhao, B.; Tao, X.; Xu, G. Promoting diesel soot combustion efficiency by tailoring the shapes and crystal facets of nanoscale Mn_3O_4. *Appl. Catal. B* **2019**, *242*, 227–237. [CrossRef]
43. Lei, J.; Chen, J.; Bai, B.; Wang, P.; Wang, S.; Li, J. Regulation of the existing state of Cu species in high water-toletant CuO_x-Mn_3O_4 for VOCs mixtrue catalytic oxidation. *Chem. Eng. Sci.* **2024**, *284*, 119455. [CrossRef]
44. Guo, H.; Liu, X.; Hojo, H.; Yao, X.; Einaga, H.; Shangguan, W. Removal of benzene by non-thermal plasma catalysis over manganese oxides through a facile synthesis method. *Environ. Sci. Pollut. Res.* **2019**, *26*, 8237–8247. [CrossRef]
45. Wang, J.; Su, J.; Zhao, G.; Liu, D.; Yuan, H.; Kuvarega, A.T.; Mamba, B.B.; Li, H.; Gui, J. A facile method for preparing the $CeMnO_3$ catalyst with high activity and stability of toluene oxidation: The critical role of small crystal size and Mn^{3+}-O_v-Ce^{4+} sites. *J. Hazard. Mater.* **2024**, *470*, 134114. [CrossRef]
46. Hu, Z.; Mi, R.; Yong, X.; Liu, S.; Li, D.; Li, Y.; Zhang, T. Effect of crystal phase of MnO_2 with similar nanorod-shaped morphology on the catalytic performance of benzene combustion. *ChemistrySelect* **2019**, *4*, 473–480. [CrossRef]
47. Perupogu, V.; Rajendiran, R.; Balla, P.; Seelam, P.K.; Aravindh, A.; Kim, S.; Pinapati, S.R.; Nakka, L. Insight into the role of metal support interface on shape dependent MnO_2 support through the synergistic effect between Ag and MnO_2 for the total oxidation of propane. *J. Environ. Chem. Eng.* **2024**, *12*, 112655. [CrossRef]
48. Shao, Q.; Cheng, Z.; Gao, L.; Li, T.; Zhang, J.; Long, C. Tuning the properties of oxygen vacancy around Cu-Mn mixed oxides on dealumination Y zeolite by P-doping to achieve ultra-efficient catalytic ozonation of toluene with low ozone consumption. *Appl. Catal. B* **2023**, *339*, 123154. [CrossRef]
49. Trinh, Q.H.; Mok, Y.S. Effect of the adsorbent/catalyst preparation method and plasma reactor configuration on the removal of dilute ethylene from air stream. *Catal. Today* **2015**, *256*, 170–177. [CrossRef]